A Guide to the Literature of Astronomy

A Guide to the Literature of Astronomy

Robert A. Seal

Libraries Unlimited, Inc. - Littleton, Colo. - 1977

Copyright © 1977 Robert A. Seal
All Rights Reserved
Printed in the United States of America

LIBRARIES UNLIMITED, INC.
P.O. Box 263
Littleton, Colorado 80160

Library of Congress Cataloging in Publication Data

Seal, Robert A
   A guide to the literature of astronomy.

   Bibliography: p. 283
   Includes index.
   1. Astronomy--Bibliography. I. Title.
Z5151.S4   [QB43]   016.52   77-12907
ISBN 0-87287-142-8

TABLE OF CONTENTS

INTRODUCTION 7

1—REFERENCE SOURCES IN ASTRONOMY 15
    Introduction 15
    Guides to the Literature 16
    Library Catalogs 19
        Card Catalogs, 19; Book Catalogs, 20
    Book Reviews 21
    Abstracts and Indexes 23
        Journal Articles, 23; Technical Reports, 30
    Translations 33
    Directories and Biographical Information Sources 34
    Dictionaries 40
    Encyclopedias and Encyclopedic Sets 42
    Handbooks, Manuals, Almanacs, and Yearbooks 47
    Atlases and Catalogs 56
    Sources of Scholarly Contribution 70
        Professional Journals, 70; Proceedings of Symposia,
        Conferences, and Workshops, 84; Publishers' Series, 88;
        Observatory Publications, 90; Review Literature, 92

2—GENERAL MATERIALS 96
    General Works 96
        For the Layman and Student, 97; For the Astronomer, 109
    Amateur Astronomy 112
        Beginning, 113; Intermediate, 118; Advanced, 126;
        Sundials, 131
    History 132
        Biography—General, 133; Biography—Advanced, 139;
        General, 143; Scholarly Works, 152
    Textbooks 156
        For the Liberal Arts Student, 158; For the Science
        Student, 164; Advanced or Supplementary Material, 168
    General Periodicals 172

3—DESCRIPTIVE ASTRONOMY 177
    Solar System 177
        General Works, 179; Sun, 184; Moon, 192; Planets, 197;
        Comets, Meteors, and Meteorites, 206; Atmospheres, 212;
        Gravity, 216; Celestial Mechanics, 218

3—DESCRIPTIVE ASTRONOMY (cont'd)
Stars 220
    General Works, 220; Structure and Evolution, 228;
    Variable Stars, 238
Milky Way and Interstellar Space 242
    General Works, 243; Interstellar Space, 246
Galaxies and Cosmology 250
    Galaxies, 250; Cosmology, 253; Quasars, 256

4—SPECIAL TOPICS 258
Instrumentation and Techniques 258
Space Science and Astrodynamics 264
Radio Astronomy 270
The New Astronomies 276

APPENDIX 283

AUTHOR-TITLE INDEX 290

SUBJECT INDEX 303

INTRODUCTION

GENERAL COMMENTS

Astronomy in the 1970s is far more than the study of the planets, the Moon, the Sun, and the stars. It encompasses dozens of related disciplines, too: physics, mathematics, computer science, chemistry, geology, atmospheric studies. The subject's multi-disciplinary aspects become evident when one examines the preparation necessary in the training of an astronomer. In the undergraduate years, the aspiring stellar scientist studies a widely varied curriculum which often includes more physics than astronomy, several math courses, one or more foreign languages, one or two computer languages, and, sometimes, related science course-work. A substantial number of social science and humanities disciplines are studied along with the science and technology. At the graduate level, aiming at the master's and Ph.D. degrees, the student finally begins concentrating on astronomy per se. Related course-work in mathematics and physics rounds out the advanced program. The result is an astronomer who has a solid, deep background in the physical sciences, a necessary prerequisite to a lifework of studying the heavens.

Astronomy's interdisciplinary nature is also apparent from its literature, the subject of this book. In this vast body of writings one finds books and journals on hundreds of topics, aimed at all levels of readership. The literature of astronomy ranges from general texts for high school and college students to pictorial works for the layman to advanced treatises for the scientist. It includes a large collection of reference works for the librarian and astronomer: catalogs, dictionaries, almanacs, bibliographies, abstracts, atlases, and ephemerides. The literature of astronomy presents, too, a voluminous serial collection to be dealt with: journals, observatory publications, review literature. It embraces countless proceedings of colloquia, symposia, and workshops. It is this wide body of information sources that this guide intends to present and explain.

Dealing with a scientific field of many subject classifications and physical forms of materials is difficult enough for the librarian and user, but another factor makes it even more difficult to cope with the astronomical literature. This phenomenon, not unique to astronomy, is the rapid rate of change and discovery in the field, which alters "what is now known" faster than ever before. Today's results

8 / Introduction

are replaced tomorrow by new information; few data can be static; the amount of fresh knowledge grows at an alarming rate. John S. Glasby, in *Variable Stars* (1968), recognized the phenomenon of change in astronomy:

> Astronomy never professes to give the last word. It grows by a tentative exploration of the great realm of the unknown by expounding theories which, however attractive and well put together, must inevitably be modified as our knowledge grows and eventually satisfy observation, for this is one criterion by which every astronomical theory stands or falls.

Certainly a prime ingredient in the great change in astronomy in recent years has been the space program. Since 1958, artificial satellites, Moon and Mars landings, orbiting observatories, and planetary flybys have contributed to the rapid flow of information and change. Hundreds of books and periodical articles detail the exciting space exploration story and its contributions to astronomical research. There is little reason to believe that change caused by space astronomy and other methods of research, such as those in non-visible light, will diminish.

The problems created by the fast pace of discovery are augmented by the information explosion, with its resultant headaches for librarians and their patrons. The overflow of scientific papers and books creates numerous problems. What does one do, for example, to provide access to the dozens of new journals that appear each year? What can be done with the increasing number of astronomy textbooks, spawned in part by the publish-or-perish syndrome, but a result in part of the information explosion itself? In short, how can the librarian cope with a broad, fast-changing body of knowledge, providing assistance to several types of potential users?

Knowing the subject matter, at least on an elementary level, helps the librarian or student to cope with this difficult situation. Much of this knowledge is achieved by osmosis, as the years go by, but reading an introductory text or taking an elementary astronomy course is a speedier way. The size and quality of the astronomy collection one has to work with play a role, too, in the kind of service one can provide. There are no easy answers. Indeed, this book, intended to help the reader along the rocky road described above, makes no promises of instant success. It is hoped, though, that it will be a good starting point for the librarian, student, or layman who wishes to gain some knowledge of the literature of astronomy and some proficiency in using it.

PURPOSE

The purpose of this book is to serve as an introduction to the literature of astronomy, presenting major and representative works from nearly all areas of subject matter. Full bibliographic data, annotations, and a brief evaluation are provided for each work. The guide is intended mainly for those individuals who are either totally unfamiliar with astronomical information, or who wish to gain a better knowledge of the literature than they currently possess. This group includes, but is not limited to, the non-astronomy librarian, the student, the amateur astronomer, the layman, and the non-astronomer scientist.

In the past there has been only one major astronomy bibliography, *Astronomy and Astrophysics: A Bibliographic Guide*, by D. A. Kemp (see entry 2 in this guide). Though one of the best bibliographies in the sciences, it is now sorely out of date. More important, however, is the fact that it is not for the novice but rather for the astronomer, the astronomy librarian, and the science bibliographer. It is so complete and detailed that it is likely to overwhelm the uninitiated reader. Dr. Kemp states in the introduction that his work is also intended for the non-scientist and librarian, but it is too comprehensive for the amateur astronomer or the newly appointed science librarian who has little or no previous knowledge. In short, a new guide is needed to supplement Dr. Kemp's excellent work—a book for the beginner as well as the experienced astronomer.

A volume on the literature of astronomy must categorize the various types of information and must explain and describe these information forms. Such an approach will help the new reader better understand the entire scheme of things by explaining examples of each type of astronomical literature. Also, the author hopes that this guide will also be of use to those who are already familiar with the basic sources of astronomy and astrophysics. The annotations, besides describing each book for the general reader, are intended to help the librarian select up-to-date items for the collection; the bibliographic information provided for each entry can be of further assistance in the ordering process. The astronomy instructor may wish to use this guide in one of three ways: as a supplement to textbook reading; as an aid in preparing a course reading list; and as a selection tool for choosing the texts themselves. The astronomer may use the book as a more up-to-date reference source; it is hoped that the guide will also be placed on the ready reference shelf alongside *Kemp*.

Summarizing, the author intends this book to be 1) an up-to-date bibliographic source in astronomy, 2) a supplement to D. A. Kemp's work, 3) a categorization and explanation of the various information forms in the field, 4) an introduction for the reader unfamiliar with the literature of astronomy, 5) a selection tool, 6) a ready-reference source, and 7) an aid to the astronomy instructor.

SCOPE

Not meant to be totally comprehensive, this book does not include every volume on astronomy ever published, nor even every work issued in the last five years. Rather, it includes a select group of items in the various subject areas of the field. An explanation of the selection criteria will help to define the scope of this guide, besides answering the question of why a particular item was or was not included.

The delimiting factor determining the scope of this volume is a definition of astronomy, chosen by this author from *The Random House Dictionary of the English Language* (New York, 1967): "the science that deals with the material universe beyond the earth's atmosphere." From this starting point, the subject can be divided into four major areas:

1. **General Astronomy.** Periodicals, general works, textbooks, history, proceedings of meetings on astronomy, reference works.

2. **Practical and Spherical Astronomy.** Time, celestial coordinates, spherical astronomy, amateur astronomy, observatories and their publications, telescopes, other special instruments, radio astronomy.

3. **Theoretical Astronomy.** Celestial mechanics, lunar and planetary theory.

4. **Descriptive Astronomy.** Astrophysics, cosmology, the solar system, the Moon, comets, meteors and meteorites, the Sun, stars, aurorae, the Milky Way, special topics (black holes, quasars, pulsars).

This guide covers each of the subdivisions mentioned above, although not in the same arrangement. It should be noted that the categories as listed above do not list every possible subtopic but instead include the most important elements of each subject, and therefore of its literature.

This volume concentrates on monographs and serials that are purely astronomical in nature—i.e., that fall into one of the four major subdivisions mentioned above. A physics text, for example, even though it might be used by astronomers and astronomy students, will not be reviewed. Nor will a geology book that contains one or two chapters on Moon rocks. Instead of listing these related items, this guide points the way to them by citing selected reference sources that give access to these materials.

A related topic that does not fall into the four categories of astronomy is aeronautics and astronautics. The space age has its own literature, in many ways closely allied to the literature of astronomy. Although some might argue persuasively that it is impossible and undesirable in this day and age to eliminate this information, such works have been omitted in an attempt to keep the book to a manageable size. Works on artificial satellites and the astronauts, for example, will not be found here. However, since there is an inevitable overlap between astronomy and these areas, a few books that reflect a dual subject emphasis have been included for illustrative purposes (for example, a book that tells of the use of spacecraft to explore the planets).

Finally, a few basic texts on space science have been added. Combining astronomy, physics, aeronautics, and other science disciplines, this subject does not cleanly fall into the astronomical scheme. But since it is closely related to astronomy, indeed emphasizes it, several appropriate references have been included here.

Individual items in each subdivision were chosen either because they are major works of a certain area or because they are good examples of a particular literature form or subject. Although the author made every effort to include all current, important volumes, a selective compilation is by its nature open to debate about the inclusion or omission of particular items. The author's discretion and bias, of course, were involved during the selection process, but there are several other factors that might account for the omission of certain items: works that appeared since the final manuscript was submitted for publication are obviously not included here, and works that were not available for examination are also omitted (the author personally examined each item included at least once).

Timeliness also played an important role in selection: in an effort to avoid rehashing what has been done before, the most current works were emphasized. Some older books are included, but generally the entries represent works published during the last ten to fifteen years.

Some secondary constraints were applied in the selection process. First, the books had to be in English, since the work is aimed primarily at an English-speaking audience. Certain works by non-English authors are included if they have been translated, and selected non-English works have been added if their importance to the literature cannot be overlooked (e.g., *Referativnyi Zhurnal: Astronomiia*).

Further, all entries are at the secondary school level and above. The volume does not include children's books. A few exceptions were made—for instance, some general, pictorial-type works can be read by younger readers as well as by adults, and a small number of works entered under Amateur Astronomy are intended for readers under high school age. But these items are the exception, not the rule.

Finally, and of great importance, the quality of the works played a factor in their selection. Some volumes are naturally better than others; the annotations attempt to distinguish between works of different quality, and to point out the good and bad aspects of each.

In summary, then, the scope of this work has been determined by the following factors: 1) a definition of astronomy and its subject matter; 2) major works in the field; 3) language; 4) timeliness; 5) age of readership; and 6) quality of material.

HOW TO USE THIS BOOK

The book is arranged both according to types of materials and according to the subject areas of the monographs. The table of contents reveals the headings used. Each entry includes full bibliographic information and a brief annotation. The form of the bibliographic information follows the style used in *American Reference Books Annual* (Libraries Unlimited, 1970–).

If more than three authors are responsible for a work, only the first author is given, followed by an indication that there are others. Main entry is under author or editor; if there is no known author or editor, the entry is under title.

Both the title and subtitle of an item are included in the bibliographic statement, unless the subtitle is extremely long, or not unique. If the subtitle is very long, it is often mentioned within the annotation, rather than in the bibliographic information.

The imprint follows the usual format: place, publisher, date. If there is more than one publisher, both are usually listed, with the original publisher cited first. For example, many British books are also published in the United States, and vice versa. If the second publisher is a distributor, rather than a dual publisher, this is indicated. The publisher of the edition examined is usually the only one listed, unless other information is available from other sources like the *National Union Catalog*.

The descriptive information has been somewhat simplified for this listing. Paging only includes Arabic-numbered pages in the main body of the book. Front matter indicated by small italic Roman numerals is not included in the paging statement. Illustrations (illus.) include photographs, diagrams, charts, sketches, plates, etc. Individual types of illustrations are not described separately, except when a volume contains primarily one type of illustration. Atlases that are mostly plates, for instance, are described by the term "plates," rather than "illus."

Special features are also cited: indexes, glossaries, references, notes, bibliographies, etc. Appendices are not mentioned in the bibliographic statement. Indexes other than general indexes are defined—i.e., subject, name, star. The terms "references" and "notes" are used interchangeably to refer to citations to related literature. "Bibliography" suggests a list of books, whether the list is termed "bibliography," "further reading," or "suggested reading."

The price of the book is given if it was readily available in such standard reference sources as *Books in Print* (Bowker). This includes both hardcover and paperback volumes. The lack of a price in the entry does not necessarily mean that the book is out-of-print and unavailable, although it is probably *not* available. Each item was carefully checked for a price in *BIP*, as well as in some publishers' catalogs, and the price given reflects what was found in those sources. If there is any doubt about price or availability, the publisher should be contacted. Whenever possible, prices are given in U.S. dollars.

Prices for items from government sources like NASA, National Technical Information Service (NTIS), Government Printing Office (GPO), and the Smithsonian Institution are often difficult to verify. *The Monthly Catalog of U.S. Government Publications* (GPO) is a help, but it gives prices only for recently published documents, and it is not completely reliable for information on availability and prices of older publications. In many cases, it is necessary to check with government offices before placing an order. NASA documents and NTIS technical reports are priced roughly according to the number of pages. The most recent issues of *STAR* (*Scientific and Technical Aerospace Reports*) and *Government Reports Announcements and Index* (*GRA & I*) should be checked for current price scales. The prices for NASA publications listed in this guide are original costs.

Individual book numbers like Library of Congress (LC) card number, International Standard Book Number (ISBN), International Standard Serial Number (ISSN), and British Library number (B or GB), are given when they could be found. When one book has two publishers, both LC and ISBN numbers are given if they could be identified. If the book has two publishers but only one number is given, then the number refers to the first publisher listed. ISBNs for both hardcover and paperback editions are listed when known, and other numbers (such as NASA and GPO document numbers) are stated when applicable.

The annotations are both descriptive and evaluative. They provide information on what is in the book, the author's purpose, when discernible, and criticisms and/or praise of the work. In-depth analyses of books can be gotten from reviews, as outlined in the Reference section under Book Reviews.

The recommendations such as "for the layman," "for the student," "for the astronomer," are usually fairly accurate, but such categorizations do not always apply. For example, some laymen are very well versed in astronomy and can handle material designated for the advanced student.

A NOTE ON UP-TO-DATENESS

All librarians and information users are painfully aware of the difficulty of keeping reference books and texts up to date. It is not easy to find the latest books on a particular subject. It is important, though, to keep in mind that "out of date" can refer to two distinctly different kinds of works: first, a work containing information that is incorrect because it has been superseded by newer information, and second, a work that lacks newer material (i.e., the information it includes is old but still valid). A book can be so out of date that its information is useless from a practical standpoint. But a work can also have good, though old, data, while lacking the latest information.

Any reference book is almost instantly out of date. It does not include items that appeared shortly after the manuscript reached the publisher, nor will it include those works which will appear, at an almost exponential rate, during the months and years after its publication. Frequent updates and new editions represent an attempt to solve the problem. Nevertheless, as a record of what was being read and consulted in astronomy and astrophysics in the 1970s, it is hoped that this guide will remain a useful bibliographic source in astronomy.

ACKNOWLEDGMENTS

The author wishes to thank the following libraries whose resources were used in the compilation of this guide: University of Virginia Library, National Radio Astronomy Observatory Library, Library of Congress, U.S. Naval Observatory Library, NASA Headquarters Library, Jefferson-Madison Regional Library (Charlottesville, Va.), Richmond (Va.) Public Library, and Martin Luther King Library (Washington, D.C.).

Special and sincere thanks go to Professor Thomas Whitby of the University of Denver, who encouraged me to undertake this work; to Sarah Martin of NRAO, who provided invaluable assistance, and to Adela, who endured three years of research, writing, and re-writing.

R.A.S.
March 1977

1

REFERENCE SOURCES IN ASTRONOMY

INTRODUCTION

Overall, the literature of astronomy and astrophysics rates high marks for both the quality and the quantity of its reference material. In its ranks are scores of handbooks, dictionaries, encyclopedias, directories, and indexes, as well as a vast number of special books called atlases and catalogs, which display and list stars, planets, galaxies, and other celestial objects. The section which follows describes the many astronomical reference sources, from works designed for quick reference to useful series to professional journals. Related material from physics, mathematics, etc., is occasionally included among the astronomy books to illustrate the importance of non-astronomical information. This is the most comprehensive and least selective section of the guide, and it includes as many current sources as possible. The only exception to this comprehensiveness is in the portion devoted to atlases and catalogs; only a small selection of these tools are described from the scores available. Librarians will wish to consult the Appendix at the end of this book for a list of essential reference works for all types of libraries; this checklist should be useful to those evaluating an existing collection and to those involved in acquisitions.

The reader should note, too, that many volumes which might be considered "reference" are not recorded in the sections immediately following, but instead are listed with their appropriate subject divisions. Examples are the handbooks on the planets and various review-type texts.

A good astronomical reference collection, of course, does not consist exclusively of books in astronomy and astrophysics. Volumes from physics, chemistry, mathematics, and other scientific fields are necessary, as well as general works like encyclopedias, dictionaries, geographic atlases, and almanacs. The librarian may need to consult other subject literature guides and standard sources like Sheehy's *Guide to Reference Books* (1976) to select certain non-astronomy works.

Since the astronomy reference collection has numerous physics books, the observatory librarian should also become familiar with the literature of that field. This guide lists several essential physics information sources that the librarian will need to know, and further mention of these items is made in the individual sections below.

GUIDES TO THE LITERATURE

The need for additional, up-to-date bibliographic material in astronomy provided the main impetus for writing this book. There are five standard sources already; only one, however, is specifically written about the literature of astronomy. All should be known to the astronomy librarian and should be on the ready reference shelf in the library. The major bibliographic source in astronomy in recent years has been D. A. Kemp's *Astronomy and Astrophysics; A Bibliographic Guide* (1970), an excellent work which comprehensively covers all areas of interest to astronomers, librarians, and students of astronomy. It lists materials by form and subject, emphasizing the latter. A standard reference volume, *Kemp* covers the literature world wide as no other work has ever done.

A related book, *Use of Physics Literature* (1975), edited by Herbert Coblans, helps the librarian cope with publications in that field and has a good, but short, section on the astrophysical literature. Both Kemp and Coblans should be required reading for librarians and astronomers.

Also cited below are three reference book compilations, one very general, Sheehy's *Guide to Reference Books* (1976) and two general scientific, Walford's *Guide to Reference Material, Vol. 1: Science and Technology,* and Malinowsky's *Science and Engineering Literature*. All three of these are quite good overall, though their coverage of astronomy is not especially good, since the space they have available for the subject is limited. Nevertheless, they are established and extremely handy bibliographic sources, necessary for nearly all library collections.

1. Coblans, Herbert, ed. **Use of Physics Literature**. London, Boston, Butterworths, 1975. 290p. refs. indexes (subject, author and title). (Information Sources for Research and Development). $18.95. LC 76-353687. ISBN 0-408-70709-7.

Useful for the astronomy librarian who wants an overview of physics information sources, this volume consists of 16 chapters written by both scientists and librarians. Beginning with a chapter of introduction on the types of materials encountered and their significance, this guide then proceeds to a discussion of the structure and control of the literature, written by Coblans. The chapter on "Science Libraries" gives the physicist an overview of the major institutions in Great Britain and the United States, the three classification systems (LC, Dewey, and UDC), card and computer-produced catalogs, microforms, and information services. Painfully brief, it is nonetheless a well-presented section. The bulk of the volume focuses on the types of physics literature, beginning with reference material and general works. The major tools are cited, with bibliographic data and brief descriptions. Next is a look at bibliographies, reviews, and abstracting and indexing services; the most important works are listed, including both primary and secondary sources. There are sections on patents and translations, history, and theoretical physics.

As it happens, chapter nine describes the literature of astrophysics, expertly written by an astronomer, A. J. Meadows, and a librarian, J. G. O'Connor. Although it's by no means comprehensive, this part does introduce most of the important astrophysical information items under the following headings: abstracting services, data compilations, ephemerides, visual aids (atlases), catalogs, journals, review journals, report publications, and monographic literature. The physicist wishing a brief but good introduction to this area need look no further. The remainder of this treatise covers information sources in mechanics and sound; heat and thermodynamics; light, electricity, and magnetism; nuclear and atomic physics; crystallography; instrumentation; and computer applications. Two appendices list acronyms and publishers of physics books. Every astronomy and physics library should have this guide, and it should be required reading for every astronomy librarian who has already read Kemp's *Astronomy and Astrophysics: A Bibliographic Guide* (1970).

2. Kemp, D. A. **Astronomy and Astrophysics: A Bibliographic Guide**. London, MacDonald Technical and Scientific; distr. Hamden, Conn., Archon Books, Shoestring Press, Inc., 1970. 584p. indexes (author and subject). glossary. (The MacDonald Bibliographical Guides). £10.00; $25.00. LC 70-505036. ISBN 0-208-01035-1 (Archon); 0-356-03011-3 (MacDonald).

Intended as a key with which to access the astronomical literature, this extensive work is already a classic in the field of bibliography. Divided into 75 sections, the book covers every possible type of astronomical book from reference volumes to review works. In all, it contains 3,642 entries from all over the globe, covering astronomy and including selected related works on physics, mathematics, geophysics, and so on. Of particular interest to the librarian and bibliographer are the first two sections, "Reference media," and "Star catalogs, ephemerides, etc." The former includes extensive lists of bibliographies, periodicals, abstracts and indexes, review literature, directories, encyclopedias, dictionaries, observatory publications, and much more. The latter is one of the most extensive checklists anywhere of catalogs, atlases, and other observing aids. The vast majority of the work, however, is composed of various subject sections of important and selected books and periodical articles. Topics include the solar system, stars, galaxies, radio astronomy, practical astronomy, celestial mechanics, cosmology, and dozens of others. A 13-page list of periodicals, their abbreviations, and country of origin is invaluable, as is the 19-page section of abbreviations used in the text.

Each of the entries contains full bibliographic information, the number of references cited (and their year span), and a concise evaluation or comment.

A large number of citations are to journal articles and published papers, especially in the topical portions of the text. The author's criterion for selecting items was the likelihood that the material would be useful to the person consulting the guide. In that regard, the author had four types of users in mind when he began work on this volume: the astronomer, the non-astronomer scientist, the non-scientist, and the librarian. Despite this, the novice should steer away from this book; it would be overwhelming to the inexperienced. In fact, one minor drawback to this fine book is that it doesn't contain much for the amateur astronomer or reader of popular materials, nor does it contain much in the way of general texts. Overall, however, this is a fine effort, *the* authoritative guide, highly recommended for all astronomy and science libraries.

3. Malinowsky, H. Robert, Richard A. Gray, and Dorothy A. Gray. **Science and Engineering Literature: A Guide to Reference Sources.** 2nd ed. Littleton, Colo., Libraries Unlimited, 1976. 368p. index. $14.00. LC 76-17794. ISBN 0-87287-098-7.

This guide describes the nature of the various fields of science and engineering, clarifies the structure of their literature and research processes, and provides annotated citations to the basic reference titles. Sources underlying all the disciplines are covered in the first three chapters, while Chapters 4 through 12 treat major fields. The last chapter covers library resources and literature searching.

The section on astronomy gives a brief introduction to the structure of the science, then lists and annotates 47 titles. Entires give full bibliographic information and are arranged under the following headings: treatises; guides to the literature; abstracts, bibliographies, and indexes; representative review journals and annual surveys; dictionaries and encyclopedias; handbooks; atlases; navigation aids; and directories.

4. Sheehy, Eugene P., comp. **Guide to Reference Books.** 9th ed. Chicago, American Library Association, 1976. 1015p. index. $30.00. LC 76-11751. ISBN 0-8389-0205-7.

Known for many years as "Winchell," after its most recent compiler, Constance Winchell, this standard work covers all subject areas in its bibliographic endeavors. *The* handbook on every information desk in public, college, and special libraries, *Sheehy* lists the world's most important reference books in the pure and applied sciences, the social sciences, and the humanities. Entries, which are relatively brief, include author, title, imprint, number of pages, whether illustrated or not, a descriptive note, and Library of Congress call number. Like most reference works, it was already sorely out of date when it came off the press. For all practical purposes, the cutoff date for entries was 1973, although there are a few 1974 titles.

Despite its authority and overall usefulness, its section on Astronomy leaves much to be desired. There are only 41 entries (33 in English); several of these are really not reference books but are semi-standard works which should not necessarily have been included. Part of the problem of the inadequacy is the omission of several important items like C. W. Allen's *Astrophysical Quantities*, J. Hedley Robinson's *Astronomy Data Book*, and *Landolt-Borstein Astronomy and Astrophysics.* The *Harvard Books on Astronomy* (texts, not ready reference material) are mentioned, but Chicago's *Stars and Stellar Systems* (standard review series) is not. There appears to have been no logic in the selection of items for this part of the work. All astronomy libraries should have *Sheehy* because of its broad coverage and general usefulness—but not for its coverage of astronomy.

5. Walford, A. J. **Guide to Reference Material: Volume I: Science & Technology.** 3rd ed. London, The Library Association, 1973. 615p. £6.00. index. LC 73-174024. ISBN 0-85365-326-7.

A kind of British equivalent to Sheehy's *Guide to Reference Books*, this work is international in scope and emphasizes English and European publications. This volume on science and technology has nearly 70 entries for astronomy, and about 50 additional citations on related topics like navigation and cartography. The third edition was a welcome addition to the literature— the older volume (1966) had become far outdated. Adequate bibliographic information and very brief annotations accompany each item. Quite a few of the references are to non-English language material, but this adds to the book's value even more.

LIBRARY CATALOGS

Card Catalogs

The library card catalog is the most basic reference source available in most libraries. It is the listing, in card form, of all materials held by the library: books, journals, newspapers, proceedings, phonograph records, filmstrips, and so on. The catalog offers the user three major access points to this information: author, title, and subject. By knowing any one of these, the library patron can fairly easily locate the book in question—if, of course, the work is held by the library. With a few exceptions, author and title are fairly straight-forward approaches to the information, and the reader can consult the catalog quite easily by using one or the other.

Subject access may be a little more difficult; some advance preparation will lead to more productive results. Those who undertake a subject-oriented search usually do not have a particular book in mind; rather, they want as many books as possible on a particular topic. Since nearly every item in the catalog has one or more "subject terms" describing what the book is about, the reader may approach a topic from more than one direction. When a book or journal has two or more subject headings, a common situation, duplicate cards are made for the catalog, one for each subject term. A particular item is often listed in the catalog under a variety of subject terms, as well as under author and/or title.

If the reader has no idea what subject words to look under in the catalog, then the library's subject headings list should be consulted first. Most libraries use a standard set of subject terms, either the *Sears List of Subject Headings* or the *Library of Congress Subject Headings*. The *Sears List* is used mainly by school and public libraries, while the *LC* list is most often used in colleges and universities and in special libraries. The reason for having lists of standard subject terms to describe the books is so that there will be some consistency within individual libraries and among different libraries. These lists of subject terms are the keys to the card catalog.

The list of subject terms is a useful source for determining which one of many synonymous subject headings is used for individual concepts. For example, will a book on asteroids be found under ASTEROIDS, PLANETOIDS, or MINOR PLANETS? Consulting the list of subject terms used in the library's catalog, will guide the user to the appropriate terms.

Both the *Sears* and the *LC* lists include good introductions and supplementary materials that explain their use. As an aid to the user of this guide, *LC* and *Sears* subject terms are given for each category of material included. It is hoped that these select terms will be useful to the reader in conducting a thorough subject search.

Book Catalogs

Card catalogs in book form are worth mentioning since the reader may encounter one in his public or special library. Book catalogs contain all the information found in the traditional card catalog file; in fact, they are often just printed pages of catalog cards. Their main advantage, besides their compact format, is that they can be duplicated fairly easily, although usually at no small cost. Thus, a library system can simply keep a copy of the book card catalog in each branch instead of separate card files, which require extensive filing and maintenance. An extremely important six-volume astronomy book catalog that is potentially of great use to astronomy librarians and researchers is the *Catalog of the Naval Observatory Library (Washington, D.C.)* (1976). This catalog provides access to one of the most important collections of astronomical literature in the United States. It is the only astronomical card catalog in printed form, and the possibilities for its use are many. A detailed description follows.

6. **Catalog of the Naval Observatory Library (Washington, D.C.).** Boston, G. K. Hall & Co., 1976. 6v. $490.00 (U.S.); $539.00 (foreign). ISBN 0-8161-0031-4.

A substantial source of astronomical book data, this set lists the holdings of one of the most extensive observatory library collections in the United States. Approximately 75,000 volumes are included in this book catalog, which is a photographic reproduction of the Naval Observatory's dictionary card catalog of authors, titles, subjects, serials, and cross-references, 21 cards to a page. A combination of locally produced cards and Library of Congress cards, this work is the only publication listing an entire astronomy library's materials. The reproduction quality is very good, the cards being reduced to 3/5 normal size. The astronomy librarian who can afford to buy this should do so, as there are many possible uses for the information included: bibliographic, verification, cataloging, reference, interlibrary loan, acquisitions, and more. It's like having an additional card catalog right in your own library. The Naval Observatory library is strong in many areas, all reflected in this catalog: history of astronomy, observatory publications, astrometry, journals, almanacs, ephemerides, and navigation. The library also has a substantial rare book section, as well as many popular and juvenile works. A selection of mathematics, physics, and other science works are included, too. Until the late 1950s, the Naval Observatory Library produced analytics for important journal articles, and cards for these appear in the catalog. A substantial number of foreign language materials can be found here, too, although there are certain areas stronger than others. Highly recommended for any observatory and university library which supports an astronomy program, this work was created under the direction of Brenda G. Corbin, librarian.

BOOK REVIEWS

Astronomy librarians, like their colleagues in other fields, are directly involved in selecting and purchasing books, journals, and other materials. Book selection is not simple, and the librarian who is not well-versed in the various subject areas of astronomy will require advice on the merits of a particular book or journal before placing an order. Advice along these lines may come from an astronomer on the staff (in many cases the staff request individual items themselves and the librarian merely places the order), or from a written source such as book reviews. Guides to the literature are also intended in part to serve as selection aids.

The best and more current book reviews are found in astronomical journals, many of which include them in each issue. Astronomy librarians regularly consult the book review sections in several periodicals of interest. The following journals contain book reviews: *Sky and Telescope, Astronomy, Astronomische Nachrichten, Astrophysics, Journal of the British Astronomical Association, Earth and Extraterrestrial Sciences, Icarus, Journal for the History of Astronomy, Mercury, The Moon, Observatory, Physics Today, Journal of the Royal Astronomical Society of Canada, Solar Physics, Soviet Astronomy,* and *Space Science Reviews.* Most or all of these journals are found in the average observatory library. Other astronomical journals contain reviews, but not always on a regular basis.

Secondary sources include *Technical Book Review Index* and *New Technical Books*, two journals that list current publications of interest to the scientific community at large. The former contains the most references to astronomical books, listing excerpts from reviews already written about the works. The latter merely lists contents and provides a descriptive note. Both contain enough information to allow easy ordering of the books. University science libraries usually subscribe to these two publications, and some astronomy libraries also include them on their reference shelves. Neither is expensive, and both are very useful for acquisitions work.

The librarian's best advice on how good a book is should come from astronomers or scientists involved in the particular subject field. Their opinion should always be sought first, if possible. Book reviews are good back-up sources for this kind of information, but in many instances, they are the primary source of selection criteria.

7. **AAAS Science Books & Films**. Washington, D.C.: American Association for the Advancement of Science. v.11– , 1975– . 4/yr. annual index.

A secondary source of book review information in astronomy, this publication includes critiques of both physical and social science books. Entries, arranged by Dewey Decimal Classification, have complete bibliographic citations, a 200-word review, and indications of intended readership, as well as a recommendation. There are usually two to six reviews on astronomy per issue, ranging from works for junior high students to professional books. Although it is relatively inexpensive, it does not have enough information on astronomical literature to make its purchase worthwhile to the observatory library. In a college-level general science library or in a public library, it would be of great use, since it covers the spectrum of books and films. It supersedes *AAAS Science Books* and *Science Books: A Quarterly Review*.

Write: American Association for the Advancement of Science, Publications Department (Dept. W3), 1515 Massachusetts Ave. N.W., Washington, D.C., 20005. Cost: $16/yr. (members); $28/yr. (non-members).

8. **New Technical Books**. New York: The Research Libraries, New York Public Library; v.1– , 1915– . Monthly (except August and September). LC 15-24011. ISSN 0028-6869.

Not to be considered a primary source of book review information for astronomy librarians, this journal only contains two or three entries per issue under Astronomy and Allied Sciences. The astronomy librarian would do better, of course, to consult the book reviews in the astronomy journals themselves, or even *Technical Book Review Index*. Nevertheless, *New Technical Books* is an excellent publication that the astronomy librarian should be aware of, for, like *TBRI*, it is a useful way to pick up titles in fields related to astronomy, items that would not normally be covered in reviews in astronomical journals.

Entries are arranged according to the Dewey Decimal Classification and contain the standard complete bibliographic information, contents (in full or abridged), and a descriptive note of 30 to 50 words. There is little or no critical comment, but the list is highly selective (evident from the small number of books included under each category) and one can assume that only the best works are listed. Each issue has a subject and author index, and there are volume/annual indices as well. This periodical should be considered by the astronomy librarian partly because it is a good aid for selection and partly because it is so inexpensive.

Write: The Research Libraries, New York Public Library, Fifth Avenue at 42nd Street, New York, New York 10018. Cost: $7.50/year.

9. **Technical Book Review Index.** New York: Special Libraries Association. v.1– , 1935– . Monthly (except July and August). LC 37-9385. ISSN 0040-0890.

The astronomy librarian who cannot find the time to locate and read entire reviews should subscribe to this excellent selection tool. Divided into five major subject areas (Pure Sciences, Life Sciences, Medicine, Agriculture, Technology), the magazine contains excerpts from book reviews for dozens of current monographs. The astronomy reviews are included in the Pure Sciences section, arranged by author; they include both general and technical entries. Each citation contains enough bibliographic data for ordering, and includes the reference to and excerpt from the review or reviews for each title. The 20 to 60 word paragraph summarizing the review hits the high points and usually emphasizes the book's strengths. The reviews are generally quite favorable, giving the reader the impression that most books listed are worth purchasing. This is probably true, and the quality of the books included indicates careful screening by the *TBRI* editorial staff, a definite strong point. In short, the journal is highly recommended for astronomy librarians involved in book selection. To obtain, write Special Libraries Association, 235 Park Avenue South, New York, New York 10003. Cost: $17.50 (U.S. and Canada): $19.50 (elsewhere).

ABSTRACTS AND INDEXES

Journal Articles

Astronomy journals can be located in the card catalog, under either title or subject, but the library user often wants an individual journal article instead of an entire volume. The keys to locating and identifying journal articles in any subject discipline, whether astronomy, psychology, or any of a dozen areas, are the so-called abstracting and indexing periodicals. Published on a regular basis, usually bi-weekly or monthly, with periodic cumulations, these tools attempt to provide selective or comprehensive indexing coverage of the world's serial literature. The major difference between an abstracting journal and an indexing journal is that the latter does not include brief summaries (abstracts) of the articles. The condensed descriptions of the articles can save unnecessary work by steering the reader away from information which he or she does not really want. Since article titles do not always reflect the contents, the importance of abstracts cannot be over-emphasized. Fortunately, most abstracting/indexing journals do provide abstracts.

"Abstracts and indexes" (as librarians refer to them) are usually arranged by subject categories, listing the latest publications in each. The subject terminology used does not coincide with either *LC* or *Sears*. Each index has its own vocabulary, which the reader must follow in order to make the most effective use of the index. In principle, the approach is analogous to a subject search in the card catalog.

Besides the subject access, the reader may also consult the author, number, or titles indexes that are often included in each issue. Each entry includes appropriate bibliographic data: title, author(s), journal title, volume, issue, date, page, etc.

Among other things, librarians use the indexes to 1) locate particular articles for library patrons; 2) compile "one-shot" or regular bibliographies on various subjects; 3) verify interlibrary loan requests in-house before sending to other libraries. The student, astronomer, or general reader uses the abstracts and indexes for the following major reasons: 1) locating articles for a term paper or research project; 2) locating articles in a given subject field that would not be included in publications regularly read; 3) locating articles that would otherwise remain unknown because of a lack of time to read or scan the appropriate journals.

Because the scientific periodicals serve as the astronomer's most important source of information, the abstracting and indexing journals, the keys to this information, should be an essential part of the observatory reference collection. This section lists ten key abstracts and indexes that are frequently consulted by students, astronomers, and librarians in the areas of astronomy, astrophysics, and their applications. The most important and best-known is *Astronomy and Astrophysics Abstracts*, a semi-annual publication covering journals, monographs, and proceedings. Its major drawbacks are its infrequency and the time lag between the articles' publication dates and their appearance in the index. This latter dilemma is not unique to *AAA*, however.

Despite the drawbacks, these indexing journals are excellent sources of information, giving the users comprehensive access to the literature of astronomy. The importance of these keys and the bibliographic data they contain cannot be over-emphasized.

In addition to the items discussed below, the following abstracts and indexes frequently contain references to astronomical periodical articles: *Mathematical Reviews, Science Abstracts, Meteorological and Geoastrophysical*

Abstracts, Chemical Abstracts, Government Reports Announcements/Index, Applied Mechanics Reviews, Readers' Guide to Periodical Literature, Applied Science and Technology Index. All provide selective reference to certain astronomical and astrophysical journals. *Readers' Guide* and *Applied Science and Technology Index* are usually found in public and college libraries, and the former is in many school libraries. The others can be found in large public libraries and university and specialized libraries. Some contain more pertinent information than the others, but all contain at least some reference to the astronomical literature.

10.   **Astronomischer Jahresbericht**. Leipzig: W. de Gruyter. v.1-68, 1899-1968. Annual. LC 7-880.

For 70 years this abstracting-bibliographical work provided exhaustive coverage of books, book reviews, articles, and observations from the world's literature of astronomy. Sponsored by Astronomisches Rechen-Institut zu Berlin, its material is international in scope, although the language of the text is mainly German, with the exception of conference papers. The author index of this yearly volume is adequate, but the subject index is not—it refers the reader only to major subject divisions as given in the table of contents. Nearly 7,000 entries per year are included, but not all have abstracts. An indispensable bibliographic source, especially for retrospective searching, it was superseded in 1969 by *Astronomy and Astrophysics Abstracts*.

11.   **Astronomy and Astrophysics Abstracts**. Berlin, New York: Springer-Verlag. v.— , 1969— . Semi-annual. indexes. About $32.00/volume. LC 71-104650.

Prepared under the auspices of the International Astronomical Union, this excellent series provides comprehensive coverage of the world's literature of astronomy. *The* index in the field, *AAA* is the continuation of *Astronomischer Jahresbericht* (1899-1968) which for seven decades was the authoritative source on astronomical and astrophysical journals, books, and proceedings. Entries are arranged according to 108 subject categories under the following general headings: Periodicals, Proceedings, Books, Activities; Applied Mathematics, Physics; Astronomical Instruments and Techniques; Positional Astronomy, Celestial Mechanics; Space Research; Theoretical Astrophysics; Sun; Earth; Planetary System; Stars; Interstellar Matter, Gaseous Nebulae, Planetary Nebulae; Radio Sources, Quasars, Pulsars, Infrared, X-Ray, Gamma-Ray Sources, Cosmic Radiation; Stellar Systems.

Of particular interest besides the usual topical entries are the following selected areas: Section 008, which lists observatories and astronomical institutes (130 in all); Section 010, which has references to reports of societies, associations, and organizations; Section 012, on proceedings; and a list of comets discovered in the past year. There's something for everyone; it's more than just an abstracting publication, since it contains much supplementary information.

Titles of papers are given in the language of the author whenever possible, and an author abstract is used when available. Ninety percent of the entries are in English, the rest are mainly German and French. Each individual reference has author/title information, publisher, citation data (if a serial), price (if a monograph), and abstract. There are many cross references to related articles, and excellent subject and author indexes accompany each twice-yearly volume. A five-year index (v. 1-10, 1969-73) was published in 1976 (v.15/16, 655p.). The only drawback to this fine source is its infrequency.

12.   **Astronomy and Astrophysics Monthly Index.** Sierra Madre, Calif.: Olivetree Associates. v.1– , 1976– . Monthly.

A need for fast indexing of the major astronomical journals and a need for an easy way to keep up with the literature were both filled by this publication, which appeared in the mid-1970s. Before *AAMI*, librarians and their users had to rely on sources that were slow (*Astronomy and Astrophysics Abstracts*) and non-comprehensive (*Physics Abstracts* and *Current Physics Index*). The only drawback of this useful publication is that coverage is limited to professional journals. Consequently, *Sky and Telescope* and *Astronomy* are missing; they can be accessed through other indexes, like *Readers' Guide*, but these tools are often unavailable in the observatory library. (No explanation is given of the omission of these journals.) *AAMI* is divided into two sections, an author index and permuted title index. The latter has mixed acceptance among librarians, since titles do not always contain pertinent subject terms, but that will not be debated here. Entries include author(s), title, journal title, volume, page, and year. In most cases there are duplicate entries for multiple-author articles, although a spot check in one issue showed several omissions. The front cover lists journals regular indexed (about 30), while the back page shows what individual issues are contained within. A periodical index of this type was desperately needed, and at this writing, the *AAMI*, after being out 10 months, has been doing a good job of fulfilling its goals. For university and observatory libraries, it is well worth the price.

Write: Olivetree Associates, P.O. Box 236, Sierra Madre, Calif. 91024. Cost: $120/yr. (domestic); $165/yr. (foreign).

13.   **Bulletin Signalétique 120: Astronomie, Physique Spatiale, Géophysique.** Paris: Centre de Documentation du C. N. R. S. v. 32– , 1971– . Monthly. abstracts. indexes. LC 74-9700. ISSN 0007-5337. Continues **Bulletin Signalétique 120: Astronomie et astrophysique; physique du globe** (v.22– . 1961–).

The third major world index to astronomical literature, this publication is entirely in French, and is not used much in the United States. (The other two are *Astronomy and Astrophysics Abstracts* [Springer-Verlag] and *Referativnyi Zhurnal* [Nauka].) Part of a series of abstracts and indexes covering hundreds of subjects, this section comprehensively covers astronomy and astrophysics, as well as space physics and geophysics. It covers all the major Western astronomical periodicals, and dozens of others throughout the world. Entries, arranged according to a heirarchical subject classification, include bibliographic citation in the original language and an abstract in French. A typical issue has 500 to 600 entries, complete with subject, author, and geographic indexes, all cumulated annually. Its coverage overlaps to a large extent with the coverage of similar publications on astronomy, and the librarian should carefully consider the budget before deciding to purchase. Large university libraries and some astronomy libraries serving foreign scientists will most likely benefit from this indexing service.

Write: Centre de Documentation, Informascience, 26, rue Boyer, 75971, Paris Cedex 20. Cost: 365 F/year.

14. **Current Contents: Physical and Chemical Sciences.** Philadelphia: Institute for Scientific Information. v.1– , 1961– . Weekly. index. ISSN 0011-3417.

The astronomer who needs a convenient way to keep up with the journal literature should leaf through this handy publication. Covering the same serials as *Science Citation Index*, including about 30 astronomical publications, this section of *CC*, like all others, consists of photographically reproduced journal title pages. The issues covered in this section include selections from chemistry, physics, mathematics, astronomy, etc. Access to information is accomplished in one of several ways: articles are arranged within subject sections (astronomy is under Space Sciences), there is an alphabetical list of journal titles, and there is a subject index. An added feature is the Author Address Directory, a computer-produced index to the authors in each issue of *CC*, which facilitates writing for a reprint. *Current Contents* is probably not used as frequently as it should be, which is an unfortunate situation for so valuable a reference tool. It is expensive, but it could possibly be substituted for several little-used serial titles in the library.

Write: Institute for Scientific Information, 325 Chestnut, Philadelphia, Pennsylvania 19106. Cost: $135/year domestic; $145/year foreign.

15. **Current Physics Index.** New York: American Institute of Physics. v.1– , 1975– . quarterly. abstracts. indexes. ISSN 0098-9819.

A secondary index to astronomical journal articles, *CPI* is a primary source for physics periodicals, covering 42 serials published by AIP. Of these titles, only three pertain directly to astronomy: *Astronomical Journal, Soviet Astronomy,* and *Soviet Astronomy Letters*. The purchase of this index, then, is questionable, especially for those astronomy libraries short on funds. On the other hand, the importance of the physics literature to astronomers would support an argument in favor of having this title in the observatory library. Its large overlap with *Physics Abstracts* (INSPEC), however, must also be considered.

One advantage of *CPI* is that it is extremely timely; a quick check of a 1976 issue revealed several entries only one month old. It is arranged according to the AIP Physics and Astronomy Classification Scheme, and entries are also accessible by an alphabetic subject index, as well as an author index. Citations include full bibliographic information, an author abstract, and miscellaneous information like Current Physics Microform number. Entries with more than one major theme are listed in the text under the most important heading, with cross references under other topics. Each issue contains around 4,000 abstracts, and readers can order copies of the papers indexed for $0.25 a page ($10.00 minimum). About nine-tenths of all United States physics journals and one-half of all Soviet publications (translated by AIP) are covered, so the scope is extensive. The fifth and sixth issue of each volume are cumulative subject and author indexes, respectively. It supersedes *Current Physics Advance Abstracts* and *Current Physics Titles*.

Write: American Institute of Physics, 335 E. 45th St., New York, N.Y. 10017. Cost: $30/year (U.S. members); $95/year (U.S. non-members). Foreign rates are slightly higher.

16. **Mineralogical Abstracts**. London: The Mineralogical Society of Great Britain, The Mineralogical Society of America, v.1– , 1920– . Four/yr. abstracts. index. LC 29-1702. ISSN 0026-4601.

Students and astronomers studying meteorites and outerspace geology should be familiar with this scholarly indexing journal, which comprehensively covers the literature. Of the 17 subject divisions, two are specifically astronomical: lunar studies and meteorites and tektites. The former frequently contains references to articles reporting on the analysis of Apollo lunar samples, theoretical studies on lunar formation, and geologic history. No doubt articles on Martian topics will soon appear in this section. The other subject division contains citations to papers on the chemical composition of meteorites and tektites, as well as their classification and origin. References to information on the origin of the solar system as related to meteoric studies can be found here as well. Data on meteorite falls and finds are included, too. Entries consist of an abstract number, full bibliographic citation (title in original language), and an English abstract of varying lengths. One can expect to find 20 to 60 abstracts per issue on astronomical subjects, drawn from journals in astronomy, geology, chemistry, and general science. Overall, the journal is concerned with petrology, mineral data, geochemistry, gemstones, physical properties of rocks, and other geologic concerns. Other features include an adequate book section, issue author indexes, and a cumulative author/subject index each year.

Write: Publications Manager, Mineralogical Society, 41 Queen's Gate, London, SW7 5HR. Cost: $50.00 (£20.00)/year.

17. **Physics Abstracts**. London: Institution of Electrical Engineers. v.69– , 1966– . Bi-weekly. abstracts. indexes. ISSN 0036-8091.

A primary source of reference in astronomical literature, this publication is one of several abstracting and indexing journals published by INSPEC. It covers physics comprehensively and includes a substantial section on astronomy and astrophysics. Covering books, reports, journals, dissertations, patents, and conference proceedings, world wide, it contains approximately 85,000 items each year in all areas. Entries include pertinent bibliographic data and a brief abstract, and each issue has a subject and author index, which are accumulated semi-annually. A further help in searching are the cross references to related citations under most subject categories. The average time delay for articles to appear in this index is about four months. This excellent journal superseded *Science Abstracts, Series A,* in January 1966, and continued its numbering.

Write: Fulfillment Manager, IEEE, 345 E. 47th Street, New York, N.Y., 10017. Cost: Paper or fiche: $430.00/year; paper and fiche: $645.00/year.

18.   **Referativnyi Zhurnal: Astronomiia.** Moscow: VINITI. 1963– . Monthly. abstracts. indexes. LC 63-55191. ISSN 0486-2236.

One of the primary worldwide sources of astronomical information, this Russian publication is just one section of a large series of abstracting and indexing journals. The coverage is quite comprehensive, and contains the following subject areas: General; Astronomical instruments; Observatories; Theoretical astronomy; Astrometry; General problems of astrophysics; Solar system; The Sun; Stars; Nebulae and interstellar matter; Galactic astronomy; and Extragalactic astronomy. Most articles indexed are from Western astronomical journals, making *Referativnyi Zhurnal: Astronomiia* potentially very useful outside the Soviet Union. Unfortunately, the number of users in the United States is small, mainly because the serial is in Russian.

It is mainly found in large university libraries and some observatories, where librarians often use it to track down obscure citations. A typical issue has about 800 to 1,000 abstracts over 100-125 pages. Each entry has the title in Russian and original language, the citation in the original language, and a Russian language abstract. A fully-translated edition would be an important contribution to the literature, even though such a publication would not be timely. On the other hand, *Astronomy and Astrophysics Abstracts,* the comparable English counterpart, is not very timely itself. There are annual subject indexes in Russian, as well as author indexes. A related journal of interest is *Referativnyi Zhurnal: Fizika* (physics literature), also published by VINITI. *Astronomiia* continues in part *Referativnyi Zhurnal: Astronomiia i geodeziia,* v.1-9, 1953-62.

Available through Les Livres Éstranger, 10, Rue Armand-Moisant, 75737 Paris Cedex 15. Cost: About $60.00/year.

19.   **Science Citation Index.** Philadelphia: Institute for Scientific Information. 1961– . Quarterly. LC 63-23334. ISSN 0036-827X.

One of the most comprehensive and powerful indexes ever published, SCI covers over 2,500 periodicals, worldwide. Divided into three sections, Source Index, Citation Index, and Permuterm Subject Index, it includes 30 astronomy and astrophysics publications in its coverage. The first section (Source Index) is a corporate and personal author guide to articles in the so-called "source journals." Anonymous papers are arranged by journal title, then article title. Each entry lists co-authors, journal title, title of article, volume number, pagination, year, and number of citations with the paper. The latter feature is a rare and highly useful item. The Citation Index is a unique guide which lists authors not by articles they wrote, but by other authors who cited the former group in later published articles. This not only shows how often a particular author and his papers were cited, but gives the searcher immediate access to related articles by different authors on the same topic. One can go from cited author to source author to cited authors to source authors, etc., in a cascading effect, thus linking dozens of citations.

Although astronomers and their librarians will make good use of the two aforementioned sections, the third part, the Permuterm Subject Index, may be the most helpful. Here, under a subject word like astrophysics or Moon, there are references to authors, who are then looked up in the Source Index. General terms list more than one subheading, for finer topical searches (e.g., Moon—Magnetism). The cost will prevent acquisition by most smaller libraries, but college libraries will usually carry the subscription. The astronomer and astronomy student should be familiar with and know how to use this excellent publication.

Write: ISI, 325 Chestnut St., Philadelphia, Pa. 19106, or ISI, 132 High Street, Uxbridge, Middlesex, England. Cost: $1800/year for parts 1 and 2; $1050/year for the subject index.

Technical Reports

The abstracting and indexing (a/i) journals go beyond the coverage of periodicals to include the technical report literature; several a/i publications cover only such documents. A technical report is a record of a research and development (R & D) project completed or in progress. These so-called tech reports are rarely published formally and are usually not refereed; they are often highly technical, detailed documents of limited distribution and interest. Technical reports are issued by private companies, government agencies, or non-profit institutions like universities. Tens of thousands of such reports are issued each year, mainly from government sources.

The National Aeronautics and Space Administration is a significant producer of technical reports and is the largest source, not surprisingly, of astronomical report literature. These documents, produced directly by NASA or by their many subcontractors, are indexed by *Scientific and Technical Aerospace Reports* (*STAR*), an a/i journal which covers aerospace information from NASA and other sources. From an astronomy standpoint, the NASA SP (special publications) series of technical reports is the most important. Geared mainly toward the public and students, this series reports on the space program and astronomical research carried out by NASA. Many of these SP's are coffee-table type books with scores of beautiful photos. Any library with an astronomy collection of any size will surely have a few from this fine group of books. They are excellent publications with well-written texts and superb illustrations.

A related publication is *International Aerospace Abstracts*, which covers books, journals, and proceedings worldwide in the aerospace fields. Although it doesn't specifically cover the technical report information, it is included here because of its close connection to the information described in *STAR*.

A more general index to United States government tech reports is *Government Reports Announcements & Index*, which cites astronomy reports from NASA and other government sources, including those that are aerospace-related as well as others. It should be considered a secondary source of astronomical reports, however.

Of the three, *STAR* would be most essential to the observatory library, and large research libraries will probably have all three. There is, unfortunately, no comprehensive government reports index, but the librarian can safely rely on the three sources above to handle most requests.

20.   **Government Reports Announcements & Index.** Springfield, Va.: U.S. Department of Commerce, National Technical Information Service (NTIS). v.74– , 1974– . Bi-weekly. abstracts.

The primary source of information on U.S. government technical reports, this index should be considered only as a secondary source of astronomical data. Research and development reports on astronomy, astrophysics, and celestial mechanics are found in each issue in Field (section) 3. Each citation includes report numbers, price, authors, title, pagination, date, and descriptors; non-NASA citations include an abstract. The major drawback to using *GRA&I* to find astronomical tech reports is that the citations are mainly to NASA documents already indexed in *Scientific and Technical Aerospace Reports (STAR)*. *GRA&I*, therefore, should be consulted only after trying *STAR*. It is doubtful that the astronomy librarian would want or need this excellent index, but university libraries surely would, for it covers all of science and technology, as well as the behavioral and social sciences. Any non-classified government R & D publication is listed in this guide, which has subject, author, and report number indexes for each issue and annually. Despite its being a secondary source for government astronomical reports, astronomers and astronomy librarians should become familiar with it, since it covers many related fields as well. Continues *Government Reports Announcements* and *Government Reports Index*, which continued *U.S. Government Research & Development Reports* and its index. Write: U.S. Department of Commerce, National Technical Information Service, 5285 Port Royal Road, Springfield, Virginia 22161. Cost: U.S.: $165/year; Canada and Mexico: $180/year; Foreign: $200/year.

21.   **International Aerospace Abstracts.** New York: American Institute of Aeronautics and Astronautics. v.1– , 1961– . Semi-monthly. abstracts. indexes. LC 65-56077. ISSN 0020-5842.

This companion to *Scientific and Technical Aerospace Reports (STAR)* covers the world's published literature in aeronautics, space science, and technology. It includes citations to periodicals, books, conference proceedings, and journal and journal article translations. It does not contain references to the technical report literature; the reader should consult *STAR* for this information. The presentation of the abstracts and bibliographic information is identical to that used in *STAR*, and the subject categories are the same, too.

An important abstracting publication, it should be held by most observatory libraries and by any other library that can afford it. References to astronomical literature are found here under Space Sciences, as in *STAR*.

Write: American Institute of Aeronautics and Astronautics, Technical Information Service, 750 Third Avenue, New York, New York 10017. Cost: $400.00/year (domestic); $550.00/year (foreign). Indexes: $300.00/year (domestic); $400.00/year (foreign). Combination: $550.00/year (domestic); $800.00/year (foreign).

22.   **Scientific and Technical Aerospace Reports (STAR)**. Washington, D.C.: NASA. v.1– , 1965– . Semi-monthly. abstracts. indexes. LC 64-39060. ISSN 0036-8741.

Although this abstracting journal consists mainly of information about technical reports in astronautics, space sciences, and supporting disciplines, it also includes references to the astronomical report literature. Entries on astronomy, astrophysics, lunar and planetary exploration, solar physics, and space radiation can be found in *STAR* under the heading "Space Sciences." Each citation contains a great deal of information describing the listed document: document number, corporate source, authors, title, availability, cost, report numbers, subject codes, contract or grant numbers, and an abstract. Each issue has a variety of indexes, and cumulations are published twice a year: subject, personal author, corporate source, contract number, and report/ accession number indexes.

STAR contains references to the following type of publications: "NASA, NASA contractor, and NASA grant reports, reports issued by other U.S. Government agencies, domestic and foreign institutions, universities, and private firms, translations in report form, NASA FEDD . . . documents, NASA-owned patents and patent applications, dissertations and theses." Also, "a separate section of information on aerospace-related *On-Going Research Projects* is inserted into each issue of *STAR*. The insert presents titles of active NASA grants and university contracts, summary portions of recently updated *NASA Research and Technology Operating Plans* (RTOP's), and notices of non-NASA research projects that were funded in the most recent or current fiscal year."

An extremely valuable source, especially for information on astronomical applications, this abstracting tool should be in all observatory libraries, as well as in many university and large public libraries.

STAR also provides a Selective Dissemination of Information (SDI) service called *NASA SCAN*. Individuals working under NASA contracts, researchers, etc., can subscribe to this service, which selectively scans each issue of *STAR* to find reports of interest on particular topics. The subscriber fills out a profile, a list of important subject terms, and this list is used to extract reports, which are then sent on a regular basis to users. Not only does this free the researcher from having to search *STAR* every week, but it also does a better job of retrieving the information than most manual searches can. More details on this computerized SDI service can be found in any issue of *STAR*. It is especially useful to astronomers working on research projects and to librarians who may be preparing bibliographies.

For a subscription to *STAR* write: Superintendent of Documents, U.S. Government Printing Office, Washington, D.C., 20402. Annual Cost: $66.90 (U.S.); $83.65 (foreign); indexes: $28.10 (U.S.); $35.15 (foreign).

TRANSLATIONS

Astronomy librarians are occasionally asked for an English translation of a foreign journal article. Fortunately, these occurrences are fairly rare because English is becoming the standard scientific language, and many foreign astronomical publications are published only or mainly in English or have an English edition (e.g., *Acta Astronomica* (Poland), *Bulletin of the Astronomical Institutes of Czechoslovakia, Publications of the Astronomical Society of Japan*). Further, three important Soviet journals are being regularly translated: *Soviet Astronomy* and *Soviet Astronomy Letters* (American Institute of Physics), and *Astrophysics* (Plenum). Consequently, a major portion of the important foreign astronomical periodicals are being translated already.

Nevertheless, occasions arise when a translation is needed, and the librarian should first consult *Translations Register Index* and *World Index of Scientific Translations*. These two publications list a handful of astronomical translations in each issue, the majority from Eastern Europe and the Far East. How to obtain these translations is spelled out in detail in these serials. The former source is much more helpful than the latter, which lists only a small number of astronomical translations. Failing to find the article listed in these indexes leaves the librarian little choice but to try to find someone who will translate the paper at a justifiably exorbitant price.

Foreign astronomy monographs are frequently translated into English, but often several years after the original publication. Many translated works are listed in this guide. Numerous Russian astronomy texts have been translated by the Israel Program for Scientific Translations (Jerusalem), and these are sometimes available through the National Technical Information Service (NTIS) in Springfield, Virginia, NASA, and John Wiley and Sons (New York).

23.   **Translations Register-Index**. Chicago: National Translations Center. v.1– , 1967– . Monthly. LC 74-16502. ISSN 0041-1256.

While not vitally important to most astronomy libraries, this publication could conceivably be of great use to certain institutions. Included here are citations to unpublished English translations of the world's literature in the natural, physical, medical, and social sciences which have been received by the National Translations Center. The "register" section of the serial includes the citations arranged by subject; a category called Astronomy and Astrophysics includes celestial mechanics. Each entry provides very brief bibliographic data, enough to identify and order, as well as availability and price. Most documents listed can be obtained from the NTC. The "index" section is in two parts: journal citation arrangement and patent citations by country. The publication cumulates semi-annually. There is not too much listed under Astronomy and Astrophysics, so it might be difficult to justify purchase. The other problem is that translating is not generally done on a systematic basis, so there is no way to predict what articles will appear in the pages of this journal. Nevertheless, an astronomy library may be able to make great use of this publication, depending on the activities of the observatory it supports. To obtain, write: National Translations Center, John Crerar Library, 35 West 33rd St., Chicago, Illinois 60616. Cost: $50.00 (U.S. and Canada); $55.00 (elsewhere).

A similar publication the librarian should know about is the *World Index of Scientific Translations*, from the European Translations Centre in The Netherlands, which lists non-Western language materials translated into Western languages. This source, however, contains very few references to astronomical translations, and its value to the observatory librarian is questionable.

DIRECTORIES AND BIOGRAPHICAL INFORMATION SOURCES

Quick access to people, places, and things is an appropriate description of the items listed in this section. Most of the books presented are not devoted entirely to astronomy, but instead are more general scientific works with a section or sections devoted to astronomical information. Many astronomical reference works are not purely astronomical, then, and there are some illustrative examples in this section.

Both current and retrospective biographical sources are included, each, of course, serving a different purpose. The retrospective sources are often useful for writing term papers on famous astronomers, while the former, the current directories of people, are useful for letter-writing, introductions of speakers, etc. Unfortunately, the biographical directories are never comprehensive, and they often include only famous scientists. What is needed is an annual biographical handbook of astronomers, including more than just scientists who are faculty members. At least one of the items below lists only faculty-type astronomers.

Directories of observatories are lacking, too. Although the guide to radio astronomy installations is updated regularly, there is no overall directory of observatories that is updated annually. A solution to this problem, and one mentioned above, would be to expand and update annually the excellent but "old" *International Physics and Astronomy Directory*. The literature of astronomy would be greatly improved if such a task were undertaken. There is, fortunately, an excellent directory aimed at the amateur astronomer and general reader, *U.S. Observatories: A Directory and Travel Guide* (1976), which provides adequate coverage of a non-technical nature.

A directory of astronomy libraries and details on their collections would be a welcome addition to the reference literature as well, at least in the eyes of the astronomy librarian. For the time being, though, the *Directory of Special Libraries and Information Centers* serves the purpose fairly adequately.

24.   **American Men and Women of Science**. New York and London: R. R. Bowker Company. 1906– . Irregular. LC 6-7326. ISSN 0065-9347. $50.00/vol.; $300.00/set. ISBN: v.1: 0-8352-0866-4; v. 2: 0-8352-0867-2; v.3: 0-8352-0868-0; v.4: 0-8352-0869-9; v.5: 0-8352-0870-2; v.6: 0-8352-0871-1; v.7: 0-8352-0872-9 (index).

An indispensable biographical reference source, this seven-volume set contains over 110,000 entries in a who's who format. Now in its 13th edition (1976), edited by the Jaques Cattell Press, the work is a gold-mine of information on astronomers and other scientists. Each entry has the following data: birthplace and date, marital status, major subject areas, educational and employment history, special interests, current position, title, address, and professional memberships. There are cross references under some names to the previous (12th) edition if new information could not be obtained for this edition. The index volume, which consists of discipline and geographic breakdowns, is especially useful. Its lengthy section on astronomy lists over 1,200 astronomers under the following subheadings according to the scientists' interests: astronomy, astrogeology, cosmology, planetary atmospheres, radio astronomy, x-ray astronomy, cosmochemistry, astrophysics, celestial mechanics, and theoretical astrophysics. Since there is no biographical dictionary of astronomers, this series becomes the best source of information for the observatory librarian. It is good but not totally comprehensive; the coverage is selective (only well-known scientists are listed), and it is also dependent on the scientists themselves, who must voluntarily supply the data for the book.

25.   **Directory of Physics and Astronomy Staff Members**. New York: American Institute of Physics. 1– , 1959/60– . Annual. $15.00. LC 72-626731. ISBN 0-88318-208-4 (17th ed.).

An essential book for both university and non-academic astronomy libraries, this directory lists approximately 18,000 physics and astronomy personnel at 2,300 institutions, including North American colleges and universities, and federally funded research and development centers. The first section of the guide includes geographic listings of institutions and faculty for the United States, Mexico, Canada, and Central America. The next two parts are alphabetical lists of academic and research center staff, and colleges and universities, respectively. Each entry gives the address and phone number of the department of physics and/or astronomy at a particular institution, the list of faculty and staff, etc. It does not include biographical data; it is merely a checklist of sorts. The appendices provide statistical data on the various types of institutions represented in the directory. This is a worthwhile purchase for observatory and academic libraries.

26.   Fisk, Margaret, ed. **Encyclopedia of Associations**. 9th ed. Detroit, Gale Research Co., 1975. 3v. LC 74-22265. v.1: National Associations of the United States. $55.00. ISBN 0-8103-0126-1. v.2: Geographic-Executive Index. $38.00. ISBN 0-8103-0131-8. v.3: New Associations (quarterly, loose-leaf). $48.00. ISBN 0-8103-0130-X.

An excellent all-purpose directory of organizations which includes ten references to major astronomical associations. Data for each entry include name, acronym, address, phone numbers, chief official, date founded, membership, staff, subgroups, descriptions, meetings, publications, sections, committees, mergers, and name changes. Arrangement is by the keyword in the title (or one assigned by the editor, if there is no word in the title describing the subject of the organization). The references to astronomical organizations, including both amateur and professional groups, are listed in section 4: "Scientific, Engineering, and Technical Organizations." Larger public and academic libraries will want to acquire this.

27.   Gillispie, Charles Coulston, ed. **Dictionary of Scientific Biography**. New York, Charles Scribner's Sons, 1970-76. 14v. diagr. bibliog. $40.00ea. LC 69-18090. ISBN 0-684-10122-X.

For reliable, well-written biographical sketches of famous astronomers of the past, this is *the* source to consult. In fact, there is nothing else in its class. Begun in 1970, this encyclopedic work is international in scope, covering those significant individuals whose contributions would fall into the present-day categories of mathematics, astronomy, physics, chemistry, biology, and earth sciences. The scholarly yet highly readable articles include descriptions primarily of the individual's contributions in science, with little or no emphasis on personal life. The excellent accompanying bibliographies, however, point to other works containing this information, as well as to other original works by the scientists.

Astronomers are well represented, and rightly so, because of their many achievements in the physical sciences. Among the "star-gazers" included are Copernicus, Ptolemy, Kepler, Brahe, Galileo, Newton, and many, many more. The depth of the pieces depends on the importance of the person and his or her contributions. There is no other work available of this magnitude and quality devoted to scientific biography, and the serious scholar and librarian will want to be familiar with it.

28.   **International Physics & Astronomy Directory, 1969-70**. New York, W.A. Benjamin, 1969. 802p. $35.00; $17.50pa. LC 73-99508. ISBN 0-8053-0377-4; 0-8053-0378-2pa.

This voluminous directory contains a wealth of vital information on a variety of subjects. According to the introduction: "The purpose of this volume is to provide, for the first time, a unified, comprehensive, and up-to-date directory containing easily accessible facts that answer most of the professional reference needs of physicists and astronomers." Although no longer up to date, the work is an excellent effort, worthy of most astronomers' reference shelves. A listing of the table of contents will indicate the comprehensiveness of the text: Academic Departments and Faculties; Faculty Index; Geographical Index of Universities and Colleges; Laboratories; International Organizations; Societies; Meetings; Meetings Calendar; Grants and Fellowships; Graduate Support; Awards; Research in Science Education; Journals; Books in Print; Directory of Publishers. Much of this information could be obtained elsewhere, but the fact that it is all in one place makes this volume quite attractive. A new edition is desperately needed, since a large portion of the information is so outdated that it is virtually useless. Revisions on a yearly basis would constitute a significant contribution to the literature of astronomy.

29.   Kirby-Smith, H. T. **U.S. Observatories: A Directory and Travel Guide**. New York, Van Nostrand Reinhold Company, 1976. 173p. illus. index. bibliog. $11.95; $6.95pa. LC 76-4448. ISBN 0-442-24451-7; 0-442-24450-9pa.

Presenting information on over 300 U.S. astronomical observatories, this fine volume is not a "directory" in the strictest sense. More than just a list of names and addresses, *U.S. Observatories* provides the reader with some fascinating stories of the great telescopes and the astronomers who used them. In fact, the first half of the guide is devoted to detailed descriptions of fifteen major observatories: Harvard College, Smithsonian Astrophysical, U.S. Naval, Leander McCormick, National Radio Astronomy, Allegheny, Yerkes, McDonald, Sacramento Peak, Kitt Peak National, Lowell, U.S. Naval (Flagstaff), Mount Wilson, Palomar Mountain, and Lick observatories. Historical background, descriptions of research and equipment, and best of all, information for visitors are given for each. Sections differ slightly with respect to the historical background and the descriptions of the physical sites. Besides the informative summaries, there is an abundance of information for the astronomical tourist; included are address, hours of visitation, tours, nearby accommodations, and nearby scientific attractions. Helpful hints (such as don't try to pull a trailer up the mountain roads to the Hale Observatories, or don't take children under five years of age to any observatory) make the text both interesting and highly useful.

The second half of the book, the "directory," contains brief entries for about 280 other institutions. Arranged by state, then city, this part mainly includes visitors' information, but historical background and descriptions of the telescopes are given for some of the more important observatories. In addition to radio and optical observatories, the author also includes planetaria, making the book even more valuable. Useful for any library, it should also be on the serious amateur's bookshelf, for casual reading, reference, and vacation planning. Nothing like it exists in the literature; a comprehensive guide for the non-scientist has long been needed.

38 / *Reference Sources in Astronomy*

30. **List of Radio and Radar Astronomy Observatories.** Washington, D.C., National Academy of Sciences, National Academy of Engineering, Committee on Radio Frequencies, 1970– . Irregular.

This pamphlet contains valuable information on the world's radio and radar astronomy installations. The booklet is divided into U. S. and foreign observatories, and gives specific information about each radio telescope: location, information officer, sponsors, type, size, height, sky coverage, collecting area, polarization, and other miscellaneous data. Additionally, there are summaries of frequencies being monitored for radio astronomy observations, as well as frequencies being used for radar astronomy observations. Useful for any astronomy library or science information center, this work provides fairly up-to-date information on radio telescopes, something not easily available for their optical counterparts. For further information, the reader is referred to Thornton Page's *Observatories of the World* and Henry Tompkins Kirby-Smith's *U.S. Observatories: A Directory and Travel Guide*.

31. **McGraw-Hill Modern Men of Science.** New York, McGraw-Hill, 1966, 1968. 2v. illus. index. bibliog. LC 66-14808.

Although slightly dated, this work is a primary source of biographical information on contemporary scientists, including 42 astronomers and astrophysicists. The majority of entries, ranging from one to four pages, were written by the subjects themselves. The articles are far more useful than the brief entries in a who's who type work, since the information describes the scientist's work in addition to providing biographical sketches. Space being limited, however, there are far fewer entries (about 700) than in *American Men and Women of Science*. The biographees here are the most important in their fields.

A kind of biographical supplement to the *McGraw-Hill Encyclopedia of Science and Technology*, this work's entries contain cross references to subject areas in the aforementioned set, useful for the individual who wishes to read more on a particular subject. Further, the work can also be used to identify scientists connected with specific concepts or general subject fields; the excellent indexes provide this access. Librarians should be well aware of this item, for astronomical biographical information as well as for general science information. This two-volume set includes a drawing of each scientist with the sketch. For all types of libraries.

32. **1976-77 Graduate Programs in Physics, Astronomy, and Related Fields.** New York, American Institute of Physics, 1977. $10.00.

Updating a 1971 edition of the same work, this book remains the authoritative guide to graduate work in astronomy and physics. Containing information on over 200 doctoral and 300 master's degree programs, it will be of great use to prospective graduate students and faculty advisors. Mainly a geographical (U.S. and Canada) arrangement of the aforementioned graduate programs, it also includes related summaries of doctoral program research specialties, Ph.D. programs in related fields, and an index of graduate programs by institution, arranged geographically. Entries include information on the graduate faculty, research specialties of staff, course requirements, financial aid, number of students, etc. Other useful information is physics and astronomy manpower statistics, research expenditures, and admission requirements. For the university and observatory library, it is a handbook worth having on the reference shelf.

33. Page, Thornton. **Observatories of the World**. Cambridge, Mass., Smithsonian Astrophysical Observatory, 1967. 41p. refs. LC 67-8792.

This brief guide concerns itself with astronomical observatories, both optical and radio, and does not include meteorological observatories, seismic observatories, and the like. The latter half of this paperbound book is a listing of the world's observatories as of 1967, arranged by country under two major headings, optical observatories and radio observatories. A good source of quick reference for the science library (and many public libraries), its entries contain 1) the name of the observatory, date founded, and city; 2) a list of telescopes and the size of mirrors and/or lenses, or size of dishes, arrays, etc. for radio instruments; 3) major observing programs. Telescopes planned for the future are listed if they were on the drawing board by January 1, 1967.

The introductory matter includes explanations of various observatory equipment and observing programs, as well as the history of observatories. Unfortunately, this material is painfully brief, since it was written as part of an encyclopedia article. The reader interested in a better treatment of the history of observatories should look elsewhere—Donnelly's *A Short History of Observatories* would be a good starting point. Certain encyclopedias like *McGraw-Hill Encyclopedia of Science and Technology* will better cover observatory equipment, as would any number of general texts.

The usefulness of this guide is seriously impeded by its age—a new edition is desperately needed. In the meantime, this edition will suffice.

34. Young, Margaret L., *et al.* **Directory of Special Libraries and Information Centers**. 3rd ed. Detroit, Gale Research Co., 1974. LC 73-3240. v.1, Directory of Special Libraries and Information Centers in the U.S. and Canada. $55.00. ISBN 0-8103-0279-9. v.2, Geographic-Personnel Index. $35.00. ISBN 0-8103-0180-2. v.3, New Libraries (quarterly, loose-leaf). $57.50. ISBN 0-8103-0281-0.

A potentially useful reference work for the astronomy librarian, this work details the locations of the major collections of astronomical literature (and other subjects) in North America. Observatory, college, public, and special libraries that have substantial astronomy collections are all listed in this guide. The overall arrangement of the book is alphabetical by name of the library; university collections of astronomy materials are listed under the name of the institution. Entries include address, telephone, librarian, number of staff, date founded, subjects, number of volumes, special collections. The subject index in this edition is a slight improvement over the old version. Under each broad category, such as astronomy (69 entries), is a list of numbers that refer to entries somewhere in the book. The new arrangement breaks down this sometimes large set of numbers by state and province, which gives the searcher much more direction than previously was given. However, a list or directory of astronomy libraries, with greater detail than appears here, is greatly needed. There are several subject divisions that include astronomical collections: astronomy (69); astrophysics (27); lunar science (2); radio astronomy (8); and others.

DICTIONARIES

Both foreign language and definitional dictionaries are represented in the literature of astronomy, although their numbers are not too great. It should be noted that there is really little difference in the definitional type works listed in this section and the so-called one-volume "encyclopedias" listed elsewhere. The definitions of terms in the encyclopedias may or may not be broader in scope, so it really doesn't make much difference which type book is consulted. More often than not, the choice of "encyclopedia" over "dictionary" or vice-versa is purely arbitrary.

35.   Chiu, Hong-Yee, ed. **Chinese-English, English-Chinese Astronomical Dictionary**. New York, Consultants Bureau, 1966. 173p. $20.00. LC 65-10966. ISBN 0-306-10739-2.

This rather specialized tool would not be appropriate for most astronomy libraries, but it would be desirable for any institution that has Chinese faculty and graduate students, or where there are studies involving the use of Chinese astronomical literature. Except for a few minor changes, the first half of this book is reproduction of the Chinese-English section of the *Chinese-Russian-English Astronomical Nomenclature* published in 1959 in Peking. This is followed by an English-Chinese part, which might be helpful for cross-reference purposes and for helping Chinese astronomers to read the English astronomical literature.

36.   Hopkins, Jeanne. **Glossary of Astronomy and Astrophysics**. Chicago, University of Chicago Press, 1976. 169p. $10.95. LC 75-14799. ISBN 0-226-35172-6.

Students and librarians will benefit most from this excellent guide to astronomical terminology, which substantially updates previous similar works (e.g., Åke Wallenquist's *Dictionary of Astronomical Terms*, 1966). Arranged alphabetically, the approximately 2,000 terms include single subject word definitions, phrases, descriptions of particular celestial objects (stars, planets, comets, etc.), astronomical and physical constants, and selected related terminology from physics. Most entries are two or three lines long (which is sufficient), but certain definitions are much longer; for example, for each of the planets there is a long paragraph listing its basic physical characteristics and numerical quantities, such as orbital period, mass, surface temperature, etc. The book is well written, but is not unique. It does, however, include many new terms missing from previous astronomical dictionaries, and therein lies its value. The astronomer might also wish to use this book when beginning work in a new, unfamiliar field. The definitions are neither too simple nor too technical, so the volume is ideal for any library with a science collection.

37.   Kleczek, Josip. **Astronomical Dictionary: In Six Languages.** Praha, Nakladatelství Československé academie věd; distr. New York, Academic Press, 1961. 972p. $48.50. LC 62-2177. ISBN 0-12-411950-6.

Not a book that defines terms, this is an equivalency-type dictionary, in which words are merely translated from one language to another. As such, this work would be extremely useful in translating astronomical literature from or to one of the following languages: English, Russian, German, French, Italian, and Czech. All terms listed were found in the literature, so it is unlikely that many words have been overlooked. A revised edition would be highly desirable, however, since the "new" astronomies (radio, x-ray, infrared, etc.) have added dozens of new terms to the vocabulary. Still, probably 90 percent of astronomical terminology is presented here. The dictionary is divided into two parts; the first is a subject arrangement (i.e., all related terms are together), and the second is an alphabetical list for each language. An essential part of any astronomical reference collection, it includes phrases from related disciplines like mathematics and physics, too. The user is also directed to volume 9 (1964) of the *Encyclopaedic Dictionary of Physics*, which is a multi-lingual dictionary that contains many astronomical terms and has Spanish and Japanese entries not found in this work.

38.   Lapedes, Daniel N., ed. **McGraw-Hill Dictionary of Scientific Terms.** New York, McGraw-Hill Book Co., 1974. 1662p. illus. $39.95. LC 74-16193. ISBN 0-07-045257-1.

Even though this dictionary is extremely comprehensive, spanning dozens of disciplines, it should not be considered a prime source of astronomical definitions. Hundreds of entries are included for astronomy, but they are painfully brief and frequently inadequate. The novice should especially beware— often terms are defined using other words that may or may not be familiar to the reader. "Quasar," for example, is described as having large red shifts. The amateur may or may not know the phrase, "red shift," and might be confused when consulting this dictionary's definition of it. In short, the dictionary is mainly for college science students, scientists, and advanced amateurs, and not for laymen. (This criticism, though true in the area of astronomy, may not be valid for other areas.)

Each entry includes a short definition and a word or phrase indicating the subject field from which a term was taken. No attempt is made to show pronunciation, derivation, or syllabication. Readers seeking good astronomical or astrophysical definitions should look elsewhere—for example, in one of the encyclopedias listed in the next section. For very, very short explanations, though, this work will suffice.

39. Wallenquist, Åke. **Dictionary of Astronomical Terms.** Garden City, New York, Natural History Press, 1966. 265p. illus. (American Museum Science Books, B12). LC 66-12201.

Though a bit outdated, this little book is an excellent source for astronomical terms. The definitions for the more than 1,700 entries are clear and concise, and many are illustrated with diagrams. Besides subject terms, the dictionary includes brief entries for famous astronomers and observatories. Well-known stars (e.g., Vega) and constellations are listed as well. Although best suited for the layman, this volume can also be used by astronomy and physics students for quick reference. One of the best available, although now out of print, this dictionary should be part of any library reference collection. Translated by Sune Engelbrektson (*Astronomiskt Lexikon*).

ENCYCLOPEDIAS AND ENCYCLOPEDIC SETS

The encyclopedias listed here are one-volume publications, with one exception, which would be suitable for all types of libraries having even the smallest astronomy collection. Although a few of the works below differ only slightly from the definitional dictionaries listed elsewhere in this guide, several go far beyond the short definition for the topics covered. Of special note are Satterthwaite's *Encyclopedia of Astronomy* and Weigert and Zimmerman's *A Concise Encyclopedia of Astronomy*, both of which include excellent illustrations, fairly long explanations, and comprehensive topical coverage.

The two general scientific encyclopedias included below are worth mentioning since they complement so well the astronomical works that share the same purpose. The one-volume *Van Nostrand's Scientific Encyclopedia* is most useful as an astronomical source in a public library that may not be able to purchase an astronomical encyclopedia, or as an "other discipline" source in an astronomy library. The *McGraw-Hill Encyclopedia of Science and Technology* has the longest articles of any encyclopedia listed, but because of its general nature, many secondary but important topics are excluded. It is the best general scientific work of its kind, however, and all astronomy libraries should have it.

The chief problem with encyclopedic works is that they go out of date quickly (like other reference works), so they lack the most current data. Fortunately, the basic information stays the same, making these handbooks useful for a number of years.

40. Bizony, M. T., ed. **The New Space Encyclopedia: A Guide to Astronomy and Space Exploration.** New York, E. P. Dutton & Co., Inc., 1973. 326p. illus. $14.95. LC 77-77915. ISBN 0-525-16629-7.

Because this volume covers both astronomy and space science, it has a wider appeal than books concentrating only on the former. Consequently, this book would be an excellent selection for many types of libraries, from secondary school to special. Subjects are arranged alphabetically and are frequently illustrated with excellent diagrams and color plates. The descriptive articles are usually short and to the point, and they provide interesting and complete explanations. Entries were written by various contributors, usually well-known scientists. A handy volume for quick reference questions.

41.   Fairbridge, Rhodes W., ed. **The Encyclopedia of Atmospheric Sciences and Astrogeology**. New York, Van Nostrand Reinhold Publishing Co., 1967. 1200p. illus. index. refs. (Encyclopedia of Earth Sciences Series, V. 2). $39.95 LC 66-26059. ISBN 0-442-15071-7.

Although the majority of the information in this massive volume pertains to meteorology, there are many entries for astronomical and astrophysical topics. These are brief but provide enough good information to explain subject matter to the uninitiated reader. The book is intended for scientists of all types, from those still in high school to professors emeriti. It would be useful but not absolutely essential for an astronomy library, and it should be in any general science or medium-sized to large public library.

42.   Flügge, S., ed. **Handbuch der Physik. (Encyclopedia of Physics.)** Berlin, Springer-Verlag, 1955-62. 54v. illus. indexes. refs. LC A 56-2942.

This well-known review series is a standard source for both physics and astronomy. Written in English, German, and French (with an emphasis on English), its scholarly papers cover spectra, radiation, particle physics, dynamics, geophysics, astrophysics, and dozens of other topics. Of particular interest to astronomers are the last five volumes: v. 50: *Astrophysics I: Stellar Surfaces and Binaries* (458p., 1958, o.p.); v. 51: *Astrophysics II: Stellar Structure* (831p., 1958, $107.50. ISBN 0-387-02299-6); v. 52: *Astrophysics III: The Solar System* (601p., 1959, $86.00. ISBN 0-387-02416-6); v.53: *Astrophysics IV: Stellar Systems* (565p., 1959, $86.00. ISBN 0-387-02417-4); v.54: *Astrophysics V: Miscellaneous* (308p., 1962, $59.40. ISBN 0-387-02844-7). Standard review format is maintained in most cases, beginning with introductory material and followed by a discussion of important research. Illustrations are frequent, mainly taking the form of line drawings and graphs, but also consisting of many black-and-white photographs. There are English and German title pages, English-German, German-English subject indexes, and occasional French topical guides. Out of date to a great extent, it is still useful to graduate students, scientists, and librarians on occasion. Its best use now is as a source of references to the literature of the period. More recent texts to be consulted are the "Stars and Stellar Systems" series (University of Chicago) and the supplement volumes of the *Encyclopedic Dictionary of Physics* (Pergamon).

43.   Galiana, Thomas de. **Concise Encyclopedia of Astronautics**. Glascow, London, William Collins Sons & Co.; Chicago, Follett Publishing Co., 1968. 294p. illus. (Collins World Reference Library). LC 68-137022. B 68-10693.

Another potentially useful reference book that should be in public or small college libraries. It desperately needs to be updated, however; the vocabulary has been rapidly changing because of the space program, and many new space missions have been carried out since 1967, when this volume was published. Nevertheless, until a new edition appears, this book will serve the purpose. It defines basic astronautics vocabulary, outlines early space missions, and lists many related terms, among them astronomical topics. Translated from the French by Dr. A. E. Roy.

44.   **McGraw-Hill Encyclopedia of Science and Technology**. 4th ed. New York, McGraw-Hill Book Company, 1977. 14v. illus. index. bibliog. $497.00. LC 76-44232. ISBN 0-07-079590-8.

This familiar, comprehensive work includes approximately 8,000 articles covering the physical, life, and earth sciences, as well as engineering. Astronomy is well represented, making this encyclopedia an excellent source of information on a variety of topics of interest for students, amateur astronomers, and non-astronomer scientists. Articles are written for readers with at least a high school education, and many appear to be on the college level. Readers who desire a brief but good explanation of most astronomical topics should be pleased with the contents of the various pieces, each written by an established scientist. Each entry is subdivided into its component sections, all prefaced with a bold-face heading for easy location of special topics. Each article contains several "see" references to guide the reader to related or similar topics, and most also contain short bibliographies. The encyclopedia's index is very good and makes subject searching rather easy. All college libraries will want this important work, as will most medium-sized to large public libraries and some high school libraries. This latest edition contains many revised and new pieces, making it a valuable, up-to-date source for the sciences. In fact, if one were to extract all articles on astronomy from this set, one would have a fine general text for the serious student.

45.   Muller, Paul. **Concise Encyclopedia of Astronomy**. Glasgow, London, William Collins Sons & Co.; Chicago, Follett Publishing Co., 1968. 281p. illus. (Collins World Reference Library). LC 68-12313. B 68-10154.

Intended for the layman, amateur astronomer, and student, this non-technical reference guide contains 850 entries of varying lengths on the aspects of our Universe. Following an alphabetical arrangement of terminology, this encyclopedia has basic definitions, historical notes, and discussions of astronomical theory. Subject coverage, though now somewhat out of date, is farily broad and encompasses stars, planets, the Moon, the Sun, comets, galaxies, and much more. Also covered are star names, capsule descriptions of astronomers, telescopes, units of measure. For the school, public, and small college library, this fine book was translated from the French by Dr. R. E. W. Maddison.

46.   Satterthwaite, Gilbert E. **Encyclopedia of Astronomy.** New York, St. Martin's Press, Inc., 1971. 537p. illus. $15.00. LC 78-26106. B 71-06737. ISBN 0-600-41106-0.

Possibly the best encyclopedic work in the field, this excellent volume includes a multitude of entries on nearly every imaginable topic in astronomy. The brief and concise entries include diagrams when they are needed to help explain the definitions. Phrases are arranged according to the most significant word, which serves to bring many related topics together. Besides covering the usual spectrum of topics, the book also includes short biographical sketches of famous astronomers, reflecting the author's historical approach to astronomy. Also included are many abbreviations and space-age vocabulary, the former being frequently neglected in works such as these. A further added attraction is information on various observatories and their histories, an uncommon but welcome feature. The variety of information, especially that usually not found in similar books, along with interesting text, good illustrations, and comprehensive coverage, make this, overall, one of the best in its class. For all types of astronomical literature collections, this encyclopedia is aimed more toward the general reader than is Weigert's *A Concise Encyclopedia of Astronomy*.

47.   Thewlis, J., ed. **Encyclopaedic Dictionary of Physics: General, Nuclear, Solid State, Molecular, Metal and Vacuum Physics, Astronomy, Geophysics, Biophysics and Related Subjects.** London, Pergamon Press, 1961-62. 7v. v.8: indexes (1963). v.9: multi-lingual glossary (1964), £24.00, ISBN 08-009928-9. Supplements: v.1: 1966, £8.00, 08-011835-6; v.2: 1967, £10.00, 08-011889-5; v.3: 1969, £10.00, 08-012447-X; v.4: 1971, £14.00, 08-006359-4; v.5: 1975, £16.00, 08-017056-0.

Librarians and scientists alike will consult this extensive set when they need a broad treatment of physics or astronomy on the advanced level. Consisting of seven basic volumes of signed review articles arranged alphabetically by subject heading, this encyclopedia also has an author-subject index tome and an excellent multilingual glossary volume (English, French, German, Japanese, Spanish, and Russian). The latter would be useful in any science library, especially in supplement to Kleczek's *Astronomical Dictionary in Six Languages* (1961). The articles, which range from one paragraph to several pages long, have illustrations, equations, tables, charts, and bibliographies. The seven encyclopedia volumes are kept up to date by the supplemental volumes, the most recent of which appeared in 1975. These books contain entries on new topics and updatings of previous articles. The subject matter is very comprehensive, and astronomy is very well represented: the solar system, stellar evolution, observations, equipment, galaxies, and much more. The supplemental books have additional data on the "new" astronomies. The set is a good review of basic, intermediate, and advanced physics, geared toward the graduate students, astronomers, and physicists. The basic eight volumes are now out of print, unfortunately, but most large science libraries have the set. It is hoped that an updated version will be published someday.

48. **Van Nostrand's Scientific Encyclopedia.** 5th ed. Princeton, N.J., Van Nostrand Reinhold, 1976. 2400p. illus. $67.50. LC 76-18158. ISBN 0-442-21629-7.

A primary general science reference work, this one-volume encyclopedia can be found in all types of libraries, since it can be used by laymen, students, and specialists. This latest edition is touted as "completely new" in a publisher's flyer, with less than 20 percent of the fourth edition carried over. An inspection shows many new topics, as expected, more illustrations, the usual extensive cross references, and the succinct, clear text typical of this standard source. The coverage of astronomy, while not as extensive as, say, that found in *Encyclopedia of Astronomy* (Satterthwaite, 1971), is quite good, with a very up to date selection of topics from the "new astronomies." The user of a non-observatory library might expect to find this work instead of one of the astronomical dictionaries mentioned elsewhere here, so it is worth mentioning, even emphasizing. The length of the articles is somewhat limited, as one might expect them to be in an all-encompassing encyclopedia in one volume, but this does not detract from its usefulness. Basically, one is presented with passages longer than dictionary definitions, but rarely more than several paragraphs. The system of cross references should be noted: words in an article that have entries of their own are printed in boldface, leading the reader immediately to important related material. Recommended for college and public libraries, it may also be used in some special libraries.

49. Weigert, A., and H. Zimmerman. **A Concise Encyclopedia of Astronomy.** New York, American Elsevier, 1968, c1967. 368p. illus. $9.95. LC 68-23260. ISBN 0-444-19796-6.

The librarian should be aware of this valuable handbook, which is a translation from the German *ABC der Astronomie*, by J. Home Dickson. The book's good points are that the definitions and explanations are generally short and always concise, and the text is easy to understand and well illustrated. Illustrations are in the form of diagrams and photographic plates that help explain particular facts and ideas. The cross references are more than adequate, so the reader can easily find the topic in question. Designed for students and laymen, this excellent encyclopedia is so complete and so well written that it should be considered for even the smallest library collection. The librarian should read this book, along with Satterthwaite's *Encyclopedia of Astronomy*, to get a good overview of the subject material that must be dealt with on a day-to-day basis. Supplementary information includes tables of stars, additional illustrations, and star maps.

HANDBOOKS, MANUALS, ALMANACS, AND YEARBOOKS

Although the books in this section might best be characterized as "ready reference," a large number of the reference works listed elsewhere in this guide could also fall into that category. Nevertheless, "quick reference" best describes these volumes, which are found in the library, in the laboratory, in the observatory, or at home. Their main purpose is to provide numerical information, tables of data, in a convenient format for fast and easy use. Included here are both professional-type works used by the astronomer and the university student, and general reference handbooks suitable for the public and college library, as well as for the home bookshelf.

The monographs below are basically of two types: those used for calculations or "theoretical" work and those used for observations or "practical" work. Of the "theoretical" handbooks, three come to mind immediately for the professional astronomer: Allen's *Astrophysical Quantities*, Lang's *Astrophysical Formulae*, and *Landbolt-Borstein Astronomy and Astrophysics*. Together, they provide an excellent source of quick information comprehensive enough to satisfy the needs of almost any astronomer. Only the last-named is substantially out of date and in need of a new edition. *Basic Astronomical Data* (Stars and Stellar Systems, v.3) is another advanced manual worth mentioning; its value lies not in its tables (it has few) but in itx explanations of the use and collection of astronomical information, an area not covered in detail elsewhere. The corresponding handbook for the amateur astronomer is the *Astronomy Data Book* by J. H. Robinson, a superb compilation of quick reference data for the serious stargazer.

In the category of so-called "practical" handbooks come the almanacs and other annually produced works that provide data useful in observational astronomy. Easily the most-used and best-known astronomical reference work is an almanac, *The American Ephemeris and Nautical Almanac*, now in its one hundred and twenty-fifth year. Best suited for the professional astronomer and student, this work, and several other national almanacs like it, provides pages and pages of data essential for carrying out observations of the Sun, Moon, stars, and planets. There are several similar almanacs for the amateur observer, one of the best being the *Handbook of the British Astronomical Association*.

The titles mentioned above are only samplings, it should be remembered. There are many other handbooks in the two categories, some listed below and some not. The ones here are currently the best known and most used, however. The annotations will help librarians choose the handbooks suited for a particular library and clientele. In any event, a large number of these should be included in the astronomy reference section, since they are among the most important books needed by the library user.

50. **The Air Almanac.** Washington, D.C., Government Printing Office. Jan/Apr 1953– . Issued three times per year: Jan-Apr, May-Aug, Sept-Dec. $6.45/vol. LC 53-61239.

Although mainly for air navigation, some of this book's tables could be used by the astronomer for observing. The major portions of the almanac are the ephemerides of the Sun, Moon, Aries, and the planets, and moonrise and moonset. In addition, there is a great deal of miscellaneous data of use to the navigator, including the sky diagrams. These diagrams show the position of selected stars and planets at a given time on a given day of the year at stated intervals of latitude and longitude. These can be used for quick reference to determine position. The principal purpose of this almanac is to provide astronomical data for air navigation in a form convenient for its users. The makers of the book have succeeded well in this goal. The explanations are clear and extensive—everything needed is in this one handy book. Formed by the union of the *American Air Almanac* and the *British Air Almanac*, it is now issued jointly by the Nautical Almanac Office, U. S. Naval Observatory, and Her Majesty's Nautical Almanac Office.

51. Akademii Nauk SSSR. Institut Teoreticheskoi Astronomii. **Efemeridy malykh planet. (Ephemerides of Minor Planets.)** Leningrad: "Nauka," 1947– . Annual. tables. 2r. 42k. LC 50-33889.

A wealth of information on the orbits and positions of asteroids, as well as related planetoidal data, can be found in this handbook for the astronomer. In particular, this ephemeris includes "a) elements of all numbered and of three unnumbered (Apollo, Adonis, Hermes) minor planets; b) dates of the oppositions; c) opposition ephemerides for . . . the current year . . . ; d) ephemerides of bright planets; e) ephemerides for some unusual planets; f) critical list of observations of minor planets by January 1 . . . " of the previous year. Besides these tables, there are lists of new orbital elements received in the past year, drawing attention to new observations and subsequent corrections. The preface is in both Russian and English (the Russian is longer and more detailed), as are the headings on all lists, so Western readers will have no trouble using this guide. A total of 1,861 asteroids were listed in the 1976 edition, arranged numerically under orbital elements, and by opposition dates in the ephemerides. Readers wishing an English-language treatment with more physical descriptive data should consult *Tables of Minor Planets* (Pilcher and Meeus, 1973).

52. Allen, C. W. **Astrophysical Quantities.** 3rd ed. London, The Athlone Press, University of London; distr. Atlantic Highlands, N.J., Humanities Press, 1974, c1973. index. refs. £6.25; $20.25. LC 74-160372. ISBN 0-485-11150-0.

Nearly every possible type of astronomical constant and numerical quantity is included in this handy volume for professional astronomers and students. The main difference between this work and Lang's *Astrophysical Formulae* should be apparent from the titles—this work contains specific data, not formulae derivation and use. The volumes should be used together, since they are complementary.

The appearance of a third edition of this excellent work was especially welcomed by astronomers and librarians, since the previous edition was greatly outdated. Several new sections have been added, including plasmas, solar wind, pulsars, Messier objects, quasars, and Seyfert galaxies. The information is arranged according to the following sections: Introduction; General Constants and Units; Atoms; Spectra; Radiation; Earth; Planets and Satellites; Interplanetary Matter; Sun; Normal Stars; Stars with Special Characteristics; Star Populations and the Solar Neighbourhood; Nebulae, Sources and Interplanetary Space; Clusters and Galaxies; Incidental Tables.

53.   Astronomisches Rechen-Institut. **Apparent Places of Fundamental Stars**. Heidelberg, West Germany, 1940– . Annual. DM 42 (about $13.00). LC 41-25670. ISBN 3-7650-0075-2 (1975 ed.).

An essential book for any observatory. The volume for 1975 contains the mean and apparent places for 1,535 stars in the *Fourth Fundamental Catalogue*, published by the Astronomisches Rechen-Institut. Information for each star includes R. A. and Dec., mean place, magnitude, spectral class, etc. Stars are arranged according to increasing right ascension. There are introductions in English, French, German, Spanish, and Russian, and there is an index arranged by constellation.

Produced annually under the auspices of the International Astronomical Union, this volume was first published in 1940 (for 1941) under the direction of H. M. Nautical Almanac Office, Royal Greenwich Observatory. In 1960 the task was taken over by the Astronomisches Rechen-Institut.

54.   Eichhorn, Heinrich. **Astronomy of Star Positions**. New York, Frederick Ungar Publishing Co., 1974. 357p. index. bibliog. $25.00. LC 73-81764. ISBN 0-8044-4187-1.

The title of this book might lead the reader to believe it is a text on spherical astronomy, but it is not. Rather, it is a guide, and a very good one, to the information contained in star catalogs. This book is both useful and unique because such information would be difficult, if not impossible, to find elsewhere. As the author points out, so much attention is being paid to the new, sensational developments in astronomy (pulsars, quasars, etc.) that the practical elements like positional astronomy are often overlooked; this book was written in response to a need for such information. This volume dwells at length on the information related to positional astronomy, covering Astronomical Coordinate Systems; The Acquisition of Astronomical Data; General Discussion of Star Catalogues; Compilation Catalogues; and Systematic Zone Catalogues. The librarian should also read this work, which is subtitled "A Critical Investigation of Star Catalogues, the Methods of Their Construction, and Their Purpose."

55.   Goldstine, Herman H. **New and Full Moons: 1001 B.C. to A.D. 1651**. Philadelphia, American Philosophical Society, 1973. 221p. (Memoirs of the American Philosophical Society, V.94). $5.00. LC 72-89401. ISBN 0-87169-094-2.

The companion to Bryant Tuckerman's work on planetary, lunar, and solar positions of the past, this book contains for each year (1001 B.C. to 1651 A.D.) the date, time, and longitude for each new and full moon in that year. The data are calculated for an observer in Babylon (equivalently Baghdad) which, by definition, puts the observer exactly 3 hours east of Greenwich. Times given are civil times, based on a 24-clock with its origin at midnight. A handy reference book for both astronomers and historians.

56.   **The Handbook of the British Astronomical Association.** London, Burlington House, 1921– . Annual. $3.00pa.

Intended for observers in Great Britain and Western Europe, this yearly guide contains a variety of handy information for the amateur observer. Information includes, but is not limited to, data on the planets' positions and visibility, eclipses, the Sun and Moon, time reckoning, lunar occultations, occulted stars, planetary occultations, information on the planets (including asteroids), a meteor diary, periodic comets, and so on. The occultation predictions list such events for Great Britain, Australia, and New Zealand. To obtain, write British Astronomical Association, Burlington House, Piccadilly, London W1V 0NL, England.

57.   Jones, Kenneth Glyn. **Messier's Nebulae and Star Clusters.** New York, American Elsevier Publishing Co., 1969, c1968. 480p. illus. glossary. bibliog. indexes. $26.50. LC 73-406500. B 69-00103. ISBN 0-444-19896-2.

A most complete and excellent work describing the nebulous objects in Messier's famous catalog of the eighteenth century. The book has so much information that at first one does not know where to begin. Fortunately, a good table of contents and three excellent indexes (name, subject, and object) solve the problem by laying out the book's arrangement. Besides listing each object and providing every imaginable detail about it, the book provides historical and biographical background about the nebulae and clusters, and about Messier and other astronomers involved. Extremely well illustrated with photos, sketches, and diagrams, it includes coordinates, star charts, and classification. An excellent aid for the observer, this volume belongs in all astronomy and most public and university libraries.

58.   Kitamura, Masatoshi. **Tables of the Characteristic Functions of the Eclipse and the Related Delta-Functions for Solution of Light Curves of Eclipsing Binary Systems.** Tokyo, University of Tokyo Press; State College, Pa., University Park Press, 1968, c1967. 341p. refs. $60.00. LC 67-30318. ISBN 0-8391-0001-9.

This volume is a candidate for the reference section of the observatory library, especially where astronomers are engaged in work with eclipsing binaries. Much useful astrophysical data can be obtained from observations of eclipsing variables, and the data in these tables will be helpful in such work. A detailed introduction explains the equations and methods used in deriving the computer-produced quantities in the book.

59. Lang, Kenneth R. **Astrophysical Formulae: A Compendium for the Physicist and Astrophysicist.** Berlin, New York, Springer-Verlag, 1974. 735p. refs. index. $78.80. LC 73-20809. ISBN 0-387-06605-5 (N.Y.); 3-540-06605-5 (Berlin).

This fine volume containing hundreds of fundamental formulae ought to be in the reference section of every astronomy library. Information is presented in textual form, a welcome divergence from the usual arrangement of mostly tables and lists of data. In general, only derivations for the simpler formulae are included—the extensive list of references point the reader to the sources of long and complicated formulae derivations. Aimed at astronomers and graduate students, this invaluable source is divided into five major sections: 1. Continuum Radiation; 2. Monochromatic (Line) Radiation; 3. Gas Processes; 4. Nuclear Astrophysics and High Energy Particles; 5. Astrometry and Cosmology. The excellent table of contents is quite detailed and complements the extensive subject and author indexes. Lang's book is an important addition to the astronomical reference collection, filling an existing need for a compendium of this nature.

60. Mason, Brian, ed. **Handbook of Elemental Abundances in Meteorites.** New York, Gordon and Breach Science Publishers, 1971. 555p. illus. refs. index. (Series on Extraterrestrial Chemistry, V.1). $56.00; $28.00pa. LC 71-148927. ISBN 0-677-14950-6; 0-677-14955-7pa.

The information in this book is the result of the activities of the Working Group on Extraterrestrial Chemistry of the International Association of Geochemistry and Cosmochemistry which gathered data and references on the various elements found in meteorites since a systematic study in this area began in 1923. The volume is arranged so that each chapter concentrates on one particular element; the research done and the research results are detailed in each section. Tables, diagrams, and a multitude of references back up the textual matter of this excellent work, one of the few of its kind.

61. Meeus, Jean, Carl C. Grosjean, and Willy Vanderleen. **Canon of Solar Eclipses.** Oxford, Pergamon Press, Inc., 1966. 749p. £17.80; refs. $44.50. LC 64-25676. ISBN 0-08-011015-0.

The remarkable book details every solar eclipse from 1898 A.D. to 2510 A.D. The data included here on 1,450 eclipses were produced by a computer, and the result is a massive volume containing a wealth of information valuable to astronomers. Included are general interest data for each event such as Julian day, date, time, Saros number, type of eclipse, and specific information on the Besselian elements and central line data. The former refers to "elements which characterize the geometric position of the shadow of the Moon relative to the Earth," and the latter refers to information on the path of the Moon's shadow. Fifty-eight maps showing those paths are an important part of the volume. This valuable book should be in every university and observatory library, as well as in selected large public libraries. The work substantially updates Oppolzer's *Canon der Finsternisse* (1887).

62. Moore, Patrick, ed. **Yearbook of Astronomy**. London, Sidgwick & Jackson Ltd.; New York, W. W. Norton & Co., Inc., 1962– . Annual. illus. bibliog. index. $9.95 (1976 ed.). LC 62-1706. ISBN 0-393-06404-2 (1976 ed.; Norton).

One of the better amateur observing handbooks available, this volume contains, in addition to the usual observing data, several articles on topics of interest to the reader. In fact, one-third to one-half of the book is made up of such chapters, written by amateur and professional astronomers, which describe interesting celestial events, like eclipses or meteor showers, or which merely explain certain phenomena, like double stars. The approach is refreshing— it keeps the book from being a dry volume of tables and charts. Standard annual sections include the Monthly Notes, brief comments on the celestial events of each month, including new and full Moon, positions of the planets, and several short paragraphs on interesting stars, constellations, and celestial occurrences for the period. There are also excellent star charts and supplementary tables which will be of use to the casual and serious observer. Finally, there is a listing of amateur astronomical societies and a section of new recent books of interest. While not the most comprehensive such handbook around, it is certainly one of the most interesting.

63. Nautical Almanac Office, U.S. Naval Observatory. **The American Ephemeris and Nautical Almanac.** Washington, D.C., Government Printing Office, 1852– . Annual. index. $10.35 (1976 ed.). LC 7-35435.

This is the one book found in every observatory, in its library and on the telscope console. The information it provides, which is extremely useful to both professional and student observers, includes tables of universal and sidereal time, daily positions of the Sun, Moon, and all the planets, sunrise and sunset, moonrise and moonset, a list of observatories, Julian date table, and much more. It is this important information that tells the astronomer which way to point the telescope to find certain celestial objects, and which assists in the calculations to determine the positions of still other celestial bodies. Other important parts of the almanac are the conversion tables of sidereal to solar time, arc to time, and vice versa for each. The latter part of the text contains an explanation of the tables and other data. Also published in Great Britain by Her Majesty's Nautical Almanac Office under the title *The Astronomical Ephemeris*; except for a few introductory pages, the books are identical for each country, and each nation contributes the data used.

There is also an *Explanatory Supplement to the Astronomical Ephemeris and to the American Ephemeris and Nautical Almanac*, which has detailed explanations of the basis and derivation of the ephemerides. It also contains historical notes and other useful information such as permanent astronomical tables not found in the ephemeris.

64. Nautical Almanac Office. U.S. Naval Observatory. **Astronomical Phenomena for the Year 19–**. Washington, D.C., Government Printing Office, 1950– . Annual. 68p. $0.80pa.

This handy pamphlet is, in part, a reprint of selected pages from *The American Ephemeris and Nautical Almanac*. It contains information on a variety of topics useful to the astronomical observer: universal, standard, and daylight times; religious calendars; civil calendars; chronological eras and cycles; a current calendar; planetary configurations; equation of time; information on Polaris; time conversion tables; eclipses; sunrise, sunset, and twilight information; moonrise and moonset; publications of the U.S. Naval Observatory. A library that does not need *The American Ephemeris and Nautical Almanac* should purchase this useful booklet, and serious amateur observers will want it, too.

65. Percy, John R., ed. **The Observer's Handbook**. Toronto: Royal Astronomical Society of Canada. 1– , 1907– . Annual. illus. $3.00pa.

Written for the amateur astronomer, this handy guide's main feature is a day-by-day calendar of astronomical events for the entire year. The handbook contains much more, however: the positions of Jupiter's and Saturn's satellites, bright star data, lunar occultations, asteroids, meteor showers, double stars, nearest stars, variable stars, star clusters, radio sources, etc. Of interest is the photometric catalogue of stars brighter than 3.55 magnitude, and the list of Canadian impact meteorite craters. Primarily for the Canadian observer, it can be used by U.S. residents, too. To obtain, write The Royal Astronomical Society, 252 College Street, Toronto, Ontario M5T 1R7.

66. Pickering, James S. **1001 Questions Answered about Astronomy**. rev. ed. New York, Dodd, Mead & Co., 1975. 420p. illus. index. $7.95. LC 75-4045. ISBN 0-396-07184-8.

Actually, there are 1,048 questions in this reference type volume suitable for school, public, and home libraries. Arranged into 15 sections on the Sun, Earth, Moon, planets, stars, comets, galaxies, and more, this fine book has the answer to nearly every inquiry that might arise in a reference situation. Neither too elementary nor too technical, this encyclopedia-type book was first published in 1958, and it has most recently been revised by Patrick Moore. Other than the major category headings, the only way to locate individual pieces of information is by using the index. Questions on similar topics tend to be clustered together in each category, but there is no pattern to this. Besides its use as a ready-reference book, this excellent volume can be read straight through or browsed, depending on the interest of the reader; it would be an ideal gift for the backyard astronomer. Of special interest are the chapter on the constellations, which tells the history and lore of the groups of stars; the up-to-date section on radio astronomy; and the chapter on telescopes and other instrumentation. Providing a wealth of information on nearly all areas of astronomy, this book comes highly recommended. It is hoped that future editions will follow.

67. Pilcher, Frederick, and Jean Meeus. **Tables of Minor Planets**. Jacksonville, Ill., F. Pilcher, Illinois College, 1973. 104p. refs. $4.00pa. (prepaid). LC 73-80379.

This remarkable book belongs in the reference collection of every observatory library where studies of asteroids are being carried out. Containing a wealth of information not easily obtained elsewhere, it lists over 1,800 asteroids and a multitude of statistical tables describing all possible aspects of each. Among the tables of minor planets are orbital elements, magnitudes, a list of discoverers, and various special tables. It is an indispensable tool for observers and students of the planetoids, and librarians should consider this for the observatory ready-reference shelf.

68. Robinson, J. Hedley. **Astronomy Data Book**. New York, Halstead Press, John Wiley and Sons, 1972. 271p. illus. $10.95. LC 72-9496. ISBN 0-470-72801-9.

A unique and highly useful handbook is the result of the author gathering a wide variety of information from dozens of sources. The introduction in this work for the student and amateur astronomer states that this book should be used in conjunction with a star atlas; this may be so, but there are many other uses as well. A glossary of very basic terms begins the book, followed by a list of important dates in astronomy. Then comes a wealth of data in the form of tables, text, and diagrams on every conceivable topic. Information on eclipses, comets, planets, stars, meteors, and other stellar and non-stellar objects abound. Highly recommended for the public and college library, it will save the user countless hours of searching.

69. Strand, K. A., ed. **Basic Astronomical Data**. Chicago, University of Chicago Press, 1963. 495p. illus. subject index. refs. (Stars and Stellar Systems, v.3). $16.50. LC 63-11402. ISBN 0-226-45955-1.

Not really a handbook, this book deals with the major types of astronomical data available from ground-based observatories. Each of the 24 chapters by various contributors stands alone, authors were given general areas to cover as they saw fit. Consequently, there is some, but not too much, overlap. Lists of data are not the norm in this volume—rather, each chapter is in the form of a bibliographic essay that discusses the data, its characteristics, and uses. Further, the work evaluates each type of information as to its completeness and accuracy. Methods of observation and equipment are not discussed here. Excellent bibliographies accompany the chapters.

Chapter titles are as follows: 1. Astronomical Reference Systems; 2. The System of Fundamental Proper Motions; 3. The Reference System of Bright, Intermediate, and Faint Stars and of Galaxies; 4. Photographic Proper Motions; 5. Proper Motion Surveys; 6. Trigonometric Stellar Parallaxes; 7. Radial Velocities; 8. Classification of Stellar Spectra; 9. Quantitative Classification Methods; 10. Spectral Survey of K and M Dwarfs; 11. Photometric Systems; 12. Interstellar Reddening; 13. Applications of Multicolor Photometry; 14. The Stellar Temperature Scale; 15. Empirical Data on Stellar Masses, Luminosities, and Radii; 16. Polarization of Starlight; 17. Surveys and Observations of Visual Double Stars; 18. Surveys and Observations of Physical and Eclipsing Variable Stars; 19. Empirical Data on Eclipsing Binaries; 20. The Calibration of Luminosity Criteria; 21. The Absolute Magnitudes of Classical Cepheids; 22. The Luminosities of Variable Stars; Appendix I, Star Catalogues and Charts; Appendix II, The National Geographic Society-Palomar Observatory Sky Survey.

For astronomers and graduate students.

70. Tuckerman, Bryant. **Planetary, Lunar and Solar Positions. Volume I: 601 BC to AD 1**. Philadelphia, American Philosophical Society, 1962. 333p. bibliog. (Memoirs of the American Philosophical Society, v.56). $6.00. LC 62-14516. ISBN 0-87169-056-X. **Volume II: AD 2 to AD 1649**. Philadelphia, American Philosophical Society, 1964. 842p. bibliog. (Memoirs of the American Philosophical Society, v.59). $7.50. LC 64-14093. ISBN 0-87169-059-4.

A wealth of numerical data is contained in this two-volume ephemeris, which lists positions of planets, Sun, and Moon for a period of 2250 years, "to an accuracy and spacing suitable for historical purposes. . . . Each page contains the (tropic) [with respect to the mean equinox of date] geocentric longitudes and latitudes, in degrees and decimal fractions, of the Sun, Moon, and naked-eye planets (except the latitude of the Sun, which is zero) for two consecutive years." The excellent introduction explains in detail the use of the ephemeris, and its symbols, design, theory, construction, and accuracy, as well as estimation of perturbation and rounding errors. Generated by an IBM 704 computer, the data could be of use to astronomers and historians.

71. Voigt, H.H., ed. **Landolt-Borstein Numerical Data and Functional Relationships in Science and Technology**. Group VI, Volume I: *Astronomy and Astrophysics*. Berlin, New York, Springer-Verlag, 1965. 711p. illus. refs. bibliog. $167.70. LC 62-53136. ISBN 0-387-03347-5 (N.Y.).

Now over 12 years old, this massive reference volume still retains much of its original value. One of a series of books on numerical data, it focuses on the basic equations and relationships of astronomical research. Page after page of tables, graphs, and equations, interspersed with text, make this work a valuable part of the astronomy reference collection. Although most of the text is in German, some is in English, as are most of the summaries and the table of contents. The presentation is both comprehensive and critical, with the contributors emphasizing the results of previous research. An updated, all-English version would make a valuable contribution to the literature. The ten major sections of the book are Astronomical Instruments; Position and Time Determination; Astronomical Constants; Abundances of Elements in the Universe; The Solar System; The Stars; Special Types of Stars; Star Clusters and Associations; The Stellar System; Galaxies.

ATLASES AND CATALOGS

Both the amateur and professional astronomer rely heavily on two special types of reference sources, the atlas and the catalog. Although these books are most frequently "star atlases" and "star catalogs," there are dozens of other types as well, ranging from general-type works covering many different celestial objects or phenomena to very special works concentrating on one type of stellar or non-stellar body.

Like geographical atlases, astronomical atlases are volumes of maps, usually of the sky, but also of the Moon, Mars, etc. The charts may be drawn by an artist, or they may be photographs with the features labeled. As a result of the recent flurry of space probes and outer space cameras, the newer atlases are mainly, or entirely, photographs (e.g., those that concentrate on the Moon). Such works help astronomers locate particular objects for observation; they are of service both to professionals with 40-inch reflectors and to amateurs with 3-inch refractors.

More important than atlases, are catalogs, the astronomer's right hand. These books are both indexes to certain atlases and separate entities in themselves. The catalog is a tabular listing of celestial objects with a variety of data for each, like position, spectral type, magnitude, and so on. Sometimes a star atlas and catalog will be combined, like many of the works for the amateur, but frequently the catalog stands alone, to be used by itself, for a variety of practical and theoretical purposes. Star catalogs were probably first developed to serve as a detailed record of observations, to be used later for reference and calculations. The catalog provides the data in a convenient form for easy access and use.

Most of the entries here are single-volume works that selectively cover all or certain portions of the sky. There have been, however, major efforts in the last century to produce the so-called "survey star catalogs," multi-volume sets that attempt to include as many stars as accurately as possible. Several of these endeavors have been carried out by many observatories, each taking a piece of the sky, painstakingly working to produce a comprehensive, exact catalog. Six of these monumental tasks are extremely important; the librarian who wants to have a good knowledge of sources in this area should be aware of them.

Briefly, they are

1) *Bonner Durchmusterung* (1860); *Southern Durchmusterung* (1886); and *Cordoba Durchmusterung* (1892). Covers the entire sky, including half a million stars collectively, down to 9th magnitude.

2) *Astronomische Gesellschaft Katalog* (AGK or AG_1) (1890-1910). Covers $-2°$ to $+80°$ in declination, stars down to 9th magnitude. Volumes for the Southern Hemisphere were only partially completed.

3) *Yale Zone Catalogues* (1925-1959). A project to re-observe the stars in the *AGK* using the photographic method, a more accurate procedure, used to establish more exact position coordinates.

4) *Astrographic Catalogue* (AC) (1903-1963). A cooperative effort of over 20 observatories aimed at producing a very accurate star catalog. A total of 141 volumes were published in the 61-year period, an effort unparalleled by anything before or since. The unusual feature of this set is that stellar coordinates are given in rectangular coordinates instead of right ascension and declination.

5) *The National Geographic Society–Palomar Observatory Sky Survey* (1949-1955). This survey produced an atlas covering -33° to +90° declination, using the 4-ft. Schmidt telescope at Mt. Palomar. A photographic atlas consisting of 936 pairs of plates taken in red and blue light, it is printed on 14x17in. heavy paper. Its 1872 photographs comprise the most comprehensive sky atlas to date. It was reprinted and re-issued for the fifth time in 1977.

6) *ESO/SRC/Atlas of the Southern Sky* (1976). Complementing the *Palomar Sky Survey*, this set of 1212 photographs on transparent film cover -90° to -20° declination. Further details were unavailable at the time of this writing.

Atlases and catalogs of the sky showing non-visible light objects are the newest and among the most interesting of this type of literature. Among these works are radio atlases, x-ray and infrared catalogs, and many more. Because these publications are often brief and are continuously being updated, they are often issued as parts of books, proceedings, journals, and journal supplements. The new astronomies are relatively young and have spawned but a few standard works. Below are six examples, the first three of which are important sources for the radio astronomer.

i) "Parkes Catalogue of Radio Sources; Declination Zone +20° to -90°," in *Australian Journal of Physics. Astrophysical Supplement* (no. 7, April 1969).

ii) "Cambridge Catalog of Radio Sources" appearing in several parts in two publications: *Royal Astronomical Society. Memoirs.* and *Monthly Notices.* See this guide's Appendix for details of volume numbers.

iii) "Master Source List" of 78,000 radio sources, available on computer tape or in printout format. Issued irregularly by the Ohio State University Radio Observatory. Formed by the merging of several radio sky surveys, it was first published in the *Astrophysical Journal. Supplement Series* (v.20, no.8, 1970) as "A Master List of Radio Sources."

iv) "List of Quasi-Stellar Objects," in *Quasi-Stellar Objects* (Burbidge and Burbidge, 1967).

v) "Catalog of X-Ray Sources," in *X-Ray Astronomy* (Giacconi and Gursky, 1974).

vi) "Catalog of Pulsars," in *Neutron Stars, Black Holes, and Binary X-Ray Sources* (Gursky and Ruffini, 1975).

The reader, especially the librarian, should keep in mind that there are many, many more than the selected examples above. The selection of catalogs and atlases in the bibliography is painfully brief. A complete list of these items would have been difficult and of questionable value in this selective, introductory work. A good but dated alternate source is Kemp's *Astronomy and Astrophysics; A Bibliographic Guide* (1970), which includes an excellent list of the most important atlases and catalogs available.

Sky atlases and star catalogs are found in libraries of all types, from secondary school to observatory. The public library will wish to collect several good amateur or general astronomical atlases and catalogs, and the observatory will lean toward more advanced titles. Among the volumes listed below are some of each type—general and specialized, elementary and advanced. Classics of past days are included along with books currently in high use. In all, the references in this section provide a good cross section of what is currently being used.

72. Abt, Helmut A., and Eleanor S. Biggs. **Bibliography of Stellar Radial Velocities.** Tucson, Ariz., Kitt Peak National Observatory, 1972. 502p. $9.00. LC 72-194110.

Included in this unusual volume are 44,000 references from 25 astronomical journals listing the radial velocities for 25,000 stars. It is unusual because there are few works that list celestial data *with* a reference to the source of that data; more efforts along these lines should be attempted. This massive compilation contains references which include the following information for each star: R. A., Dec., visual magnitude, spectral class, HR and HD numbers, and average radial velocity. Additionally, each reference contains journal, volume, page, and year, and a special notation for variables, double stars, etc. The authors have scanned every volume of each of the 25 journals to gather their data. This book is a must for every observatory reference collection.

73. Alter, Dinsmore, ed. **Lunar Atlas.** New York, Dover Publications; distr. Gloucester, Mass., Peter Smith, 1968. 343p. illus. 154 plates. $12.00. LC 67-28175. ISBN 0-8446-1531-5 (Smith).

Probably one of the last atlases to include exclusively Earth-based photographs of the Moon, the book provides an excellent view of all aspects of the lunar surface. Included are pictures of the various lunar phases, as well as plates illustrating the various landscape features: maria, mountains, craters, etc. Although more detailed, clearer photos now exist because of the space probes and manned landings, this volume is highly recommended, both for its historical value and because it includes good photographs of the Moon at a distance.

This Dover edition is an unabridged re-publication of the 1964 edition published by the North American Aviation, Inc. The plates in this reprint are about one-third smaller than the original issue, however. The introduction states that the main intention of the atlas is to "aid lunar research by presenting the surface of the moon in accurate photographic and descriptive detail, from a fundamental selenographic and astronomical point of view." More detailed photographs have obviously been obtained since.

74. Bečvář, Antonín. **Atlas Australis 1950.0.** Praha, Nakladatelství, Československé, Akademie Věd; distr. Cambridge, Mass., Sky Publishing Corp., 1964. 24 color plates. refs. $14.00. LC 65-43677.

This atlas "contains all stars in the zones south of declination -30°, which are included in available catalogues and whose precise positions are well known. Their photovisual magnitudes are expressed in steps of 0.5 magnitudes."

A total of 104,045 stars are included in this fine book, which shows stars as faint as 13th magnitude and variables as faint as 10th magnitude. Spectral type is shown in color, and the size of the stars indicates magnitude. This scheme makes the atlas both very attractive and easy to use. Although star names are not given, boundaries of constellations are shown. An excellent atlas to be used in conjunction with an adequate catalog like the Smithsonian Astrophysical Observatory *Star Catalog*.

75. Bečvár, Antonín. **Atlas Borealis 1950.0**. Praha, Nakladatelství, Československé, Akademie Věd; distr. Cambridge, Mass., Sky Publishing Corp., 1962. 24 color plates. refs. $19.50. LC 63-38077.

A duplicate of the *Atlas Australis 1950.0* in form and purpose, this volume contains only stars north of declination +30°.

76. Bečvár, Antonín. **Atlas Coeli 1950.0**. Praha, Nakladatelství, Československé, Akademie Věd; distr. Cambridge, Mass., Sky Publishing Corp., 1962. 16 color plates. refs. LC 64-28084.

A standard basic sky atlas containing stars down to +7.75 apparent magnitude and galaxies as faint as 13th magnitude. Unlike many similar works, it uses color to indicate star clusters, nebulae, galaxies, and clouds of interstellar gas and dust. The volume shows boundaries of constellations and indicates their principal stars, and gives NGC and IC numbers of galaxies brighter than 12th apparent magnitude (and includes those as faint as 13th). The boundaries of the Milky Way are given, as well as the paths of the celestial equator and ecliptic, and the work also includes various radio sources and multiple star systems. Explanations and the table of contents are in Czech, Russian, English, and German. An attractive, important atlas—for all observatory collections.

77. Bečvár, Antonín. **Atlas Coeli—II. Katalog. 1950.0**. Prague, Czechoslovakia, Academy of Sciences Publishing House; Cambridge, Mass., Sky Publishing Corp., 1964. 369p. bibliog. tables. NUC 65-64985.

The companion volume to the *Atlas Coeli* star atlas, this handy book contains numerical data on stars down to visual magnitude +6.25 and all other celestial objects depicted in the atlas. Stars are arranged by GC number (from the *Boss General Catalogue*), each containing right ascension and declination, apparent and absolute magnitude, spectral type, parallax, radial velocity, proper motions, constellations, and notes. Also included are: 1) an index of named stars arranged by constellations, 2) a correlation index between GC and HD catalog numbers, 3) a list of double and multiple stars, 4) tables of visual and spectroscopic binary elements, 5) variable stars and novae index, 6) a list of open star clusters, 7) planetary nebulae index, 8) a table of bright diffuse nebulae, 9) a list of galaxies, 10) a Messier catalog, and 11) supplementary tables. A valuable volume to have, with or without Bečvár's beautiful atlas.

78. Bečvář, Antonín. **Atlas Eclipticalis 1950.0.** Praha, Nakladatelství, Československé, Akademie Věd; distr. Cambridge Mass., Sky Publishing Corp., 1964. 32 color plates. refs. $16.50. LC 65-52523.

This work is strictly a star atlas, including all stars listed in the Yale Zone Catalogues between declinations -30° and +30° with no limiting magnitude. Color is used to indicate stellar spectral type. The book also includes variables down to 10th magnitude and multiple systems with combined magnitudes down to 10. Explanations in Czech, Russian, English, and German. Another quality work from the Czech astronomer.

79. Boss, Benjamin. **General Catalogue of 33,342 Stars for the Epoch 1950.** Washington, D.C., Carnegie Institution, 1937; reprinted by Johnson Reprint Corp., 1962. 5v. tables. refs. (Carnegie Institution. Publications. 468) $105.00. LC 38-4623. ISBN 0-384-07699-8 (Johnson).

A predecessor to more complete and more accurate works like the SAO *Star Catalog*, this publication provides very precise positions and proper motions for approximately 33,000 stars over the entire sky. Including all stars down to 7th magnitude, and more as faint as 9.0, the catalog filled a need at the time for a precise set of values of stellar coordinates. The familiar GC numbers from this work still appear in cross references in other catalogs, and this volume, itself a classic, is still used from time to time. The entire first volume is given to explanations, appendices, descriptions, and historical background. The appendices include systematic corrections in right ascension and declination, ephermerides, and peculiar proper motions. The catalog proper contains GC number, HD number, magnitude, spectral type, coordinates (1950), epoch, proper motion, probable errors, variations in coordinates, remarks, and more. Although not as complete as the SAO *Catalog*, it is easier on the eyes, and for many projects it has a substantial enough number of stars with a high degree of accuracy.

80. Bowker, David E., and J. Kenrick Hughes. **Lunar Orbiter Photographic Atlas of the Moon.** Washington, D.C., NASA, 1971. 41p. 675 plates. illus. bibliog. $19.95. LC 70-607341. NASA SP-206.

This massive volume combines photographs taken by five Lunar Orbiter spacecraft. The entire surface of the Moon is included, and each photograph includes the picture's center coordinates, north deviation, Sun angle, spacecraft altitude (km), and scale of the photograph. Besides the photographs there are three useful tables: Summary of Lunar Orbiter Missions; Support Data; and Lunar Features. The extensiveness of the coverage and the quality of the photographs makes this large volume a worthwhile purchase for the observatory and university. The only drawback to this atlas is that lunar features, such as craters, are not named.

81. Brown, Basil J. W. **Astronomical Atlases, Maps and Charts: An Historical and General Guide.** London, Dawsons of Pall Mall; distr. New York, International Publications Service, 1968. 200p. illus. index. bibliog. $30.00. LC 74-512209. ISBN 0-7129-0131-0.

Well illustrated with pictures of old astronomical maps and charts, this beautiful book is a reprint of the original 1932 work. The author tells interesting stories of the history of star charts, sky atlases, and star catalogs, and provides a wealth of detail on the individual items discussed. The book not only tells of the maps themselves, but also of the mapmakers, many of whom were famous astronomers and cartographers. While star atlases, charts, and catalogs are the main thrust of the book, other topics are covered as well: solar and spectroscopic charts, the Moon, the planets, and other subjects related to celestial mapping. Besides providing valuable information to scientific historians and astronomers, this work is just plain fun to look at and browse through. A modern-day version of this work, including the revolution in celestial mapmaking caused by space probes, would be interesting.

82. Cannon, Annie J. and Edward C. Pickering. **The Henry Draper Catalogue.** Cambridge, Mass., Harvard College Observatory, 1918-1924. 9v. tables. (Astronomical Observatory of Harvard College. Annals. v. 91-99). LC 18-18382.

The basic catalog of stellar spectral classifications, this massive work lists data for approximately 225,000 stars. The tables include HD number, corresponding number in the Bonn, Cordoba, or Cape Durchmusterungs (survey star catalogs of the latter half of the nineteenth century), right ascension and declination for 1900, photometric magnitude, photographic magnitude, spectral class, photographic intensity, and remarks. Explanatory material on the system of spectral classification, including sample stars for each category, precedes the tables of data. Still used by astronomers, this classic catalog is primarily the work of Miss Cannon, who toiled six years in its completion. A tenth volume, *The Henry Draper Extension*, 1925-36, was issued to include almost 50,000 additional faint stars in certain areas of the Milky Way which were not included in the main catalog. The information in the *Extension* includes H. D. and B. D. numbers, R. A. and Dec. (1900), magnitude, and spectral class. In 1949, a book of the charts of the *Extension* stars was published in memory of Miss Cannon.

83. Davis, Robert J., and William A. Deutschman, and Katherine L. Haramundanis. **Celescope Catalog of Ultraviolet Stellar Observations.** Washington, D.C., Smithsonian Institution, 1973. 248p. index. $4.50.

Subtitled "5068 objects measured by the Smithsonian equipment aboard the Orbiting Astronomical Observatory (OAO-2)," this is the first catalog of stars as seen in the ultraviolet, and therefore is a unique and important reference work. Beginning in 1968, four telescopes aboard the OAO-2 took more than 8,000 television pictures of stars; these photographs are the essence of this catalog. The following data are recorded for each star included: HD and DM numbers, R.A. and Dec., apparent visual magnitude, B-V and U-B magnitudes, spectral characteristics, remarks, and references. Indexes include numerical and author keys to the references. For the observatory library.

84.   de Vaucouleurs, Gerard, and Antoinette de Vaucouleurs. **Reference Catalogue of Bright Galaxies**. Austin, University of Texas Press, 1964. 268p. notes. refs. (University of Texas Monographs in Astronomy, no. 1). LC 64-22391. ISBN 0-292-73348-8.

Updating work done by Shapley and Ames in the Harvard "Survey of Galaxies Brighter than the 13th Magnitude" (1932), this catalog reflects 15 years of enlarging and revising existing data on galaxies. Detailed numerical and descriptive information on 2,600 star systems is included. Tables are laid out across two pages under the following headings: NCG, IC, and A (anonymous) identification numbers; right ascension and declination (1950); precession; old and new galactic coordinates; type and color class of Yerkes lists; Hubble-Sandage type; diameter information; magnitudes; colors; velocities; radio and integrated flux; and references to best available photographs in the literature. The extensive introductory material includes maps of the distributions of various types of galaxies projected onto a sky map. There are also lengthy notes for most galaxies, which give description and/or sources of reference. Several appendices are included in this standard source: Integrated Magnitudes in B System; Integrated Colors in (B-V) System; Integrated Colors in (U-B) System; and Listing [of Galaxies] by Right Ascension. To be used in conjunction with its successor, *Second Reference Catalogue of Bright Galaxies*, 1976.

85.   de Vaucouleurs, Gerard, and Antoinette de Vaucouleurs, and Harold G. Corwin, Jr. **Second Reference Catalogue of Bright Galaxies**. Austin, University of Texas Press, 1976. 396p. notes. refs. (University of Texas Monographs in Astronomy, no.2). $50.00. LC 75-44009. ISBN 0-292-75507-0.

Containing 70 percent more entries than its first edition, this standard source should be used in conjunction with the earlier work. Made necessary because of the rapid growth of extragalactic studies since 1964, this catalog follows the same guidelines for inclusion of galaxies—i.e., objects must be brighter than 16th magnitude, have diameters greater than 1.5, or have redshifts less than 15,000 km/sec. A total of 4,364 star systems are indexed, in two-page-wide tables using slightly different headings from those used in the *Reference Catalogue of Bright Galaxies* (1964). Selected data given for each galaxy are NGC, IC, or A numbers; right ascension and declination; precession; new galactic coordinates (the "old" are omitted); supergalactic coordinates; morphological type; source of plate material and author of classification; type and color class in the Yerkes lists; several diameter data; magnitudes; surface brightness; color indexes; radio continuum flux density; radial velocities. The introductory material is much more extensive than that provided in the old version, with detailed explanations of derivations and sources of error. In most cases, the data given in the 1964 edition have been revised and given one more order of magnitude, increasing the numerical accuracy. This version has a type face that is 40 percent smaller, and there are about 10 more entries per page, but it is still fairly easy to read. There are 10 appendixes of numerical data of a supplementary nature. A sorely needed work, it will surely be extensively used by astronomers, students, and librarians.

86. Gutschewski, Gary L., Danny C. Kinsler, and Ewen Whitaker. **Atlas and Gazetteer of the Near Side of the Moon.** Washington, D.C., NASA, 1971. 538p. illus. refs. notes. $15.00. LC 73-607342. NASA SP-241.

This atlas contains 404 photographs taken by the Lunar Orbiter IV with craters, plains, and other lunar features clearly marked on each. The gazetteer also provides additional useful information: an index to lunar names (B&M and LPL reference numbers, longitude and latitude, photo number), a Lunar and Planetary List (LPL), Blagg and Muller List (B&M), an index to Lunar Orbiter coverage (arranged by lunar features), and notes. A useful reference tool for the lunar observer; astronomy, university, and medium-sized to large public libraries will want to acquire it as well.

87. Hoffleit, Dorrit. **Catalogue of Bright Stars.** 3rd rev. ed. New Haven, Conn., Yale University Observatory, 1964. 415p. tables.

A prime source of numerous data on 9,000 stars brighter than 6.5 magnitude, this standard work is an extremely important and useful reference book. The following information is included on the stars in this catalog, which is arranged in ascending right ascension: HR-BS number, name (Bayer or Flamsteed designation, if any), five star catalog numbers (including HD, GC, and DM), double star designation, variable star designation, R.A. and Dec. (1900), galactic coordinates (1900), R.A. and Dec. (2000), R.A. and Dec. changes (2000-1900), visual magnitude, B-V color index, spectral class, proper motion, parallax, radial velocity, double star data, and remarks. There are also two appendices: the identification of Bayer and Flamsteed stars, and an index to star names. Essential for the astronomy library, this handy catalog will be of great use to student and astronomer alike.

88. **International Auroral Atlas.** Edinburgh, Edinburgh University Press; distr. Chicago, Aldine Publishing Co., 1963. 20p. 56 plates. illus. (IAGA Publication, no. 18). ₤2.75; $6.95. LC 64-421. ISBN 0-85224-107-0 (Edinburgh); 0-202-2201-X (Aldine).

This unique atlas is intended "to serve as a standard text, providing basic definitions and descriptions, designed to classify auroras in the manner best adapted to promote their scientific study . . . the atlas should not be an observer's handbook providing details of techniques and systems of observations . . . ." The excellent introductio includes information on notation, form, qualifying symbols, structure, condition, brightness index, and colour class. Each classification of aurora is illustrated with a black-and-white or color plate. The volume was published for the International Union of Geodesy and Geophysics by the International Association of Geomagnetism and Aeronomy, Auroral Committee. No similar work is available.

89. Kopal, Zdeněk. **A New Photographic Atlas of the Moon.** New York, Taplinger Publishing Co., 1971. 311p. 214plates. illus. glossary. index. $20.00. LC 72-125480. ISBN 0-8008-5515-9.

An impressive collection of 214 plates of the lunar surface, the photographs were taken with earth-based cameras and telescopes, as well as cameras on spacecraft, and by astronauts walking on the Moon. Each plate is briefly described, including origin of photograph, and a description of the lunar feature photographed. An excellent text precedes the plates, and introductory material describes the lunar landscape and surface structure. There are no maps, drawn or otherwise, showing and naming lunar features—this is strictly a collection of photographs. A fine book for the college and public library, it is typical of many recent works inspired by the space program.

90.   Kuiper, G. P. **Photographic Lunar Atlas**. Chicago, University of Chicago Press, 1960. 23p. 230 photos, index. LC 60-2602.

Based on selected plates taken at Mount Wilson, Lick, McDonald, Yerkes, and Pic du Midi Observatories, this work was, at one time, the best available attempt at comprehensive lunar photography. Instead of the usual book arrangement, this atlas consists of loose photographs in a large box. It was, and possibly still is, the best atlas of the Moon using Earth-based photography. An accompanying pamphlet includes background information as well as tabular data on various lunar features. Four excellent supplements were subsequently issued:

1. Kuiper, G., ed; D. W. G. Arthur, and E. A. Whitaker, comp. *Orthographic Lunar Atlas.* Tucson, University of Arizona Press, 1960. Chiefly photos. refs. $16.50. LC 60-53621. ISBN 0-8165-0283-8.

2. Whitaker, E. A., *et al. Rectified Lunar Atlas.* Tucson, University of Arizona Press, 1963. 1v. (chiefly illus.) (Contributions of the Lunar and Planetary Laboratory, No. 3). $35.00. LC 63-17721. ISBN 0-8165-0077-0.

3. & 4. Kuiper, Gerard P., *et al. Consolidated Lunar Atlas.* Tucson, Lunar and Planetary Laboratory, University of Arizona, 1967. 24p. 226 photos. LC 68-63335.

91.   Kukarkin, B. V., *et al.*, eds. **Obshchii Katalog Peremennykh Zvezd. (General Catalogue of Variable Stars.)** 3rd ed. Moscow, "Nauka," v.1-2: 1969; v. 3: 1971. tables. bibliog. indexes. v.1-2: 4r 04k. v.3: 2r 49k. NUC 72-98928.

The source of numerical and positional variable star data, this work is the result of an on-going program of observation being carried out in the Soviet Union. Reflecting thousands of painstaking observations over many years, this set includes data on over 20,000 variables, including pulsating, eruptive, and novae-type stars, as well as eclipsing binaries. The arrangement of tables in the first two volumes is by constellation, with stars listed by alphabetic notation. Data for each star include name, 1900.0 coordinates (including precession), galactic coordinates, bibliographic information, type of variability, maximum and minimum magnitudes, type of stellar magnitude (UBV, photographic, visual, red, infrared), epoch, period, spectrum, etc. A lengthy introduction in both Russian and English provide the reader with guidance for using the mountains of data within. An extensive bibliography rounds out the first two books.

Volume three provides auxiliary tables on "extragalactic Supernovae; galactic Supernovae and Novae observed in the antique and medieval ages; pulsars;" there are also tables of "optically variable quasars and nuclei of galaxies." The remainder of the third book includes a list of variables arranged by right ascension (1900.0 coordinates), a list of variable stars by type, and several minor tables. Three regular supplementary volumes (1971, 1974, 1976) were issued to include data on newly discovered variables and changes in data made since the first two volumes. A fourth special supplement issued in 1972 lists the stars in vol. 1-2 by right ascension (1950.0 coordinates). Limited copies of all volumes of this extremely important work are available from the American Association of Variable Star Observers, 187 Concord Avenue, Cambridge, Mass. 02138. A must for the observatory library, it will continue to be updated in the future.

92.   Luyten, Willen J. **White Dwarfs**. Minneapolis, University of Minnesota, 1970. 124p.

This volume is a catalog of white dwarfs and degenerate stars found by the author in his Proper-Motion and Faint-Blue-Star Surveys. There are 2,934 entries, each containing 1950.0 coordinates, apparent magnitude, spectral type, and proper motion. The book should be in every observatory reference collection since no comparable work is available.

93.   Massachusetts Institute of Technology. **Wavelength Tables**. Rev. ed. Cambridge, Mass., The MIT Press, 1969. 429p. $40.00. LC 73-95288. ISBN 0-262-08002-8.

An indispensable volume for the astronomer studying stellar spectral lines, this book substantially updates and corrects the 1939 edition edited by George B. Harrison. The full title of the book explains the exact character of the data in this reference work: "Wavelength Tables with Intensities in Arc, Spark, or Discharge Tube of More than 100,000 Spectrum Lines Most Strongly Emitted by the Atomic Elements under Normal Conditions of Excitation between 10,000 Å and 2,000 Å Arranged in Order of Decreasing Wavelengths." Supplementary information includes keys to the symbols used, tables of corrections, tables of 500 sensitive lines, and other data and keys. Essential for both astronomy and physics libraries.

94.   Meinel, Aden B., Anthony F. Aveni, and Martha W. Stockton. **Catalog of Emission Lines in Astrophysical Objects**. Tucson, Optical Sciences Center and Steward Observatory, University of Arizona, 1968. 195p. bibliog. $4.00pa.

The contents of this "catalog" are the consolidation of emission line information from a variety of sources in the literature for the period 1930-1966. In addition to stellar and non-stellar astronomical sources, the authors have included contamination sources (airglow, city lights, aurora, and lightning) that can adversely affect observations of celestial objects. There are two main tables (A & B), and three supplementary tables (C, D, E): A. Observed Wavelengths (principal emission lines observed in individual classes of emission objects);

B. Identifications and Intensities (for items in Table A, arranged by wavelength); C. Object Identifications; D. Wavelength range and instrumentation for each reference cited; and E. Bibliography. The inclusion of so much information in one handy volume makes this a reference book worth having in any observatory.

95. Moore, Patrick. **The Atlas of the Universe.** London, Mitchell Beazley Ltd; distr. Chicago, Rand McNally and Co., 1970. 272p. illus. index. glossary. £11.75. $35.00. LC 71-550003. B 70-28399. ISBN 0-540-05153-5 (Beazley); 0-528-83041-4 (Rand).

Containing a multitude of excellent illustrations and dozens of astronomical topics, this work is difficult to describe adequately. Moore has divided this fine atlas into five major parts, each including historical material, photographs from space probes, and countless astronomical facts. These five sections are Observation and Exploration of Space; Atlas of the Earth from Space; Atlas of the Moon; Atlas of the Solar System; and Atlas of the Stars. This superb volume cannot be recommended enough; more than an atlas and more than a picture book, it is also an excellent text that is destined to become a standard part of the public, school, and college library.

96. Moore, Patrick. **Color Star Atlas.** London, Mitchell Beazley Publishers, Ltd; distr. New York, Crown Publishers, 1973. 112p. illus. index. notes. £2.95. $7.95. LC 73-88403. ISBN 0-517-51403-6 (Crown); 0-7188-2084-3 (Beazley).

The aim of this beautiful atlas is to give a general introduction to the stellar universe, and Moore succeeds quite well in meeting his goal. Extremely well illustrated, the book begins with an introduction to the stars, followed by a chapter on the types of stars and then a discussion of galaxies. The color star maps, interspersed with beautiful color photographs, are the best part of the volume. The effectiveness of these superb maps is slightly decreased because they are bound into the center margin, thus obliterating some stars. A chapter on observing follows, complete with constellation maps. A good book for the beginning amateur and interested layman; recommended for the public library.

97. Nilson, Peter. **Uppsala General Catalogue of Galaxies.** Uppsala, Sweden, Uppsala University, 1973. 456p. refs. bibliog. (Uppsala. Universitet. Astronomiska Observatoriet. Annaler. Bd.6). ISBN 91-554-0064-7.

Based on the Palomar Sky Survey prints, this catalog lists all galaxies north of declination -2° 30′, 12,921 in all. Galaxies were included based on the criteria of a limiting diameter of 1.′0 and apparent magnitude of 14.5 and brighter (including those smaller than 1.′0 in diameter). The objects are arranged according to ascending Right Ascension (epoch 1950.0). Extensive information accompanies each entry: R.A. and Dec., new galactic coordinates, various identification numbers from several catalogues, diameters, luminosity class, photographic magnitude, radial velocities, etc. Listed after the main catalog are extensive notes for a large number of the entries. A supplement* issued a year later includes over 400 galaxies not in the first work because their apparent diameters were too small

---

*Nilson, Peter. *Catalogue of Selected Non-UGC Galaxies.* Uppsala, Uppsala Astronomical Observatory, 1974. 16p. Observatory Report no. 5.

or because they were located below the UGC declination of -2° 30´. Appropriate mainly for the observatory collection, this is an important part of the sky catalogue literature.

98. Rahe, Jürgen, Bertram Donn, and Karl Wurm. **Atlas of Cometary Forms, Vol 1: Structures Near the Nucleus.** Washington, D.C., NASA, 1969. 128p. illus. index. $2.25. NASA SP-198.

There are all sorts of atlases these days, so why not an atlas of comets? This volume lives up to the usual high quality material from NASA. Reproductions of drawings made from visual observations of nineteenth and twentieth century comets are followed by a description and illustrations of two comets for which both visual and photographic observations were made; this makes an interesting comparison. The third and largest section of the book takes an in-depth look at three of the brightest comets of this century: Comet Morehouse 1908 III, Comet Halley 1910 II, and Comet Humason 1962 VIII. The final part of the atlas consists of photographs of six additional comets, in which the authors "illustrate the effect of telescopic exposure time and of plate and filter characteristics upon the appearance of a comet." Consequently, an historical perspective on cometary research can be gained from this book as well.

As far as possible, pertinent facts about each photograph or drawing are given: observatory, observer, telescope, plate, and filter; mid-exposure time, duration of exposure, heliocentric distance, geocentric distance, phase angle (sun-comet-earth), enlargement factor, and scale of reproduction. If this first volume of a series of books on comets is any indication of things to come, there is much to look forward to. The amount of deatil and the number of illustrations make this atlas a worthwhile purchase for nearly any library with an astronomy collection.

99. Rükl, Antonín. **Maps of Lunar Hemispheres.** Dordrecht-Holland, D. Reidel Publishing Co., 1972. 24p. 6 maps. refs. (Astrophysics and Space Science Library, v. 33). $22.90. LC 77-179896. ISBN 90-277-0221-7.

Subtitled: "Giving the views of the lunar globe from six cardinal directions in space," this fine book is not really a book, but rather a set of six lunar maps, accompanied by an index and some introductory material. Using photographs from earth-launched lunar satellites, the author's "six maps present to the reader the Moon as it would look from six mutually perpendicular cardinal directions in space—its near and far side, its eastern and western hemispheres, as well as the hemispheres seen from both poles." The introduction, titled "Maps of Lunar Hemispheres," concentrates on four topics: 1. Coordinates of Places on the Lunar Surface, 2. Orientation of the Six Maps of the Lunar Hemisphere and Their Representation, 3. Construction of the Maps, and 4. Lunar Nomenclature. The excellent maps should be studied by any serious lunar scholar.

100. Sandage, Allan. **The Hubble Atlas of Galaxies.** Washington, D.C., Carnegie Institute of Washington, 1972, c1961. 50p. illus. (Carnegie Institute of Washington. Publication 618). $12.50pa. LC 60-16568. ISBN 0-87279-629-9.

This work is an atlas of photographs whose purpose is to illustrate the Hubble classification system for galaxies. The book begins with historical background on galaxies, then moves to an explanation of the classification scheme. The bulk of the atlas is a section of photographs with text illustrating the various galactic types—spirals, barred spirals, elliptical, irregular, etc. A short but important volume for observatory and university libraries.

101.   Seitter, Waltraut Carola. **Atlas for Objective Prism Spectra**. Bonn, West Germany, Ferd. Dummlers Verlag, 1970. 56p. 65plates. refs. $55.00. (boxed). LC 70-565777. ISBN 3-427-70152-2 (English ed.): 3-427-70151-4 (German ed.).

Because of its high potential use by astronomers, this volume should be in every observatory library. The atlas is a compilation and discussion of objective prism spectra of standard stars taken with three different resolutions. The introduction, in English and German, includes a list of the standard stars used and a catalog of elements. The plates are photographic positives packaged loosely in a box, arranged by spectral type. All spectral lines are identified for each star on the photographs. The catalog of elements mentioned above is especially good: it discusses each element's importance in stellar spectral classification, emission and absorption lines, etc. The second part of this set (ISBN 3-427-70161-1) was published in 1973, but was not examined by the author of this guide.

102.   Smithsonian Astrophysical Observatory. **Star Atlas of Reference Stars and Nonstellar Objects**. Cambridge, Mass., The MIT Press, 1969. 13p. 152 plates. bibliog. $20.00. NUC 72-30373. Available from Publications Division, SAO, 60 Garden Street, Cambridge, Mass. 02138.

Probably the most accurate and most used star atlas available, this professional work has 152 individual (loose) star charts covering the entire sky. Including nearly all stars down to 9th magnitude, and some as faint as 11.0, the atlas includes normal stars, multiple systems, variable stars, stars with unknown proper motions, and non-stellar objects (e.g., galaxies, planetary nebulae, and star clusters). Of very high quality, this work could be used by both serious amateur and professional, but it would be more useful to the latter. There is a lack of specific star designations because of scale, but these can be determined with the use of various star catalogs, like the SAO. Useful information is included: an explanation, limits of charts, constellation abbreviations, and star names. Highly recommended; and for the price, you can't beat it.

103.   Smithsonian Astrophysical Observatory. **Star Catalog**. Washington, D.C., Smithsonian Institution, 1966. 4v. refs. bibliog. tables. (Smithsonian Institution. Publications. no. 4652). $37.50. LC 65-62534. Available from the U.S. Government Printing Office.

Reprinted in 1974, this popular star catalog contains positions and proper motions of 258,997 stars for the Epoch and Equinox of 1950.0. Available both on magnetic computer tapes, and in book form, this work was compiled because of the need for an accurate, computer-accessible star catalog needed in the development of a system for tracking artificial satellites in the late 1950s. The catalog is divided into four volumes or "parts," each covering a portion of the sky. Information used in constructing this work was obtained from various earlier catalogs, which are listed in the introduction. Data for each star include photographic and visual magnitudes, coordinates, proper motion, spectral type, and much more. An indispensable tool for the working astronomer.

104.   Sulentic, Jack W., and William G. Tifft. **The Revised New General Catalogue of Nonstellar Astronomical Objects.** Tucson, Arizona, University of Arizona Press, 1973. 384p. refs. (available from Sky Publishing Corp., Cambridge, Mass.). $10.50. LC 73-83378. ISBN 0-8165-0421-0.

A valuable, high-use item, this book is an updated version of J. L. E. Dreyer's *New General Catalogue of Nebulae and Clusters of Stars* (1888), which itself was a revision of John Herschel's *General Catalogue* (1864). In producing this much-needed work, the authors have greatly revised, corrected, and enlarged the *NGC*, itself a standard catalog for many years. The result, the *RNGC*, contains the following useful information for each entry: NGC identification number; type of object; R.A. and Dec. (1975); galactic coordinates; x, y position of object on the Palomar Sky Survey; magnitude of object; source of magnitude; visual appearance of object transcribed from Dreyer; Palomar Sky Survey description of each object; and cross references for *RNGC* objects. An essential part of the observatory reference collection, this work is *the* catalog of celestial objects other than stars.

105.   Vehrenberg, Hans. **Atlas of Deep-Sky Splendors.** 2nd ed. Düsseldorf, Treugesell-Verlag; distr. Cambridge, Mass., Sky Publishing Corp., 1971. 215p. illus. refs. $22.00. LC MAP 67-12.

This superb atlas should be seriously considered by anyone who does a lot of observing, whether amateur or student astronomer. The objects included in Charles Messier's catalog are the main thrust of this photographic work, which presents these nebulae, galaxies, clusters, etc., in a manner not previously attempted. The author has gathered excellent photos of the star fields containing the Messier objects, long a favorite of neophyte and experienced observers, and has arranged them in a format ideal for use by the amateur. Reproduced on a uniform scale, each plate is accompanied by a detailed description of the objects in the field, historical notes, R.A. and Dec., and references. A finder chart is included for each area to assist the observer in locating the celestial objects. An appendix contains a list of the galaxies and other objects by NGC (*New General Catalogue*) number. Finally, three beautiful mosaic, photographic maps of the Milky Way are inserted at the end. Translated from the German (*Mein Messier Buch*), this book is greatly recommended for public, personal, and astronomy libraries.

106.   Zwicky, Fritz. **Catalogue of Galaxies and of Clusters of Galaxies.** Pasadena, California Institute of Technology, 1961-68. 6v. illus. LC 67-7599.

Unmatched by any similar work, this impressive catalog consists of two distinct parts: "(a) a list of individual galaxies, designed to be complete to apparent magnitude m_p = +15.5 and (b) a list of clusters of galaxies to the limit of the 48-inch Schmidt telescope on Palomar Mountain. Unlike many other catalogues, it does not attempt to cover the entire sky all in one piece, but is split into sections of 36 square degrees, each covering the area of an individual print of the National Geographic-Palomar Observatory Sky Survey."

The following data are included in this excellent set: 1) for galaxies: 1950.0 coordinates, apparent photographic magnitude, NGC/IC numbers, radial velocities, remarks; 2) for clusters: 1950.0 coordinates of the centers, populations, estimated distances, diameters, and character. Additionally, there are charts for each section of the catalog showing locations of galaxies and clusters, and the galaxies' brightnesses and the clusters' contour lines are indicated. The excellent introduction defines terms and includes a general description of the atlas and its construction. A fine compilation, recommended for the observatory collection.

SOURCES OF SCHOLARLY CONTRIBUTION

Professional Journals

It should go without saying that journals are the most important, up-to-date source of information for the professional astronomer. It is through the many current astronomical and astrophysical periodicals that scientists keep up with the latest developments in their field. Naturally, the astronomer reads not only serial publications strictly defined as astronomical, but many related journals as well, including those in physics, mathematics, computer science, and so on.

This guide lists about 40 journal publications, most of which fall into the category of professional magazines. The list is by no means complete, nor does it include any related periodicals (with one exception). The reader should seek out such journals in other subject bibliographies. The section below deals only with the so-called professional journals, those which mainly include reports of original research of an advanced nature. The less technical serials are listed under the heading of "General Periodicals" in Chapter 2. Some astronomy magazines have material for both the scientist and general reader, but since such items are usually not as technical as the professional journals, they, too, are in Chapter 2 with the more popular publications.

In this section, three journals worth emphasizing are *Astrophysical Journal, Astronomical Journal,* and the *Monthly Notices of the Royal Astronomical Society*. The three most important and prestigious professional publications, they are the ones the astronomer most often reads and hopes to get published in. A newcomer that is gaining in popularity and distinction is *Astronomy and Astrophysics,* a European journal formed several years ago by the consolidation of several lesser-known publications. Articles in these four journals, like the other professional serials, are technical reports of research results, abstracted, heavily referenced, and full of mathematics.

The subject coverage of the periodicals in this section is quite broad. Some publications are general in nature, like the four previous titles, covering every possible topic of interest in astronomy. On the other hand, there are journals that focus on one area of the field, specifically limiting coverage to that area. Examples are *The Moon* and *Solar Physics*. Further, there are journals that can be categorized as publications of astronomical societies. *The Monthly Notices*, already mentioned, is one such example; others are the *Journal of the British Astronomical Association* and the *Publications of the Astronomical Society of Japan*. The society publications have both broad-based and specialized subject matter, but the emphasis is on the former. They are intended sometimes for professional astronomers and occasionally for amateur astronomers.

The astronomer also regularly reads three general science periodicals that frequently include articles and primary reports on astronomy: *Science* and *Scientific American* (U.S.) and *Nature* (Great Britain). These sources of current information are not to be overlooked; indeed, it was in *Nature* that one of the most significant astronomical discoveries in recent years was announced, the identification of pulsars. These three titles are almost always included in observatory libraries and in many respects are as important as the journals that concentrate on astronomy. Besides containing articles on astronomy and astrophysics, they keep the scientist aware of developments in related fields.

The observatory and university astronomy library will want to have as many journals as it can afford, in order to support the research effort. The journals, after all, are the astronomers' prime information source. Most astronomy libraries have all or most of the titles listed below, as well as a few related titles in physics and some general scientific titles like the three mentioned in the previous paragraph. Bibliographic control of the journal collection is handled using the *Union List of Serials, New Serials Titles, Ulrich's International Periodicals Directory*, and similar aids. The latter work is especially good for ordering a subscription, while the former two are useful for locating materials on interlibrary loan and for bibliographic verification. *New Serials Titles* is especially helpful for locating recently published astronomical serials. The *Subject Guide to New Serials Titles, 1950-1970* (R. R. Bowker, 1975, 2v.) is a handy aid for isolating astronomical serials of all types. Arranged by Dewey Decimal Classification, this work lists 675 astronomical publications—from journals to proceedings to observatory publications.

Each entry in this section gives the publisher, dates of publication, frequency, subscription information, and a brief evaluative note. Readers should keep in mind that the prices of journals change constantly, so the figures presented here may be out of date by the time this guide is published. It should be mentioned, too, that personal and institutional subscriptions for professional journals differ greatly in prices; libraries are often charged several times the personal subscription cost.

107.   **Acta Astronomica.** Warsaw: Panstwowe Wydawnictwo Naukowe. v.1– , 1925– . Quarterly. illus. refs. abstracts. volume author and subject indexes. LC 61-46388. ISSN 0001-5237.

Articles in this journal are written for professional astronomers; like similar publications, it covers a wide variety of topics in astronomy and astrophysics. The papers range from fewer than five pages to more than forty, and mainly cover astrophysical problems, spectroscopy, and observation of individual types of celestial objects. The articles in this secondary source, which are reports of original research from Eastern Europe, follow standard technical paper format: abstract, introduction, diagrams, references, etc. The text is in English, with an occasional French paper. Like many of the lesser-known and -used journals, this periodical is not essential to the observatory library's collection, but it should be purchased if funds permit. Published by the Committee of Astronomy of the Polish Academy of Sciences, it was issued in three parts (A, B, & C) from 1925 to 1955; in 1956 these were combined into one publication, continuing the number of series A and C (v. 6– , 1956–).

Write: Foreign Trade Enterprise, "ARS POLONA–RUCH", 7 Krakowskie Przedmieście, 00-068 Warszawa, P.O. Box 1001, Poland. Cost: $8.00/year.

108.   **American Astronomical Society. Bulletin.** New York: American Institute of Physics. v.1– , 1969– . Four/yr. (2 parts/issue). indexes. ISSN 0002-7537.

Not the typical astronomical journal containing reports of research or review articles, this publication is a record of the activities of the major U.S. astronomical organization, the American Astronomical Society. The bulk of the seven issues each year consists of abstracts of papers presented at the meetings of the Society and its divisions, with additional notes and reports of interest to the astronomical community at large appearing from time to time. The other major feature of the *Bulletin* is the observatory reports, which appear mainly in the first issue each year. These annual summaries of the activities of American observatories is a valuable record of the current state of astronomical research in the United States. This feature alone makes the price worthwhile.

Intended for members of the AAS and other interested persons, including astronomers, students, and laymen, this publication is normally received through membership in the Society. Not all libraries will want this serial, but most astronomical libraries will.

Write: American Institute of Physics, 335 E. 45th St., New York, N.Y. 10017. Cost: $15.00/year.

109.   **Astronomical Institutes of Czechoslovakia. Bulletin.** Prague: Publishing House of the Czechoslovak Academy of Sciences. v.1– , 1950– . Bi-monthly. illus. refs. abstracts. annual author and subject indexes. ISSN 0004-6248.

Articles in this Eastern European journal are scholarly, technical descriptions of original research. While not devoted entirely to any one astronomical area, the decided emphasis of the papers here is on solar and solar system research, with a large portion devoted to meteoritic studies. The text is mainly

English, but some French, German, or Russian appears from time to time; abstracts and indexes are in English and Russian. Papers are usually less than 10 pages long and follow the format of other similar professional publications. Occasional book reviews are included.

Write: Academic Press, Inc., 111 Fifth Avenue, New York, N.Y., 10003, or Academic Press, 24-28 Oval Road, London NW1 7DX England. Cost: £15.00/year.

110. **Astronomical Journal.** New York: American Institute of Physics. v.1– , 1849– . Monthly. illus. refs. abstracts. annual author and subject indexes. ISSN 0004-6256.

On a par with *Astrophysical Journal (ApJ)* in quality and importance, this journal is less frequently published and less voluminous. Like *ApJ*, it emphasizes stars and galaxies, but there are more frequent articles here on the solar system; papers on celestial mechanics and the planets are not rare. A large number of the articles are devoted to radio astronomy, a trend that is hardly surprising, considering the growth of that field. An abstract precedes each paper, but there are no subject headings or key words with it as in *ApJ*; this is unfortunate for the librarian but probably of little consequence for the astronomer. Editorial guidelines require that each paper have an introduction and various subheadings, which makes reading and searching a little easier. Like most professional periodicals, illustrations are mainly diagrams and occasionally plates. Extensive references accompany each paper. Length of articles is usually 10 pages or less, but infrequently a longer paper will appear, usually when tables of data are the main theme. The subject indexes that accompany each volume are especially useful; the librarian will appreciate the time taken to produce them. Aimed at astronomers and graduate students, this journal, published for the American Astronomical Society by the American Institute of Physics, is the oldest U.S. astronomical publication. A 50-volume cumulative index (1849-1944) is available.

Write: American Institute of Physics, 335 East 45th Street, New York, N.Y., 10017. Cost: $40.00/year (paper or microfiche).

111. **Astronomical Society of Japan. Publications.** Tokyo: Japan Publications Trading Co., Ltd. v.1– , 1949– . Quarterly. illus. refs. abstracts. annual author and subject indexes. ISSN 0004-6264.

Like the other professional journals in astronomy, this periodical publishes reports of original research on dozens of topics. But whereas other journals emphasize the observational results of such work, the Japanese seem to be more concerned with theoretical studies than their Western counterparts. Consequently, many papers here are concerned with models of stars, galaxies, planetary conditions. And although many of the articles use mathematics heavily to prove a point, a great many rely mainly on words to explain their point of view. Some papers even begin to resemble review-type articles. The author will discuss previous research done in his area enroute to setting up his model situation. The papers clearly show a great interest in stellar and galactic evolution, and many authors discuss models for stars at various stages of the evolutionary scale. Other frequent topics are radio observations, solar studies, and the interstellar medium. Papers follow standard format for professional journals and are usually 10 to 15 pages long. Shorter communications or comments on research on previous papers, called "Notes," are included on a regular basis. While not a primary souce, it should be considered for the astronomy library if funds are available.

Write: Maruzen Co. Ltd. (Export Department), Box 5050, Tokyo International, Tokyo, Japan. Cost: $53.25/year.

112.   **Astronomical Society of the Pacific. Publications.** San Francisco: Astronomical Society of the Pacific. v.1– , 1889– . Bi-monthly (even months). illus. refs. annual author and subject indexes. ISSN 0004-6280.

In accordance with the goal of keeping Society members up to date with astronomical research, the editors include both technical and non-technical papers, including historical matter, in each issue. Articles are fairly short (2 to 10 pages) when compared to similar papers in other journals, but this is of little consequence. The articles here are as good as any to be found. Among them are "refereed research reports, invited topical reviews, and abstracts of papers given at scientific meetings." Most articles concern stellar studies: observations, photometry, spectroscopy, etc., but many other subjects are covered as well. Tables of observational data are not uncommon, and occasionally these lists are quite long. Papers are written for professional astronomers but could be read and appreciated by advanced amateurs; there is not usually an overwhelming amount of mathematics and equations to wade through. *PASP* would be appropriate for any observatory library and many university collections as well.

Subscriptions are available through membership in the Society. Write: Astronomical Society of the Pacific, 1244 Noriega St., San Francisco, California, 94122. Institutional memberships are $40.00/year and include both publications (*PASP* and *Mercury*). Individual memberships: $25.00/year (both journals included); $12.50/year (*Mercury* only).

113.   **Astronomische Nachrichten.** Berlin: Akademie der Wissenschaften zu Berlin. v.1– , 1821– . 6 issues/yr. illus. refs. abstracts. annual author and subject indexes. LC 13-11169-73. ISSN 0004-6337.

The articles in this professional journal are scholarly, technical articles of original astronomical research. The text is in English, French, or German, with summaries in English and German. Unfortunately for English speaking/ reading astronomers, articles are frequently not in English, and the book reviews and subject indexes are in German. Topics cover the spectrum of astronomical subjects from stellar evolution to celestial mechanics to planetary studies. Papers are usually less than 10 pages long. Not to be considered a primary source of astronomical information (compared to *Astronomical Journal, Astrophysical Journal*, etc.), it is nevertheless an important secondary source.

Write: Akademie-Verlag GmbH, Leipziger Str. 3-4, 108 Berlin, East Germany. Cost: DM 72/year (about $22.50).

114.   **Astronomy and Astrophysics; a European Journal.** Berlin: Springer-Verlag. v.1– , 1969– . Semi-monthly. 8 vol/yr. illus. refs. abstracts. annual subject and author indexes. LC 74-220573. ISSN 0004-6361.

Formed by the merger of several astronomical journals (*Annales d'Astrophysique, Arkiv für Astronomi, Bulletin of the Astronomical Institutes of the Netherlands, Bulletin Astronomique, Journal des Observateurs, Zeitschrift für Astrophysik*), this publication reduces the scattering of research results in many periodicals, and promotes European cooperation. The continental counterpart of *Astronomical Journal* and *Astrophysical Journal*, it contains technical papers on all aspects of astronomy and astrophysics, written mainly by European astronomers. The only astronomical journal to include a standard table of contents, which guides the overall format and causes a balancing of topics, *Astronomy and Astrophysics* is subdivided as follows: I, Stars and Stellar Evolution; II, Galactic Structure, Stellar Dynamics, Interstellar Matter; III, Galaxies; Cosmology; IV, The Sun; V, Physical Processes and Astrophysical Plasmas; VI, Planetary System; VII, Celestial Mechanics and Astrometry; VIII, Instruments and Data Processing. The latter three sections are usually given less emphasis than the first five, but this can vary according to the individual issue. Research results based on radio studies are frequent.

The format of the papers is consistent with that used in similar journals: a summary (abstract) with keywords, introduction, subdivisions, references; length is 10 to 30 pages. Features include an occasional letter-to-the-editor and research notes, both in the form of very brief papers. The text is in English, French, or German, with English abstracts; the norm is English, however.

Write: Springer-Verlag, 175 Fifth Avenue, New York, N.Y. 10010. In Europe: Springer-Verlag, Berlin 33, Heidelberger Platz 3, Germany. Cost: DM 80 (about $35.50/year) for a personal subscription; DM 175/vol; DM 1400/year for institutions. Prices do not include postage and handling.

115.   **Astronomy and Astrophysics; a European Journal. Supplement Series.** Berlin: Springer-Verlag. v.1– , 1970– . Monthly. Four vol/yr. illus. refs. abstracts.

An average of 5 to 10 papers comprise each issue of this publication, which includes primarily "detailed results of interest to a small number of specialists, especially long tabular material and atlases of basic observational data." These papers are usually not too long, but they do emphasize numerical data, in keeping with the journal's purpose. Articles follow the format of the present journal and cover the spectrum of astronomical topics, but the emphasis is on stellar research.

Write: Springer-Verlag, 175 Fifth Avenue, New York, N.Y., 10010 for subscription information. Cost: DM 75/vol for institutions; free to individuals if already subscribing to *Astronomy and Astrophysics.*

116.   **Astrophysical Journal: An International Review of Astronomy and Astronomical Physics.** Chicago: University of Chicago Press. v.1– , 1895– . Semi-monthly (two parts each issue): 8 vol/yr. illus. refs. abstracts. annual and volume author indexes. annual subject indexes. LC 17-24351. ISSN 0004-637X (paper); 0091-8768 (microfiche).

If astronomers could have a choice of which journal to publish a paper in, this would surely be the periodical most frequently chosen. The most prestigious serial in the astronomical world, this publication consists of scholarly articles by and for professional astronomers. Labelling the subject matter as merely astrophysics papers would be both too general a statement and an injustice. The thrust is mainly stars and galaxies, with a heavy smattering of other topics ranging from the interstellar medium to the solar system. Since articles are reports of observations and research, the topics are usually specific: the study of spectral lines in a particular type of star, the evolution of a certain type of galaxy, the solar wind, etc. Reports include both optical and radio studies, the latter occupying more and more space each year.

Papers include an abstract (with subject headings at the end!) and are subdivided with appropriate paragraph headings, including an introduction and such topics as observations, results, analysis, conclusions, etc. There are many references as well. It is obvious that there are very strict editorial guidelines.

Each semi-monthly issue consists of two parts: the regular papers and the "letters-to-the-editor" section. The purpose of the latter category is to provide faster publication of research results. The length of papers in the journal section is rarely more than 20 pages, (most are 6 to 10 pages); the "letters" section articles are generally 2 to 6 pages long.

Published for the American Astronomical Society (AAS) by the University of Chicago, this journal has three "general" indexes covering vols. 101-135, 136-145, and 146-165. Annual subject indexes for the *Journal* and *Supplement Series* began in 1973.

Write: University of Chicago Press, 5801 Ellis Avenue, Chicago, Ill. 60637. Cost: $130.00/year (paper or microfiche editions); $234.00/year (both). Members of the AAS receive cheaper rates.

117. **Astrophysical Journal, Supplement Series**. Chicago: University of Chicago Press. v.1– , 1954– . Irregular. illus. refs. abstracts. volume author and subject indexes. LC 56-37588. ISSN 0067-0049 (paper); 0093-187X (microfiche).

An offshoot of *Astrophysical Journal (ApJ)* this series provides an opportunity to publish extensive investigative information. Papers are longer than the ones usually included in *ApJ* (often 20 to 50 pages), and usually one such report occupies an entire issue. The issues comprise two volumes per year with a total of about 500 pages. Format includes an abstract with subject headings or keywords, an introduction, subheadings, and references. Articles contain extensive tables of data—for example, stellar positions and magnitudes, information about theoretical stellar models, etc. Subject matter is similar to *ApJ* papers: stellar studies, galactic research, etc., both optical and radio astronomy.

Write Astrophysical Journal Supplement Series, University of Chicago Press, 5801 Ellis Avenue, Chicago, Illinois 60637. Cost: $16.00/vol (paper or microfiche); $28.80/vol (both).

118.   **Astrophysical Letters**. New York: Gordon and Breach. v.1– , 1967– . Monthly. 3 vol./year. illus. refs. abstracts. volume author and subject indexes. ISSN 0004-6388.

Very brief papers on research in progress or small projects, usually observational in nature, make up this atypical journal for the astronomer. These short communications of 2 to 10 pages give astronomers an opportunity to publish results more quickly than in a journal such as *Astronomical Journal*, where articles are much longer. The shortness of papers, however, is the only distinguishing factor here. Papers follow standard format: abstracts, etc., and cover the usual stellar and galactic research with both optical and radio observations. Includes book reviews.

Write: Gordon and Breach Science Publishers Ltd., One Park Avenue, New York, New York, 10016. In Europe, write: Gordon and Breach, 41/42 William IV Street, London WC2 England. Cost: $12.00/year (£10.00) (individuals); $55.00/year (£25.50) (libraries).

119.   **Astrophysics**. New York: Plenum Publishing Corp., v.1– , 1965– . Bi-monthly. 4 issues/vol. illus. refs. abstracts. vol. author indexes and table of contents. ISSN 0004-6396.

Astronomers should be aware of this journal, which is the English translation of *Astrofizika*, a publication of the Academy of Sciences of the Armenian SSR. Like its Western counterparts, it includes scholarly technical articles reporting research in the various fields of astronomy and astrophysics. Topics emphasize studies of stars and star clusters, galaxies, and the interstellar medium. Spectral observations of numerous galactic and extra-galactic objects are common, as are radio studies of star and star-like objects. Papers, usually no longer than 10 pages, contain an abstract, introductory statement, and bibliography. So-called "Brief Communications," two or three pages long, are included in each issue, reporting on research in progress. The journal began to include selected review-type articles in the January-March 1974 (Soviet date) issue.

The lag time between the publication of the Russian version and the translation ranges from six to nine months, so the articles are not as up to date as could be wished. This is the journal's only drawback.

Write: Plenum Publishing Corp., 227 W. 17th Street, New York, N.Y., 10011. Cost: $145.00/4 issues.

120.   **Astrophysics and Space Science: An International Journal of Cosmic Physics**. Dordrecht-Holland: D. Reidel Publishing Co. v.1– , 1968– . Monthly. 7 vol./yr. illus. refs. volume author indexes. ISSN 0004-640X.

It is difficult to define clearly the subject emphasis of this technical publication; as its title indicates, it covers two broad areas, the latter of which is a quite varied field. The papers are studies of the stars and the interstellar medium, and the relationship between the two. The physical aspect of these subjects, rather than descriptive astronomical features, is the underlying theme. One can find reports of original research on stellar evolution, the solar wind, cosmic dust, spectroscopic binaries, radio studies of H_{II} regions, etc. Hundreds of different topics are explored, most based on observations, some purely theoretical and highly mathematical. Because of the diversity of subject matter, there is something for everyone, so this journal would be a good purchase for any astronomy library. Papers are 10 to 30 pages long and follow standard format.

To place a subscription one may use an agency as many libraries do, or write: D. Reidel Publishing Co., 38 Papeterspad, Box 17, Dordrecht-Holland. Cost: $68.00/vol; $476.00/year. A personal subscription is $24.00/vol.

121. **Australian Journal of Physics.** Melbourne: Commonwealth Scientific and Industrial Research Organization (CSIRO). v.1– , 1948– . Bi-monthly. volume index. ISSN 0004-9506.

One of the few regular sources of astronomical information in the Southern Hemisphere, this scholarly journal's main purpose is to present original research in all branches of physics. About one-fourth of the 125-page publication is devoted to astronomy and astrophysics, covering every imaginable topic, both radio and optical astronomy. Since there is no Australian astronomical journal, this source becomes even more important. Each issue contains papers about six to ten pages long, short communications that report on research in progress. The format of the articles is typical for an advanced serial: an abstract, text, illustrations, and references. An irregular supplement publishes cumulations of astronomical data like radio astronomy catalogues, sky surveys, and summaries of observations. The *Astrophysical Supplement* issues have abstracts, short textual sections, and references when appropriate. Their emphasis in the recent past has been on radio studies, including updates and additions to the *Parkes Catalogue of Radio Sources*. Not a primary source world wide, it is nevertheless a periodical that the librarian and astronomer should get to know.

Write: CSIRO, Box 89, East Melbourne, Victoria 3002. Cost: $A25/year.

122. **British Astronomical Association. Journal.** Cambridge: British Astronomical Association. v.1– , 1890– . $10.07/yr. illus. refs. index. LC 10-5073. ISSN 0007-0197.

Although this periodical is mainly of interest to members of the BAA and to British astronomers in general, it contains a great deal of information useful and worth reading outside England. In addition to reports of meetings and other business of the BAA, it also contains short articles of general interest to both amateur and professional astronomers, as well as letters to the editor and book reviews. The articles include treatments of historical subjects, current popular astronomical topics, and aids for the amateur astronomer. About one-half the journal is designed for practicing amateur and professional observers, with short articles reporting actual observations, arranged under such headings as Meteor, Radio Astronomy, Variable Star, Solar, Lunar, and more. It is one of the few astronomical publications that publish results from non-professional astronomers. Other features of note are the classified ads, notes from other journals, a crossword puzzle, obituaries, and meeting notices.

Write: British Astronomical Association, 39 New Road, Barton, Cambridge CB3 7AY, England. Cost: $12.75/year. Special rates for younger and older astronomers.

123. **Celestial Mechanics: An International Journal of Space Dynamics.**
Dordrecht-Holland: D. Reidel Publishing Co. v.1– , 1969– . 4 issues/vol. 2 vol./yr.
illus. refs. abstracts. volume author indexes. LC 40-230002. ISSN 0008-8714.

The study of the motions and gravitational attractions of the planets and other bodies of the solar system is probably the most rigorous and highly mathematical area of astronomy. It should not be surprising, then, that this journal, which considers these problems, is one of the "heaviest" in the literature of astronomy. Papers include both highly theoretical approaches to celestial mechanics and the more practical considerations related to the movements of artificial satellites. The major difference between this publication and other professional journals should be obvious—articles in *Celestial Mechanics* deal not with observational research, but with theoretical considerations. From astrodynamics to geodesy to particle dynamics, the range of topics is great, considering the motions and interactions of hypothetical bodies and the planets, their satellites, comets. Papers include abstracts, references, etc., and vary in length from 5 to 25 pages.

The articles are in English, French, German, and Russian, but most are in English. Each issue also contains "abstracts of forthcoming papers," to preview future articles. Physics, mathematics, and astronomy libraries may wish to subscribe to this publication which has material of interest in all three areas.

Place a subscription through an agency, or write: D. Reidel Publishing Co., P.O. Box 17, Dordrecht-Holland. Cost: $24.00/volume ($48.00/year) for individuals; $70.00/volume ($140.00/year) for institutions.

124. **Comments on Astrophysics: A Journal of Critical Discussion of the Current Literature.** New York, London, Gordon and Breach. v.1– , 1969– . Bi-monthly. illus. refs. index. (Comments on Modern Physics: Part C). LC 76-7519. ISSN 0010-2679.

The short papers in this unusual publication can be described as falling somewhere between editorials and review articles. Written for professional astronomers and students, the articles provide scientists a chance to offer constructive criticism and general observations about research carried out in astrophysics and its applications. The papers are 5 to 10 pages long, include references, and do not report on original research. Although the treatment of topics is general, the level of intellectual content is high. Laymen with little background in astronomy will probably not want to read this journal, so it is unsuited for most public libraries. Subject coverage is as broad and varied as the title indicates. There are four papers per issue.

Write: Gordon and Breach Science Publishers, Inc., One Park Avenue, New York, N.Y. 10016; or Gordon and Breach Science Publishers Ltd., 42 William IV Street, London WC2, England. Cost: $21.00/year for individuals (£7.00); $47.00/year for institutions (£19.00).

125. **Earth and Extraterrestrial Sciences: Reviews, Conference Reports and Professional Activities.** New York: Gordon and Breach. v.1– , 1969– . Irregular. 8 issues/vol. illus. refs. abstracts. LC 70-23378. ISSN 0070-7902.

Although the title indicates coverage of both earth science and astronomy, the decided emphasis through 1975 was on the latter. Each issue of this compact publication contains two or three papers (of 20 to 40 pages) highlighting some of the current problems in the field. Included mainly are reviews/summaries of important conference topics, but there are also general reviews and book reviews from time to time. The papers are abstracted and contain references in the standard professional paper format. Individual subjects have included (or will include) cosmology, astronomy, astrophysics, geology, geophysics, geochemistry, meteorology, oceanography, and space research. Articles are scholarly but not overly technical, and the non-specialist can read these papers to learn about current research in related fields without being overwhelmed. Not a primary source, but an important one, this journal is one of the few in astronomy that condenses and reports on professional meetings. Unfortunately, the journal's lack of regular publication and brevity of issues does not provide comprehensive coverage of topics.

Write: Gordon and Breach Science Publishers, 1 Park Avenue, New York, N.Y. 10016; or Gordon and Breach Science Publishers Ltd., 42 William IV Street, London WC2, England. Cost: $75.00/vol (U.S.); £32.00/vol (G.B.); £35.00/vol (elsewhere).

126. **Earth and Planetary Science Letters: A Letter Journal Devoted to the Development in Time of the Earth and Planetary System.** Amsterdam: North Holland, v.1– , 1965– . Monthly. 3 vol/yr. illus. refs. abstracts. volume, author index. LC 66-9932. ISSN 0012-821X.

Only a few of the scholarly articles in this journal are related to astronomy; the great emphasis is on the earth sciences. Papers are fairly brief (5 to 15 pages) and cover research in a wide variety of astronomical, geological, and related disciplines. Articles are in English, German, and French, with English dominating. The lack of a subject index is a disappointing drawback. As might be expected, the papers on the planets other than Earth are mainly concerned with lunar geology, meteorites, and planetary atmospheres. Definitely a tertiary source for astronomy collections, it should be considered where planetary studies are being conducted.

Write: North Holland Publishing Co., Box 211, Amsterdam, The Netherlands. Cost: $140.00/year (institutions); $40.00/year (individuals).

127. **Icarus: An International Journal of Solar System Studies.** New York: Academic Press. v.1– , 1962– . Monthly. 3 vol/yr. illus. refs. abstracts. volume author and subject indexes. LC 64-4524. ISSN 0019-1035.

The only journal entirely devoted to technical papers on the solar system, this periodical should be subscribed to by all astronomy libraries with even the smallest program in planetary studies. Topics run the gamut of possibilities:

the planets and their satellites, comets, meteors, and asteroids. Optical and radio observations of our locale in space, as well as infrared, photometric, spectroscopic, and ultraviolet readings are discussed. There are occasional papers describing data gathered from various spacecraft, but the prime emphasis is earth-based observations. Besides the "standard" subject matter, there are also some "exotic" topics like exobiology, cosmogony, etc. Regular lesser features are announcements of relevant meetings, bibliographies, book reviews, and meeting reviews. Papers follow standard format and are usually less than 10 pages long.

Write: Academic Press, Inc., 111 Fifth Avenue, New York, New York 10003. Cost: $45.00/vol; $135.00/year. $49.50/vol. outside the United States.

128.   **The Moon: An International Journal of Lunar Studies**. Dordrecht-Holland: D. Reidel Publishing Co. v.1– , 1969– . Monthly. 4 issues/vol; 3 vol/yr. illus. refs. abstracts. volume, author indexes. LC 75-13570. ISSN 0027-0903.

Before Apollo 11's triumphant voyage to the Sea of Tranquility, lunar studies held a respectable but out-of-the-limelight position. In 1969, however, all that changed, and now Moon research is much more prominent. Along with an increase in the study of our satellite came this professional journal, devoted entirely to such research. It remains the only publication of its kind in English. Papers cover countless subject areas, many non-existent before 1969, ranging from studies of the interior to atmospheric research. They include, but are not limited to, geology, seismology, particle studies, lunar history, topography, chemistry, and mathematics. In short, a multitude of interdisciplinary subjects are covered, making this the most important current source of lunar information. Papers vary greatly in length, from less than 10 to 40 or 50 pages, and follow the standard format. Most are in English, but some are in Russian, French, or German. Some issues are devoted to conference proceedings, but most consist merely of submitted papers.

Write: D. Reidel Publishing Co., P.O. Box 17, 38 Papeterspad, Dordrecht-Holland. Cost: $25.50/vol (individuals); $68.00/vol (institutions).

129.   **Planetary and Space Science**. Oxford: Pergamon Press, v.1– , 1959– . Monthly. illus. refs. abstracts. volume author and subject indexes. LC 60-3632. ISSN 0032-0633.

This technical periodical emphasizes the study of atmospheres, interplanetary space, magnetic fields, and related topics. These reports of original research employ data gathered from earth- and satellite-based observations, the latter becoming more and more important and frequent. The astronomer working the area of solar-terrestrial physics and astronomy will benefit most from this publication, which contains articles on the solar wind, the ionosphere, radio bursts, etc. Maximum length of papers is 15 to 20 pages, and each has an abstract, references, etc., in the standard professional journal format. In addition to the regular papers, there are short articles called "research notes" in the form of letters, and book reviews.

Write: The Circulation Manager, Pergamon Press, Headington Hill Hall, Oxford OX3 OBW, England. Cost: $20.00/vol. (individuals); $125.00/vol. (institutions).

130. **Radio Science: Journal of the United States National Committee, International Union of Radio Science.** Washington, D.C.: American Geophysical Union. v.1– , 1966– . Monthly (except August). illus. refs. abstracts. volume author indexes, and volume table of contents. LC 66-60016. ISSN 0048-6604.

Although there are no journals yet that are specifically for radio astronomers, this publication will certainly be useful to those scientists who "listen" to the stars. Actually, the radio telescope technician will more likely be interested in *Radio Science*, since the scholarly articles here deal with the study of radio waves and the equipment used to detect and send them. There are no papers on radio astronomy in this journal—the reader will find articles on this topic in the standard astronomical journals, like *Astrophysical Journal* and the *Monthly Notices of the Royal Astronomical Society*. The articles on radio communications and research are highly technical and are 5 to 15 pages long. There are abstracts and references, and the format follows that of any professional technical journal. For the radio astronomy observatory and electrical engineering library, this journal would not necessarily be suitable for all astronomy libraries.

Write: American Geophysical Union, 1909 K Street N.W., Washington, D.C., 10006. Cost: $15.00/year (personal); $40.00/year (institutional).

131. **Royal Astronomical Society. Monthly Notices.** Oxford: Blackwell Scientific Publications Ltd., v.1– , 1827– . 3 no/vol; 4 vol/yr. illus. refs. abstracts. volume author and subject indexes. LC 1-6399. ISSN 0035-8711.

One of the major professional astronomical journals in the world, *MNRAS* ranks with *Astrophysical Journal* and *Astronomical Journal* in importance. Articles take two forms: standard papers averaging 10 to 15 pages in length, and "short communications" (mini-papers, 2 to 6 pages long), which relate progress on research projects. Both include abstracts, introductory statements, illustrations, and references. The overall format is not unlike that of *ApJ* and equivalent publications. These reports of original research include stellar studies, galactic investigations, radio astronomy, and much more, but the emphasis is on astrophysics research. The oldest continuous astronomical publication also includes papers dealing with cosmology, space research, astrometry, and instrumentation, as well as some of the "new astronomies" such as x-ray, ultra-violet, and infrared. A cumulative index (v.1-91) is available.

Write: Blackwell Scientific Publications Ltd., Osney Mead, Oxford OX2 OEL, England. Cost: $95.00/year (includes *Memoirs*, a supplement series) (£120.00); $75.00/year (individuals outside the United Kingdom) (£30.00).

132. **Royal Astronomical Society. Quarterly Journal.** Oxford: Blackwell Scientific Publications Ltd., v.1– , 1960– . Quarterly. refs. abstracts. annual indexes. LC 64-50164. ISSN 0035-8738.

This journal could be classed as a non-technical professional periodical, since its main purpose is not to impart the results of research. Instead, it is a vehicle for communicating the news of the RAS, as well as including articles and review papers aimed at the non-specialist. These articles often are biased toward one view or another, a refreshing and unique occurrence in astronomical serial literature. There is no particular subject emphasis—articles range from the historical to current topics of interest. The majority of the journal, however, is concerned with the news of the RAS, including meetings, observatory reports, obituaries, library accessions, etc. Not all libraries will want this publication, since its main theme (RAS news) may or may not be of interest; but many will want it. Laymen and students would have no trouble reading the articles, many of which are intended for non-specialists.

Write: Blackwell Scientific Publications Ltd., Osney Mead, Oxford OX2 0El, England. Cost: $24.00/year (£8.00);

133.   **Royal Astronomical Society of Canada. Journal.** Toronto: Royal Astronomical Society of Canada, v.1– , 1907– . Bi-monthly (even months). illus. refs. volume indexes. cumulative index. (v.1-25; 26-60). LC 11-23485. ISSN 0035-872X.

Comparable to the *Journal of the British Astronomical Association* in that it is intended for both professionals and amateurs, this publication has more for the former group than the latter. Articles are sometimes reports of original research, but more often than not they are general interest papers with varying levels of technical content. Some explore historical subjects—from time to time there are treatises on Canadian astronomers and observatories—but most are concerned with current research projects on the stars, planets, etc. The maximum length of papers is between 30 and 40 pages, but most are around 5 to 10 pages. Besides regular articles, there are reports of the Canadian Astronomical Society meetings with abstracts of papers given, notes from observatories, book reviews, and a regular column called "Variable Star Notes," written by the director of the American Association of Variable Star Observers. Also, inserted in the middle of each issue is the *National Newsletter*, a report of the activities of the RASC. Observatory and university libraries will wish to consider this publication, which has something for all types of astronomers; larger public libraries may want it as well.

Write: Royal Astronomical Society of Canada, 252 College Street, Toronto, Ontario, M5T 1R7. Cost: $12.50/year membership fee (includes journal).

134.   **Solar Physics: A Journal for Solar Research and the Study of Solar Terrestrial Physics.**Dordrecht-Holland: D. Reidel Publishing Co., v.1– , 1967– . Monthly. 2 issues/vol. 6 vol/yr. illus. abstracts. refs. volume author indexes.

The Sun and its effects on the Earth are the main theme of this professional journal, which covers the corona, flares, interior, surface, magnetic field, sunspots, spectrum, wind and other emissions, radio and other non-visible light observations, research carried out by the orbiting solar observatories, and much more. Solar astronomers can best keep up to date by regularly reading this publication. Besides the regular papers (10 to 20 pages long), there are shorter features, including research notes, abstracts of forthcoming papers, and book reviews. The text is usually in English, with occasional articles in French, German, and Russian. Like the other Reidel astronomy journals, this has no subject indexes, which is both frustrating and disappointing.

    Write: D. Reidel Publishing Co., Box 17, Dordrecht-Holland. Cost: $14.00/vol or $84.00/year (individual); $64.00/vol or $384.00/year (institutional).

135.   **Soviet Astronomy.** New York: American Institute of Physics, v.1– , 1957– . Bi-monthly. illus. refs. abstracts. volume indexes. ISSN 0038-5301.

    Along with *Astrophysics* (Plenum) and the new *Soviet Astronomy Letters* (AIP), this translation journal helps provide Western astronomers with news of research being carried out in the USSR. The equivalent of such important periodicals as *Astrophysical Journal*, *Astronomy and Astrophysics*, etc. *Soviet Astronomy (Astronomicheskii Zhurnal)* is published by the Academy of Sciences of the USSR. No particular subject emphasis emerges—the coverage is as diversified as the research itself. Papers on astrophysics, spectroscopy, radio astronomy, galactic studies, and more, average less than 10 pages in length. The articles in any translated publication are slightly dated by the time they reach American readers, but this is nevertheless an important journal to have in any astronomy library. In addition to the regular papers, there are "Brief Communications," which report on-going projects, and short pieces on new Soviet astronomy monographs.

    Write: American Institute of Physics, 335 E. 45th St. New York, N.Y. 10017. Cost: $140.00/year. Microfilm (16mm or 35mm) is available.

Proceedings of Symposia, Conferences, and Workshops

    Astronomers, like other scientists and professional people, meet regularly at conferences, symposia, and workshops to discuss the latest developments in their fields. These meetings are sponsored by professional organizations like the International Astronomical Union (IAU), government agencies such as NASA, universities, and other scholarly and educational groups. The conferences are frequently attended by scientists from around the globe.

    The proceedings of the gatherings, the transcripts of the papers presented and of the discussions that follow, are a valuable astronomical information source, as important as the professional journal. In fact, the papers included in the volumes of proceedings bear a remarkable resemblance to the periodical article in format and content. Unfortunately, conference papers are not as up to date as journals since it is often months before the volumes of proceedings are issued. Such volumes come about according to a fairly standard procedure. After a conference is over, a pre-chosen proceedings editor reviews the papers from the meeting, edits them, and arranges them by topic or chronologically. Sometimes the papers are submitted in advance of the conference and the published volume is ready when the participants arrive, but this happens infrequently.

Despite the importance of conference proceedings, librarians and users should be aware of two hindrances to their use. The first, the lack of an index (whether subject, author, or both) is something that little can be done about. In their haste to get the papers published as soon as possible after the symposium, the editors often neglect to inlcude what should be the most important part of the book. Any scientific book without an index is difficult to use, but a conference volume is even worse. Such works are often over 1,000 pages long— just try to isolate a small bit of information without some guidance. Happily, many editors of proceedings have been wise enough to include a subject and author index. Books without indexes can be approached only through the table of contents, which may or may not be helpful, depending on how detailed the desired information is.

The second problem is that conference proceedings are often impossible to find, verify, or order. So many meetings take place that are directly or indirectly related to astronomy and other scientific fields that it is a real problem to index and bibliographically identify them all. Consequently, it is not uncommon for a librarian to be unable to locate a volume of proceedings either in the card catalog of the library (assuming a fairly large collection) or through the available bibliographic tools.

The *Anglo-American Cataloging Rules* (ALA, 1967) state that a conference proceedings volume should be entered in the catalog under the name of the conference; for example, *Symposium on Binary Stars* or *International Conference on Astrophysics.* The patron who knows the exact title or name of the conference will probably be able to find the book in the card file. But, if the precise name of the conference is not known, and this is not an uncommon situation, the user may have to look under a variety of headings to locate the volume (e.g., "Conference on . . . ," "Workshop on . . . ," "International Symposium . . . ," and on and on). Even if the patron knows the pre-conference title, the title used for the final published volume may be completely different. If it is at all possible, then, the correct title should be determined, using various bibliographic reference books like *Astronomy and Astrophysics Abstracts, Directory of Published Proceedings, Books in Print,* etc. The problem with this approach is that there is often a lengthy delay before a particular conference volume shows up in these sources.

Ordering and purchasing conference proceedings can be especially trying. The vast majority of them are not listed in standard acquisitions sources like publishers' catalogs, *Books in Print,* and *Publishers Trade List Annual.* And the little-known, older volumes are even more difficult to handle. An example of a hard-to-find volume of proceedings would be a book from a university-sponsored conference published by the university press. Standard reference sources often do not cover such volumes. Further, many proceedings go out of print so fast that it is very hard to buy a copy even one or two years after the meeting.

Fortunately, there are two excellent, specialized bibliographic aids that greatly assist the librarian who is trying to locate and/or place an order for a volume of conference papers. The most commonly used is the *Directory of Published Proceedings,* usually referred to as "InterDok," which is the publisher.

A newer tool is *Scientific, Technical, and Engineering Societies Publications in Print, 1976-77,* which provides further coverage. The librarian should consult these two books first when ordering proceedings.

Frequently, a patron wants only a particular paper from such a volume of proceedings. In that case, the librarian should consult several of the abstracting and indexing journals (see the heading "Abstracts and Indexes"), since many routinely cover the major conferences in their subject areas. The articles may then be obtained either as an interlibrary loan request (if the article was reprinted in a journal, or if it can be determined that another library holds the volume of proceedings itself) or on microfiche, say from NASA. In order to obtain the article on interlibrary loan, the librarian consults *New Serials Titles* and/or the *Union List of Serials* (Library of Congress) to determine which library has the proceedings in question. Though they do not comprehensively cover the location of proceedings volumes, these two works do cover transcripts of the major meetings.

The single most important source of proceedings in astronomy is the International Astronomical Union (the IAU), a worldwide organization of professional astronomers. This group holds frequent conferences, symposia, and colloquia throughout the world, and in most cases publishes, or arranges the publication of, the papers presented at these meetings. The D. Reidel Publishing Co. currently issues many of the IAU publications. There are some 70 IAU volumes of symposia proceedings, and dozens of volumes of colloquia proceedings. The *Transactions* of the IAU which accounts for further proceedings books, is issued in two parts, *Reports* and *Proceedings*. These voluminous publications reflect the current state of the worldwide astronomical research effort, so they are an extremely important source of information.

Bibliographically, these scores of IAU volumes are difficult to control. Since they have been issued by several publishers over the years, they are not always easy to locate or identify in indexes. There have been scattered efforts at lists of these publications, but no "grand" index yet exists. Such a work, a monumental but needed task, would be a very significant contribution to the literature of astronomy.

There is no separate listing of proceedings in this guide. Rather, there are many individual items, including IAU monographs, cited under the various subject categories here. It should be remembered that there are many, many more and that those listed here are only a small sampling of what's available.

136.   **Astronomical Society of Australia. Proceedings.** Sydney: Sydney University Press. v.1– , 1967– . Irregular. index. ISSN 0066-9997.

An infrequent information source on Southern Hemisphere astronomical research, this conference volume contains papers from the yearly meeting of the ASA. Each paperbound issue is about 75 pages long, containing 25 to 35 papers, called contributions, about one to three pages long. Other items include a summary of Society business, three or four invited papers, and other miscellanea. In general, the short contributed papers are reports of research in progress. Unfortunately there are no abstracts and no subject index. Not a fast, up-to-date source, these proceedings provide only a good summary of Australian astronomy, with an emphasis on radio studies. This publication is cited as an example of a conference volume issued on a fairly regular basis.

Write: Treasurer, CSIRO, Division of Radiophysics, Box 76, P.O., Epping, New South Wales 2121. Cost: $A3.00/issue.

137. **Directory of Published Proceedings**: Series SEMT (Science/Engingeering/Medicine/Technology). Harrison, N.Y.: InterDok Corp. v.1– , 1964– . Monthly. 10/yr. (September to June). ISSN 0012-3293.

This important bibliographic tool lists "preprints and published proceedings of congresses, conferences, symposia, meetings, seminars, and summer schools which have been held worldwide from 1964 to date." This directory makes it easier to find and identify elusive conference proceedings. Citations include place, title of meeting, sponsor, availability and source, cost, paging, editor, LC and ISBN numbers, etc. The arrangement is chronological, and there are three indexes in each issue: editor, location, and subject/sponsor. Issues are cumulated into an annual volume four months later; indexes are cumulated annually, until a five-year volume is completed. InterDok (the frequently used name for this serial) also serves as a clearing house or acquisitions center for the proceedings listed in the directory, subject to availability. An order form accompanies each current issue.

The publication is helpful to astronomy librarians, who will find citations to the publications of the International Astronomical Union (IAU) and American Astronomical Society (AAS) listed in most issues. "InterDok" should be used along with the R. R. Bowker publication, *Scientific, Technical, and Engineering Societies Publications in Print* for the most complete bibliographic coverage.

Write: InterDok Corp., 173 Halstead Avenue, Box 326, Harrison, New York, 10528. Cost: $92.50/yr. (U.S. and foreign); Annual cumulative volume: $45.00; Cumulated index supplement: $60.00.

138. **International Astronomical Union. Transactions of the General Assembly**. Dordrecht-Holland, D. Reidel Publishing Co., v.1– , 1922– . Irregular. English, some French. LC 30-10103.

Divided into *Reports on Astronomy* and *Proceedings of the General Assembly*, this serial provides a substantial record of astronomical research worldwide. The former is a thick volume of about 700 pages, containing summaries of the 48 IAU commissions, their committees and working groups. These reports are often lengthy and contain fairly complete bibliographies of the recent literature in a particular field. Published every three years, the *Reports* include recent discoveries, summaries of projects, and news of Commission activities, over the spectrum of astronomical topics.

The *Proceedings* are reports of the triennial General Assembly meetings (two in one year) including the gatherings of the individual commissions. Strictly a summary of the business meetings of the assembly at large and things like reports on finances, etc., these volumes also include useful information like membership, committee and commission structure and constituency, and miscellaneous material like the list of lunar and Martian nomenclature in volume 15B (1973). Along with the IAU's *Highlights in Astronomy* (the technical papers presented at the General Assembly—see under General Works), the *Transactions* provides a comprehensive record of the Union's activities, accomplishments, and aspirations. The IAU *Information Bulletin* (semi-annual) helps members keep up to date between issues of the *Transactions*.

139.   Kyed, James M., and James M. Matarazzo, eds. **Scientific, Engineering, and Medical Societies Publications in Print, 1976-77**.New York, R. R. Bowker, 1976. 509p. indexes. $19.95. LC 76-26086. ISBN 0-8352-0898-2.

A helpful aid for locating elusive conference proceedings and similar publications, this work helps the librarian identify, verify, and order important materials for the astronomy and general science collection. More than double the size of the first edition (1974-75), this version lists books, journals, and non-print media issued by 369 organizations worldwide. Its value lies in the fact that it contains more than just proceedings, therefore making it a useful supplement to the *Directory of Published Proceedings* (InterDok). While the first edition had only limited coverage of astronomical societies, being limited to the United States, this latest volume has expanded geographically to include important international groups like the *International Astronomical Union* and the *Royal Astronomical Society*. The inclusion of the IAU publications is of special interest, since they represent the bulk of astronomical society information, both quantitatively and qualitatively. The IAU entries cannot be considered a complete checklist, however, since only in-print materials are included.

Arranged alphabetically by name of society, entries include ordering information (price, address, availability), bibliographic data (author, title, imprint), and form (journal, monograph, newsletter, pamphlet, film, series, etc.). Three useful indexes (author, subject, and periodical title) make searching very easy. Occasionally entries are not complete, and some information is missing, but on the whole this effort is an excellent one, useful for the astronomy librarian and reference librarian alike.

Publishers' Series

A publisher's series, for those unfamiliar with the terminology, is a group of books in a particular topical area (e.g., astronomy), issued by one publisher over an extended period of time. An example of a well-known series is "Time-Life Books," which consists of several series—on nature, history, etc. In the literature of astronomy there are several excellent series, some of which the librarian should be aware of.

For the general reader and amateur astronomer, two series come to mind immediately, both distributed by Sky Publishing Corporation. The first is the "Sky and Telescope Library of Astronomy" (McGraw-Hill), a collection of nine books that contain reprints of articles from *Sky and Telescope* magazine, the best general astronomy publication available. Each volume is concerned with one particular subject (e.g., the solar system, stellar evolution, telescopes, galaxies and cosmology, etc.) Most are still in print and are moderately priced. Edited by Thornton Page and Lou Williams Page, these excellent works are now somewhat outdated; they lack the newer material since publication. Despite this, they are still of great value, providing some historical perspective on the development of astronomy in the middle third of the twentieth century.

Another series of interest for the lay reader is the "Harvard Books on Astronomy" (Harvard University), a group of about ten or twelve volumes aimed at the average non-scientist. Included are books on the Sun, solar system, stars, telescopes, astrophysics, and more. Collectively, they are the best group of astronomy monographs ever published. For quality, comprehensiveness, and lasting value, there are none better. Most are still available, and they are relatively inexpensive. Both series are suitable for the public library and university collections, as they provide good coverage of the very basic topics in the field.

Of particular interest to astronomers and laymen alike is "Dover Books on Astronomy and Astrophysics," a reprint series which has contributed greatly to the literature by providing re-issues of classic astronomy books. The works of Schwarzchild, Eddington, Chandrasekhar, Aitken, Jeans, Hubble, Webb, and many more, mostly from the late nineteenth and the first third of the twentieth century, are made available again in unabridged form. From a historical point of view they are invaluable, but from an educational view they are even more so: much of the basic astronomy in today's texts first appeared in these reprinted books. Mainly paperback and reasonably priced, this series of two dozen or so items is a very important one.

Two of the most important series aimed at the astronomer and university student are "Stars and Stellar Systems" (University of Chicago Press) and "Astrophysics and Space Science Library" (D. Reidel Publishing Co.). The former is a set of nine volumes, eight of which have been published as of 1976. The titles of these works should be mentioned since this is currently the most important group of books in astronomy: I. *Telescopes* (1960); II. *Astronomical Techniques* (1962); III. *Basic Astronomical Data* (1963); IV. *Clusters and Binaries* (cancelled); V. *Galactic Structure* (1965); VI. *Stellar Atmospheres* (1960); VII. *Nebulae and Interstellar Matter* (1968); VIII. *Stellar Structure* (1965); IX. *Galaxies and the Universe* (1976). Each book contains a series of review articles written by astronomers who are experts in their fields. Subtitled "Compendium of Astronomy and Astrophysics," this series is found in every astronomy library and most large science collections. Since the volumes are a bit outdated, it is hoped that they will be revised someday. All are still in print, and are fairly expensive.

The other major professional series, "Astrophysics and Space Science Library," is an extensive group of over 50 volumes dealing with current problems in astronomy and its applications. This series is comprised of general texts for the advanced student and astronomer, as well as selected proceedings of astronomical conferences. All are high-quality works, suitable for astronomy libraries. Most are still available, and they are highly priced, as many technical works are.

There are, of course, many other astronomical and astrophysical series. The five mentioned here, however, are among the most important and are easily the best in terms of quality of individual items within the series. Many of the series volumes are discussed in this guide.

Observatory Publications

Another important source of astronomical information is the observatory publications, reports of activities from the scores of astronomical research centers of the world. These materials take the form of reports, circulars, bulletins, newsletters, and books, all of which relate news of research, important discoveries, and changes in facilities and staff. The documents can be preprints and reprints of journal articles written by observatory astronomers and staff, reports of meetings held at the institution, or copies of research proposals. Some observatory publications are formally published and can be purchased for the library or for personal use; a good example is the *American Ephemeris and Nautical Almanac*, issued by the U.S. Naval Observatory. It should be apparent that there is no consistency in form or purpose to this group of items, then, and that the librarian can have a difficult time in dealing with them.

Bibliographic control (keeping tabs on location, availability, etc.) of observatory publications is a most difficult job. The greatest problem is that there is no general index or guide to observatory reports, such as an abstracting and indexing journal. Locating an individual item of an observatory series is frequently impossible. Fortunately, there is an excellent volume that is of great help in controlling these documents. The *Bibliography of Non-Commercial Publications of Observatories & Astronomical Societies* lists over 200 observatories and their serial publications (current and ceased). Although individual titles and authors are not cited, this work goes a long way toward solving the bibliographic control of these elusive publications.

Other aids for dealing with observatory series are two familiar standard sources, *New Serials Titles* and the *Union List of Serials*, published by the Library of Congress. These listings include most of the observatory publications that are issued on a regular basis. These "tools" tell who holds a particular publication series and give some indication as to what pieces are held by the library. *Astronomy and Astrophysics Abstracts* is another reference source that can be consulted for help in tracking down a particular citation. The problem, however, is that there are just too many observatory materials—and too many that are irregular, ephemeral, and elusive—to be covered in one index.

The librarian should also keep in mind that there are several related reference books that can be used when dealing with observatory publications. Among these are the *List of Radio and Radar Astronomy Observatories*, published by the National Academy of Sciences, and *Observatories of the World*, by Thornton Page, both of which list the major astronomical research centers. A more recent work, mainly intended for the amateur, is a good reference source for librarians: *U.S. Observatories: A Directory and Travel Guide* (1976).

One of the best and most up-to-date observatory information sources is the *Bulletin of the American Astronomical Society* (see entry 108 under the heading "Professional Journals"). In the first issue of each year, and to some extent in succeeding issues, are capsule, annual reports (two to five pages) of most U.S. observatories, optical and radio. Included in each report is information on staff, equipment, facilities, and descriptions of major research projects. Some reports list recent staff publications and, although this coverage is spotty, the information could help the librarian in search of an observatory document. Other journals carry brief news of observatory activities; two of note are *The Observatory* (Great Britain) and the *Journal of the Royal Astronomical Society of Canada*.

In most cases, one does not subscribe to these publications in the usual sense of subscribing to a journal. One can place a "standing order" with certain observatories, but because of the irregular frequency of many of these bulletins, it is difficult to predict when they will arrive. One may get on the mailing list of certain observatories and receive their publications free or at a nominal cost. Astronomy libraries usually trade their observatory's publications on an exchange basis, thus building up a sizable collection at little cost. A new trend, though, is for certain observatories to exchange lists of their publications, rather than the items themselves. One can then request these documents free or at a stated price. This tactic prevents a library from receiving a lot of materials that it doesn't need or have room to store. Further, libraries that exchange publications may not be able to afford to send everything on their lists automatically to everyone.

In-house management of observatory publications is usually handled by the card catalog or the computer printout. Since the latter is easy to update and provides a variety of access points, it is becoming more popular, at least where computer facilities are accessible by the library and staff.

Card catalog arrangement varies from library to library, but most frequently titles are entered on cards under the name of the institution with which the observatory is connected, with cross references from the name of the observatory, if this is different from the university or institution with which it is a part. An example would be *Harvard University. Observatory. Reprints*. Each card includes dates, frequency, and other descriptive information. Usually immediately following this card is a serial record card showing issues or parts held by the library. In other words, these special serial items are treated like journals. Other libraries prefer to use the Kardex record for recording issues. In any event, individual items in a series are not cataloged or indexed. This would be too time-consuming and of questionable value. Special monographs issued by observatories, however, would probably be cataloged separately. A separate card file not connected with the main card catalog is another alternative.

The computer print-out or on-line access to a computer data bank is still another way to achieve bibliographic control. Many computer programs for handling serial publications provide multiple access for the user; that is, the information is presented more than once, arranged by title, place, subject, etc. There could conceivably be a printout of the information arranged by institution, name of observatory, location, type of research, etc. The method of control is determined by the librarian after considering resources, staff, and use of materials.

140.   **Bibliography of Non-Commercial Publications of Observatories & Astronomical Societies.** Rev. ed. Utrecht, The Netherlands, "Sonnenborgh" Observatory, 1973. index.

Prepared under the auspices of the IFLA/FIAB section of astronomical libraries and the IAU Commission 5 (documentation), this book lists 202 observatories and astronomical institutes and their publications. An invaluable tool for any astronomy librarian, it contains information not easily available elsewhere, helping in the difficult control of observatory publications. Each entry in this typewritten, mimeographed volume contains address of the institution, current and discontinued series (including titles, numbers, and dates), and availability of documents. There is no listing, however, of individual titles, which would be an insurmountable task. Arrangement is by city, with a cross reference index by observatory name. The major drawback is that the book only lists those observatories on the "Sonnenborgh" exchange list, thereby missing a few important institutions. The vast majority, though, are included. Further, a certain percentage of the holdings listed are incorrect (i.e., some publications are *not* available or numbers are incorrect), which is not surprising, since most serials lists have the same problem, due to either typographical errors, carelessness, or outdatedness. Nevertheless, there is nothing like this volume. It should be published commercially, after updating and correcting errors—it would then be available to the many librarians who would use it on a regular basis.

Review Literature

The literature of astronomy has six (five current) titles that contain excellent review articles on topics of current interest to astronomers and students. Written in bibliographic essay form, these articles examine in depth the work done by several or many scientists in a particular field of study. Since the papers are overviews rather than reports of specific research, their text is generally not too technical, although some essays are much more "advanced" than others. The reader should not expect an elementary treatment of topics, however; these pieces are definitely written for scientists and advanced students. This is not to say that the layman could not read and enjoy many of the volumes in these series.

The articles cover the spectrum of astronomical topics, and most volumes contain a number of different subjects, although some occasionally devote an entire monograph to one topic or the work of one astronomer. From historical treatises to astrophysics to radio astronomy, the reader will find a variety of subjects to consider. The importance of these works, it should be remembered, is that they provide overviews to work or research already completed. The astronomer who wants a quick introduction to a particular topic should first read any review articles pertaining to his interest.

Two of the current titles are part of the "Annual Review" series: *Annual Review of Astronomy and Astrophysics* and *Annual Review of Earth and Planetary Science*, the latter having a small part devoted to astronomy. Two works which have produced fine review articles in the past are *Advances in Astronomy and Astrophysics* (none published since 1972) and *Vistas in Astronomy* (which became a review journal in 1975). Another review-type journal is *Space Science Reviews*, a publication dealing with the highly interdisciplinary field of astronomical applications. All should be in the astronomy library, except possibly for the series devoted mainly to the earth sciences. Many university and larger public libraries will want these volumes as well. Several other journals carry short review papers from time to time: *Earth and Extraterrestrial Science, Comments on Astrophysics* and *Journal for the History of Astronomy*, to name a few. Finally, the *Index to Scientific Reviews (ISR)*, first published by the Institute for Scientific Information in 1976, provides a good coverage of review materials in general. The articles that appear in the issues and volumes of the astronomy books and journals mentioned above can be found in the *ISR* under author, title words and phrases, and organizations.

141.   Beer, Arthur, ed. **Vistas in Astronomy.** Oxford, Elmsford, N.Y.: Pergamon Press, Inc. v.1-18, 1955-75. Irregular. illus. index. refs. LC 56-4213.

No longer a review series, this continuation became a journal in 1975. During the 21 years of its publication, *Vistas in Astronomy* included numerous reviews of technical and historical topics. In fact, it was the only review series in astronomy to include historical treatises. The topics ranged from a computer program used to solve an astronomical problem to a pair of conference proceedings volumes on Kepler and Copernicus. Editor Beer's original intent of keeping astronomers as up to date as possible has been adhered to over the years; as many topics as time and space have permitted have been in the pages of this series. Several volumes have been devoted entirely to one subject—for example, history of astronomy, stellar evolution, and the work of Henry Norris Russell. Occasionally non-English articles were included, but this was the exception, not the rule. *Books in Print* should be consulted for current availability and prices.

142.   **Index to Scientific Reviews.** Philadelphia: Institute for Scientific Information, 1976– . Semi-annual. LC 75-1738. ISSN 0360-0661.

Identical in format to *Science Citation Index, ISR* is basically a subset of the former, providing a much needed source of data on review-type literature in the sciences. Including standard sources like monographic review literature and review journals, it also covers articles from scholarly periodicals of a research nature. Astronomy is represented well, with about 30 sources regularly scanned. Citation, source, patent, and permuterm subject indexes provide multiple access. If one subscribes to *Science Citation Index*, it is not really necessary to purchase *ISR* as well, since it is mainly a duplication of part of *SCI*. See description of *SCI* (under Abstracts and Indexes) for more detailed information on the workings of this excellent source.

Write: Institute for Scientific Information, 325 Chestnut St., Philadelphia, Pennsylvania 19106. Cost: $250/year; $675/3 years.

143.   Kopal, Zdeněk, ed. **Advances in Astronomy and Astrophysics.** New York: Academic Press. v.1-9, 1962-72. Irregular. illus. indexes (author and subject). LC 61-18299.

Similar to *Annual Review of Astronomy and Astrophysics*, this series tends to be slightly more technical and less bibliographically oriented. Each volume contains current review articles on dozens of topics, but occasionally one topic will occupy an entire volume, as did the 1971 edition on the Moon. The papers tend to be fairly lengthy (often more than 50 pages), so there are fewer articles per volume than in the other review literature. Although the series covers a wide range of subjects, the emphasis is on the Sun and Moon and the solar system. This is not surprising since the editor is a well-known lunar scholar. As of 1976 most volumes were still available; *Books in Print* should be consulted for the latest information.

144.   Goldberg, Leo, David Layzer, and John G. Phillips, eds. **Annual Review of Astronomy and Astrophysics.** Palo Alto, Calif.: Annual Reviews Inc. v.1– , 1963– . Annual. illus. indexes (subject and author). LC 63-8846.

The papers in this review series are in the style of bibliographic essays, with an emphasis on words and not equations. Articles are all of a technical nature, and there are no pieces on historical topics. Of all the review titles, this one provides the best balance of the subject matter; there is no particular slant one way or the other. Stars, planets, the Sun, the Moon, galaxies, etc. all get equal billing. Articles average 20 to 30 pages in length, so there are many different essays in each monograph. Like the others, it is aimed at the astronomer and advanced student. Unfortunately, the volumes go out of print rapidly.

145.   Donath, Fred A., ed. **Annual Review of Earth and Planetary Sciences.** Palo Alto: Annual Reviews, Inc., v.1– , 1973– . Annual. illus. refs. author indexes. $12.00. LC 72-82137. ISSN 0084-6597.

Although the major emphasis of this review series is earth science, there are also occasional articles on the planets besides our own. Astronomical topics in the first few volumes have included the interiors of Jupiter and Saturn, geophysical data and the interior of the Moon, as well as articles on meteorites, the Martian atmosphere, and the outer planets. Like the other annual reviews, this contains papers in the bibliographic essay style surveying work done by various researchers in particular fields. Not all astronomy libraries will want this review series, but those specializing in planetary studies will.

146.   **Space Science Reviews.** Dordrecht-Holland: D. Reidel Publishing Co. v.1– , 1962– . 9 issues/yr. 6 issues/vol. illus. refs. volume author indexes. LC 65-69362. ISSN 0038-6308.

This journal departs from the usual reports of original research by publishing review-type articles that survey various topics of interest in space science. The subject matter included in these papers for the professional astronomer is best explained by the journal's definition of space science: "scientific research carried out by means of rockets, rocket-propelled vehicles and partly also by stratospheric balloons and at observatories on the Earth or Moon." Mainly concerned with the purely scientific aspects of the subject, this publication discusses the instrumental and technical aspects, too. There are usually two or three long papers (40 to 120 pages) in each issue, as well as a handful of book reviews. Like most review articles, these pieces include extensive lists of references, often several pages long; such bibliographies are often more important than the papers they accompany, for they lead the reader to original sources. The text of this journal is in English, French, German, and Russian, with English most common.

  Write: D. Reidel Publishing Co., Box 17, Dordrecht-Holland. Cost: $34/vol (individuals); $94/vol (institutions).

147. **Vistas in Astronomy: An International Review Journal.** Oxford, New York: Pergamon Press. v.19– , 1975– . 4 issues/yr. illus. annual subject and author indexes. refs. ISSN 0083-6656.

  The continuation of the popular review series in journal form has been most welcome. The former was issued on a rather irregular basis, and it is hoped that the promise of four regular issues per year will be kept by the publishers. Like its predecessor, this journal contains bibliographic essay review articles, presenting a survey of up-to-date astronomical research. The first two issues contained five articles each, averaging 22 pages. Topics covered included x-ray astronomy, the Milky Way and related research, stellar rings, theoretical astronomy in the Soviet Union, current knowledge of Venus and Mars, and more. Astronomy libraries that subscribed to the former series will surely want to continue this journal. Perhaps the price will not be increased too much too soon.

  Write: Subscription Fulfillment Manager, Pergamon Press Ltd., Headington Hill Hall, Oxford OX3 0BW, England. Cost: $50.00/year; $95.00/2 years.

2

GENERAL MATERIALS

GENERAL WORKS

The books in this section are best categorized by the type of reader for whom they are intended—books for the general reader (by far the largest group) and those for the astronomer and scientist. The books for the layman and student, the first section, are further subdivided into picture books and general texts, two distinctly different types of material.

The emphasis in the former is on the beauty of the Universe, and the volumes are filled with many excellent photographs and paintings, often in color. As a result of the recent space probes and the close-up photos of the Moon and planets, there has been a flood of books of this type. Their numbers peaked in the early 1970s during the Moon landings, and have leveled off somewhat since then. Also called "coffee table books," these fine volumes are excellent acquisitions for all types of libraries as well as the home. They appeal to all ages and make thoughtful gifts.

The more general text also provides a good introduction to astronomy, but usually on a more advanced level. These works on the stars and planets tend to be aimed at non-science college graduates or students. They often emphasize the latest developments in astronomy (currently, topics like black holes, pulsars, etc.). Sometimes the author of such a book concentrates on one topic, like cosmology. These volumes have fewer illustrations, since the text itself is usually emphasized. General texts, like picture books, are suited for any type of library.

General books for the astronomer are few and far between, since professional observers and theoreticians are concerned with highly specialized topics. Nevertheless, there are some non-specialized works for the scientist, a few of which are included here as samples.

Card catalog subject headings: *Library of Congress:* ASTRONOMY; ASTRONOMY—PICTORIAL WORKS; ASTRONOMY—POPULAR WORKS. *Sears:* ASTRONOMY; ASTRONOMY—PICTORIAL WORKS.

For the Layman and Student

Picture Books

148.   Alter, Disnmore, Clarence H. Cleminshaw, and John G. Phillips. **Pictorial Astronomy**. 4th rev. ed. New York, Thomas Y. Crowell Co., 1974. 328p. illus. index. glossary. $12.50. LC 73-15577. ISBN 0-690-00095-2.

The fourth edition of this popular book is bigger and better than ever. The most obvious change is the inclusion of the latest information and photographs gathered by Apollo astronauts and the unmanned probes to the planets. As the title indicates, the book's strongest point is its beautiful illustrations, and for that reason, this work will be enjoyed by all types of readers, especially the young. The book is divided into 60 short chapters in nine major sections; though the emphasis is on the solar system, there are sections on stars and space science as well. Recommended for the home and public library.

149.   Bergamini, David. **The Universe**. New York, Time Incorporated; distr. Morristown, N.J., Silver Burdett Co., 1969. 192p. illus. index. bibliog. (Life Nature Library). $9.32. LC 62-18337. ISBN 0-8094-0619-5 (T-L).

There are many popular-type astronomy books for general readers, full of beautiful photographs and pictures, enjoyable to look at and browse through. This familiar book is one such volume—informative, handsomely illustrated, and downright fascinating. It is frequently found in school, public, and home libraries because of its clear, easy-to-understand text and its educational value. The topics covered are fairly standard—the planets, the Sun, the Moon, comets, meteorites, stars, etc. There is a strong emphasis on the Universe, however; the author explores cosmological questions like "Is the Universe expanding?" and "What is the shape and size of the cosmos?" Other topics of particular interest are the birth and death of stars, telescopes, and historical background. It would be nice to see a new edition of this fine volume in a couple of years that would include some of the discoveries of the late 1960s and early 1970s. In the meantime, this book in its present form is still quite good and highly recommended.

150.   Friedman, Herbert. **The Amazing Universe**. Washington, D.C., The National Geographic Society, 1975. 199p. illus. index. glossary. bibliog. $4.25. LC 74-28806. ISBN 0-87044-179-5.

Books on the wonder and glory of the heavens continue to be published at an incredible rate, almost faster than the growth of the science itself. And here is still another "gee-whiz" volume, displaying the color and grandeur of the skies—but more successfully than most similar works. What makes this book so special is not the text, for the words are basically the same—fantastic discoveries (quasars, pulsars, x-rays, etc.) described in clear layman's language. The uniqueness here is the style of those words, the familiar, comfortable *National Geographic* style, taking the reader on an armchair tour of the Universe. Color photographs galore immediately capture one's eye, and this too, of course, is the *Geographic* style; one is tempted to skip the text and read only the captions. The astronomers themselves come alive here as in no other work, as the text includes their stories and photographs. The opening chapter entices the reader to continue with a pictorial and textual description of why astronomy has captured—and continues to capture—the fancy of mankind throughout the ages. The author next describes the known size of the Universe and how the distance to its edge has changed over the years. Chapter three looks at the closest star, the Sun, explaining and illustrating what we know about it, especially in light of recent studies in the non-visible portion of the spectrum. Stars in general, including their evolution and characteristics, follow. Chapters on pulsars and black holes, galaxies, cosmic theories and the currently "hot" topic of life in the Universe round out this fine book. Highly recommended for any type library; it is hoped that the National Geographic Society will publish more books on astronomy.

151. McCall, Robert, and Isaac Asimov. **Our World in Space**. Greenwich, Conn., New York Graphic Society, Ltd., 1974. 176p. illus. $25.00. LC 73-78567. ISBN 0-8212-0434-3.

This beautiful volume is both an astronomy book and a book on space exploration by manned and unmanned probes. It is the story of the progress in the study of our solar system and what might lie ahead in the next century. Satellites, lunar landings, voyages to Mars and beyond, and other fascinating topics are discussed in an excellent text by Isaac Asimov, renowned science fiction and science author. Equally important to this work, however, are the many strikingly beautiful paintings by Robert McCall. Any astronomy and public library will want this fine book, and it is hoped that this author-artist combination will produce other works.

152. Menzel, Donald H., and Ching Sun Yü (illus.) **Astronomy**. rev. ed. New York, Random House, 1975. 320p. illus. glossary. index. $17.50. LC 70-127542. ISBN 0-394-41564-7.

The old expression, "a picture is worth a thousand words," is an apt description of this general astronomy book. Beautifully illustrated with dozens of color and black-and-white photographs, the volume displays well the beauty of the Universe in all its glory. The first four chapters are concerned with the history of astronomy, and the author makes this section especially good reading by including sketches of many important early astronomers. Chapter 5, an introduction to the remainder of the book, gives a panoramic look at the Universe as a whole. The majority of the volume concentrates on the most of the familiar topics in an introductory astronomy work: the Sun, the Moon, the stars, the planets, the Milky Way, galaxies, quasars, etc. The clearly written text and the fine illustrations make a book that is sure to please all types of readers, from the casual backyard observer to the non-specialist. A noteworthy feature is the 24 sky maps drawn by Ching Sun Yü, which show the constellations for all parts of the sky.

153. Moore, Patrick, and David A. Hardy. **Challenge of the Stars.** Chicago, Rand-McNally, 1972. 63p. illus. $6.95. LC 71-18909. ISBN 0-528-83045-7.

Like *Our World in Space* (McCall & Asimov), this is a fascinating book about man's future exploration of the planets and beyond. Prolific Patrick Moore has written excellent text about space stations, moon bases, and planetary exploration, and artist David Hardy has provided beautiful illustrations of his ideas. The work combines fact with prediction in an interesting look at the future. Would be an excellent purchase for the public library and personal book collection.

154. Nicolson, Iain. **Astronomy.** New York, Grosset & Dunlap, 1971. 159p. illus. index. (All Color Guides). LC 71-120607. ISBN 0-448-00858-0.

Several popular handbooks of astronomy are available, and this book is one such guide. Well illustrated, and not at all technical, it should be part of the home library, since it will appeal to all members of the family, young and old alike. Like similar works, this short volume gives concise yet informative explanations of a wide spectrum of topics, including the Earth, Sun, Moon, planets, stars, etc. A special emphasis is put on the constellations, with excellent drawings and lively, informative text. There is also a good chapter on amateur astronomy which very briefly introduces to the neophyte what can be observed by the casual astronomer. Public libraries will want to acquire this volume, which can be read straight through or browsed on a rainy day. Also available from Bantam Books as Volume 20 of the "Knowledge Through Color" series (paperbound).

155. Rohr, Hans. **The Beauty of the Universe.** New York, Viking Press, 1972. 87p. illus. bibliog. (A Studio Book). $10.00. LC 77-164990. ISBN 0-670-15340-0.

The title of this fine book is an apt description of its contents. The author has prepared a short, but extremely beautiful volume displaying the stars, galaxies, and nebulae of the Universe in all their glory. An excellent candidate for the coffee table and public library, the book contains 77 black-and-white and color photographs of the heavens. An enlightened and descriptive text accompanies the photographs. The volume was translated from German (*Strahlendes Weltall*) by Arthur Beer.

Texts

156. Abetti, Giorgio. **The Exploration of the Universe.** London, Faber & Faber; distr. New York, American Elsevier Publishing Co., Inc., 1968. 178p. illus. index. LC 68-19351. ISBN 0-571-08688-8.

A typical example of a general text for the layman, this book is actually a collection of articles written by the author over several years. Because of this arrangement, the chapters do not always proceed logically from one to another, but they are nevertheless logically arranged in subject categories. Although the topical coverage is not comprehensive, Professor Abetti has hit the high points of astronomy and presents them readably. Particularly good is the section on the Sun, the author's area of expertise, in which he details vividly two solar eclipses he personally observed. The other sections of the book include discussions of the Earth, the planets, comets, stars, galaxies and the Universe. Translated from the Italian (*Esplorazione dell'universo*) by V. Barocas, this somewhat outdated volume might be enjoyed by the layman seeking a general introduction to astronomy.

157.   Abetti, Giorgio. **Stars and Planets**. London, Faber & Faber; 1966. 341p. illus. bibliog. £4.00. LC 66-71992. B 66-11379. ISBN 0-571-06702-6.

The layman with some scientific inclination and/or background would enjoy this volume, which was written for readers with no previous knowledge of astronomy. The author's approach is simple and direct—he describes in detail the most familiar objects in the sky, the planets and stars, nothing more. It is a logical beginning for the novice, since it starts with the very basics instead of trying to teach him everything at once. The first two-thirds of the book is an in-depth look at the stars, their spectra, physical characteristics, types, systems, etc. The latter part of the work is a planet-by-planet description of the solar system, including the Moon, comets, meteors, and meteorites. Like many books of the last decade, this volume was written in response to the interest in astronomy due to fantastic discoveries (quasars, etc.) and the space program. Translated from the third Italian edition (*Le stelle e i planeti*) by V. Barocas.

158.   Bova, Ben. **The New Astronomies**. New York, St. Martin's Press; New York, New American Library, 1972. 214p. illus. index. bibliog. $7.95. (St. Martin's); $1.95pa. (NAL). LC 70-184553.

Another of the "Look what's happening in astronomy!" books, this volume zeroes in on the third major era of astronomical research, the period from the mid-twentieth century to the present. Historically, the first great era was from antiquity to the early seventeenth century, during which mankind observed the skies with the unaided eye. The second era began in 1609, when Galileo first used the telescope in astronomical observations, and ended somewhere in the middle of this century.

After brief introductions to the first two eras, the so-called "optical" eras, the author spends the majority of the book exploring the great discoveries of the third era, the age when radio telescopes, infrared telescopes, etc., made great advances beyond the visible spectrum. Like other "current" books, familiar subjects show up: x-ray astronomy, radio astronomy, space flights, quasars, black holes, and so on. The author concludes by pointing out that the third era has proven the Universe to be a violent one, a view remarkably different from that of the first two periods.

159.   Corliss, William R. **Mysteries of the Universe**. New York, Thomas Y. Crowell Co., 1967. 216p. illus. index. bibliog. $5.95. LC 67-23019. ISBN 0-690-57117-8.

Like many books for the general reader, this work examines a smattering of astronomical topics and does not concentrate on any one area. As a result, the reader can selectively read chapters of interest without losing continuity, and a list of references with each section points the way to related sources. While somewhat outdated, it discusses problems that are still "mysteries," that are thought-provoking and highly adaptive to discussion. Among the chapters are cosmology (is there a beginning or end?); quasars, the age of the universe; checking out Einstein's Theory of Special Relativity; how stars work; the sunspot cycle and its effect on the solar system; Jupiter's Red Spot; the Martian canals; the "missing planet"; lights on the Moon; and the search for life beyond the Earth.

160.   Flammarion, Gabrielle Camille. **The Flammarion Book of Astronomy.** New York, Simon & Schuster, 1964. 670p. illus. index. $29.95. LC 64-15354. ISBN 0-671-26181-9.

A descendent of the 1880 *Astronomie Populaire*, a classic in its own right, this beautifully illustrated volume provides excellent explanations and treatises on the many, varied subjects of astronomy. The work is the result of several contributions by French astronomers, under the direction of Gabrielle Camille Flammarion and André Danjon, and is divided into eight "books:" The Earth; The Moon; The Sun; The Planets; Comets, Meteors, and Meteorites; The Sidereal Universe; The Instruments of Astronomy; and Artificial Satellites and Space Vehicles. Combining factual historical data with excellent and complete detail, the book lives up to its reputation as one of the finest astronomical reference works. This classic was translated from the French (*Astronomie Populaire*) by Annabel and Bernard Pagel.

161.   Glasby, John S. **Boundaries of the Universe.** Cambridge, Mass., Harvard University Press, 1971. 296p. illus. index. $11.00. LC 76-162638. ISBN 0-674-08015-7.

Most of the author's books have concentrated on variable stars, a field in which he is a recognized authority. His expertise, however, goes beyond this subject, and he has also written this excellent volume on general astronomy. Aimed at the educated layman, the book describes the "current" state of knowledge about the Universe from the Moon to relativity and cosmology. The theme of the work, as might be surmised from the title, is that as new astronomical discoveries are made, and as man learns more about his Universe, the dimensions of that Universe expand. Approximately one-half of the book is concerned with stars and stellar evolution, which is not surprising, considering Mr. Glasby's experience. The text, as usual, is clearly written and well presented, and the illustrations are good. This volume would be a fine addition to the university, public, or astronomy library.

162.   Goldsmith, Donald, and Donald Levy. **From the Black Hole to the Infinite Universe.** San Francisco, Holden-Day, Inc., 1974. 330p. illus. index. questions. $7.95pa. LC 73-86412. ISBN 0-8162-3323-3.

On the surface, this paperback looks like a typical astronomy book; the outline of a spiral galaxy appears on the front of the volume against a dark blue background. But don't judge a book by its cover! Inside, one finds astronomy, modern physics, *and* a science fiction novelette. Each chapter of this volume begins with a portion of the sci/fi story about an interstellar insurance investigator.

The approach is different, and the result is a work which is enjoyable and informative at the same time. The authors intend their work to be used as a text for a one-term college course in astronomy or physics, but it could also be used by the layman who wishes an introduction to the topics explored. Among these subjects are black holes, energy and matter, forces, protons, the expanding universe, space and time, and several others. Each chapter contains a summary and questions at the end for study and review.

163.   Hawkins, Gerald S. **Splendor in the Sky**. rev. ed. New York, Harper and Row, 1969. 292p. illus. index. bibliog. LC 69-17283.

If one spends an evening in a local observatory during an open house, one quickly learns the interests of the general public with regard to our Universe. The visitors' questions range from simple inquiries about the Moon and planets to queries about radio astronomy. It's safe to say that the astronomer on duty in the dome learns almost as much from the visitors as they do from him. It was these visitors and their interest that inspired the author, who was the decoder of Stonehenge, to write this fine book on general astronomy. A variety of topics are discussed, all of which were frequent topics of conversation between astronomer and guest: the solar system, the Moon, the stars, the Milky Way, comets, meteors, galaxies, and radio astronomy. The emphasis in this work for the general reader is on recent astronomical discoveries.

164.   Hodge, Paul W. **Concepts of the Universe**. New York, McGraw-Hill Book Co., 1969. 125p. illus. index. $5.95; $3.95pa. LC 69-18715. ISBN 0-07-029132-2; 0-07-029130-6pa.

The study of the observable universe and beyond is the theme of this short but well-written book. Dozens of excellent illustrations, both diagrams and photographs, fill the pages of this work, which examines, among other things, the distances between planets, stars, and galaxies. The newcomer to astronomy will no doubt be intrigued by these numerical figures and the "accepted" distances to the edge of our "seeable" universe. Written for the layman, the book will certainly arouse the interest of even the most casual reader. Mathematics is minimized in this excellent look at the properties of our Universe as a whole; for the public library.

165.   Hodge, Paul W. **The Revolution in Astronomy**. New York, Holiday House, 1970. 189p. illus. index. glossary. bibliog. LC 70-102430. ISBN 0-8234-0094-8.

Revolution means change and, indeed, that's what this book for the layman is all about. Change in astronomy has been rapid and great in the past decade, and the author has summarized well the most significant developments in his work on this astronomical revolution. Admittedly the subject matter is not unique—there are several other works on these same topics. Nevertheless, Dr. Hodge's style and clarity make this volume above average, and therefore a prime candidate for the interested reader. For the record, the subjects discussed and illustrated so well are quasars, radio astronomy, pulsars, ultra-violet and x-ray astronomy, the Moon, exploration of the planets, gamma rays, neutrinos, and many others.

166. Hoyle, Fred. **From Stonehenge to Modern Cosmology**. San Francisco, W. H. Freeman and Co., 1972. 96p. illus. index. LC 72-10836. ISBN 0-7167-0341-6.

The four chapters in this thought-provoking book are tied closely together by the author's contention that the motivation of scientific research is basically a religious one, a kind of belief in the elegant laws of the Universe. The first chapter discusses the relationships between science and society today, and here the author propounds his predictions for civilization. He next explores Stonehenge at some length. The motivations of the builders of the ancient astronomical observatory were religious, says Hoyle, like the scientists of the twentieth century. He agrees with Gerald Hawkins (*Stonehenge Decoded*) that the structure was an astronomical observatory, but he disagrees with him on how the so-called eclipse predictor was to have worked. The chapter details Hoyle's verification of Hawkins' basic contentions, and explains in depth the disagreement on the eclipse predictor.

The final two chapters relate what could be considered scientific research with religious motivation in the present—cosmological theories. Hoyle's discussion of the recent developments in cosmology provides a good look at what he considers the key question of cosmology to be: "whether or not there is an important interrelation between the structure of the Universe and the laws of physics." A very fine volume for the layman and scientist alike.

167. Hoyle, Fred. **Highlights in Astronomy**. San Francisco, W. H. Freeman and Co., 1975. 179p. illus. index. questions. $10.00; $5.50pa. LC 75-1300. ISBN 0-7167-0355-6; 0-7167-0354-8pa.

For the reader who is new to astronomy and would like a very brief overview of the subject, this volume comes highly recommended. The author, a renowned astrophysicist and cosmologist, has written an interesting and worthwhile book up to his usual high quality. Although there is nothing particularly unique about the presentation here, it is one of the latest such works available, so it mentions some of the newest, headline-grabbing discoveries. The pictures are superb and well chosen for the eight chapters, which are equally divided between the solar system and the stars and galaxies. The reader who is not in need of an introduction to astronomy may nevertheless wish to spend a few moments in the library leafing through the pages just to look at the color photographs. This is not a "gee-whiz" book in the strictest sense (although the illustrations are the high point), because it is written on a serious, adult level for the layman. An added feature not usually found in such works is the inclusion of discussion questions with each chapter. The book should not be considered a text, though; it is far too brief. For college, public, and personal libraries, this volume is not quite worth the hardcover price; but, then, few books nowadays are. Buy the paperback.

168. Jastrow, Robert. **Red Giants and White Dwarfs**. 2nd ed. New York, Harper and Row, 1971. 190p. illus. index. $8.95; $1.50pa. LC 79-108939. ISBN 0-06-012182-3.

Not really an astronomy book, *per se*, this well-known work is one of the relatively few science works to have a paperback as well as a hardbound edition. In short, the author tells the story of creation and evolution in the Universe, drawing heavily upon astronomy along the way. The birth and death of stars is just one portion of this exciting tale which Jastrow tells so well. A particularly good part of the book involves the theories of how life began on Earth. This second edition is well illustrated, like the first, and reflects some of the changes in thinking about the Universe brought about by Apollo and Mariner. Highly recommended for layman, student, and scientist.

169. Kienle, Hans. **Modern Astronomy: An Introduction**. London, Faber & Faber; distr. New York, Thomas Y. Crowell Co., 1969. 141p. index. bibliog. ISBN 0-571-08725-6.

Not comprehensive by any means, this work is, in a sense, a personal narrative of an astronomer's research in the first half of the twentieth century. Astronomical knowledge at the turn of the century is described in the first chapter, titled "The Year of the Return of Halley's Comet" (1910). Using the theories of that time as a starting point, the author begins to describe the changes in our concepts of the Universe as he discusses topics like telescopes and their use, astronomical coordinates, the discovery of galaxies, the inner processes in stars, and several others. It is a different approach to "introductory" astronomy— one that does not attempt to cover all bases and explain all things, but rather is a discussion of selected important topics done in an extremely interesting, readable way. Translated from the German (*Einführung in die Astronomie*) by Alex Helm.

170. Kopal, Zdeněk. **Man and His Universe**. New York, William Morrow and Co., Inc., 1972. 313p. illus. index. glossary. $2.95pa. LC 74-166343. ISBN 0-688-05014-X.

The unprepared beginner should steer away from this general treatise on astronomy for the layman and serious amateur. Though the book is not technical, the reader should at the very least, be familiar with the terminology of astronomy and also of physics. The author, a well-known astronomer and astronomical writer, has divided his work into three major parts: The Building Blocks of the Universe—The Stars; The Solar Family and Our Terrestrial Cradle; and The Universe at Large. The first section may be a bit too advanced for this type of book, but the prepared reader could handle this material on stellar evolution. Otherwise, this is an excellent volume, and Kopal's skill as writer is apparent as usual as he discusses the following topics: the planets, the Earth and Moon, the comets and meteors, our galaxy, other star systems, and the Universe. The final chapter explores an increasingly popular topic: is there life elsewhere in the Universe?

171. Lattin, Harriet Pratt. **Star Performance**. Philadelphia, Whitmore Publishing Co., 1969. 238p. illus. index. refs. LC 68-54698. ISBN 0-8059-1363-7.

The main emphasis of this remarkable volume is on planetaria, their history, and operation, but the overall theme is a "history of demonstrational astronomy." This work is unique (there are other books on planetaria, such as

The Planetarium and Atmospherium by Norton) because it delves deeply into other teaching aids in astronomy as well, and takes a look at their history. The author vividly describes globes, diagrams, astrolabes, armillary spheres, planetary instruments, and many clocks, all used in astronomical education. The beginning chapters give the reader some useful background information, including delightful descriptions of the constellations, early beliefs about the Universe, and early graphic depictions of the heavens. A person with any interest at all in this area should not miss this fine book.

172. Levitt, I. M. **Beyond the Known Universe: From Dwarf Stars to Quasars**. New York, Viking Press, 1974. 131p. illus. index. $10.00. LC 73-5232. ISBN 0-670-16107-1.

Fifteen years ago, around 1960, things were fairly quiet on the astronomy homefront. Suddenly, incredible discoveries were made, one after another, and astronomy has never been the same since. "Exotic celestial objects" like quasars, pulsars, black holes, white dwarfs, and neutron stars quickly became household words. The man on the street suddenly became aware of astronomy and its new vocabulary. This short volume presents an overview of these topics and more, in an attempt to explain the newly discovered phenomena to the layman. The text is easy to read and is interspersed with just the right number of excellent color illustrations by John Gorsuch. Since quite a few of the "new" topics are related to stellar evolution, the author precedes their discussions with a chapter on the development of stars. Read this fine book and learn, among other things, why a pint of the material making up a relatively small star called a white dwarf weighs 50 tons.

173. Lovell, Sir Bernard, *et al.* **The New Universe**. New York, Rand McNally and Co., 1968. 127p. illus. bibliog. LC 68-22239.

The lead article in this collection of eight essays gives an overview of the recent rapid development and change that has taken place in astronomy and describes how these new ideas have changed our concepts of the Universe. Following this are discussions of our galaxy, other stellar systems, and related topics. Radio astronomy has played an important role in this astronomical research concerned with the Universe as a whole, and the authors frequently refer to advances made with radio telescopes.

The book is now somewhat outdated, but the information is still valid, and the essays themselves make good reading. Each writer (all astronomers and/or physicists) has geared his presentation toward the layman with a science background, and toward the non-specialist. Quasars, radio galaxies, neutrino astronomy, and gravitation are but a few of the areas covered. A good choice for a public or college library.

174. Moore, Patrick. **Suns, Myths and Men**. rev. ed. New York, W. W. Norton & Co., Inc., 1968. 236p. illus. index. (The Amateur Astronomer's Library). $7.95. LC 68-27145. ISBN 0-393-0364-X.

Although the dust jacket proclaims this volume to be "a history of astronomy from the cave men to modern times," it is really a survey of astronomy which devotes only one-third of its text to history. Patrick Moore, the amateur astronomer's friend, gives an overview of the past, present, and future of the science, comparing the ideas of each period, and exposing some of the myths which for years were accepted fact. "The Past," section one, is a survey of the science from antiquity to the early part of the seventeenth century, concentrating on man's view of the Universe, making the transition from Ptolemy's Earth-centered system to a Sun-centered concept to the present view. "The Present" is a non-comprehensive view of current research, a look at observatories and telescopes, astronauts, stars and planets, etc. Finally, Moore lets speculation take over in "The Future," presenting his ideas on the directions astronomy will or should take, the search for life elsewhere, outer-space exploration, and finally the future of the planet Earth. Recommended for the layman and public library, it would make good casual reading, but nothing more.

175. Murchie, Guy. **Music of the Spheres**. New York, Dover Publications, 1967. 2v. (644p.). illus. index. Vol 1: $2.75pa. Vol 2: $3.00pa. LC 67-22255. ISBN 0-486-21809-0 (v.1); 0-486-21810-4(v.2).

Subtitled "The Material Universe—From Atom to Quasar, Simply Explained," this popular work has been revised and enlarged in the latest edition. The book on astronomy and related subjects is divided into two volumes, covering The Macrocosm: Planets, Stars, Galaxies (Volume 1) and The Microcosm: Matter, Atoms, Waves, Radiation, Relativity (Volume 2). The author's style is refreshing; this is not a dull, unimaginative explanation of the Earth, Moon, Sun, planets, and stars, in that order. In this book the astronomers of the past come alive, and their theories become more than equations and treatises. It is a different and fascinating account of the Universe—read it and enjoy! For the layman and anyone else who might be interested.

176. Norton, O. Richard. **The Planetarium and Atmospherium; An Indoor Universe**. Healdsburg, California, Naturegraph Publishers, 1967, c1968. 176p. illus. index. refs. glossary. $5.95; $2.95pa. LC 68-31924. ISBN 0-911010-73-4; 0-911010-72-6pa.

The layman or student with more than a passing interest in planetaria should examine this book, one of the few on the subject. A variety of topics are considered, including a look at the development of some of the various plaetarium projectors. The most interesting area of the work should be, at least for the inexperienced, the description of how the planetarium works. Along these lines there is a chapter giving a sample planetarium show, including details such as appropriate background music. Also described in this book are the atmospherium (the day-time sky projector), accessories to the planetarium, and the Pre-planetarium Period. For the public and planetarium libraries.

177. Peltier, Leslie C. **Starlight Nights: The Adventures of a Star-gazer**. New York, Harper and Row Publishers, 1965. 236p. illus. LC 65-20992.

The dedicated amateur observer will likely be inspired by reading this autobiography of one of the best-known and most accomplished non-professional astronomers. The author's first-person narrative traces the development of his star-gazing days from his boyhood on an Ohio farm in 1905 to 1965, when an observatory was dedicated on a mountain named in his honor. While reading this delightful account of the adventures of an amateur astronomer, one becomes familiar with the author's friends and family, as well as his "friends" in the sky, the stars. Among the author's accomplishments are the discovery of several comets. All public libraries should have this fine book.

178. Saslaw, William C., and Kenneth C. Jacobs, eds. **The Emerging Universe: Essays on Contemporary Astronomy**. Charlottesville, University of Virginia Press, 1972. 195p. illus. index. bibliog. $7.95. LC 72-188526. ISBN 0-8139-0397-1.

A collection of 10 lectures in essay format comprise this work for the layman and non-specialist. The important discoveries of the past decade are the basis of this volume whose introductory chapter correctly contends that our basic concepts of the Universe have been radically changed by these revelations. The collection begins close to home with a discussion of thermonuclear processes in "Why Does the Sun Shine?" and concludes with four viewpoints of cosmology, a look at the Big Bang Hypothesis and other theories. Between these topics lie thought-provoking discussions of pulsars, interstellar material, life in the Universe, and others. An excellent bibliography for each essay tops off this fine collection of current topics in astronomy.

179. Schatzman, Evry. **The Origin and Evolution of the Universe**. New York, Basic Books, 1966, c1965. 288p. illus. index. LC 65-10692.

A discussion of the more recent and some older theories of the creation of the various parts of our Universe. After a chapter on the "present" state of knowledge about the solar system, stars, and interstellar matter, the author discusses the origin and evolution of stars and stellar systems. Next is a chapter on extragalactic nebulae and cosmology, followed by a look at theories on the origin of solar system. Translated from the French (*Origine et évolution des mondes*) by Bernard and Annabel Pagel, this book would appeal to students, professors, and well-prepared amateur astronomers.

180. Schatzman, E. L. **The Structure of the Universe**. New York, McGraw-Hill Book Co., 1968. 253p. illus. index. bibliog. (World University Library). $4.95; $2.95pa. LC 67-22981. ISBN 0-07-055173-1; 0-07-055172-3pa.

An overall view of "current" astronomy and cosmological thought is the basis of this fine volume for the general reader and beginning college student. Generously illustrated with dozens of photographs of the heavens, it is concerned with the large-scale characteristics of astronomy including a description of the Milky Way; a discussion of time and its importance in the study of the evolution of the Universe; an interesting look at the distribution of galaxies in space; and finally, a grand look at the Universe as a whole. Physical laws and phenomena as they relate to the aforementioned topics are an integral part of the text. The emphasis in the book is on current (at least at that time) astronomical thought and knowledge, and not on the historical development of those ideas.

181. Sciama, Dennis W. **The Physical Foundations of General Relativity**. Garden City, N.Y., Doubleday and Co., 1969. 104p. illus. notes. index. (Science Study Series). $1.45pa. LC 68-14202. ISBN 0-385-02199-2.

Written by an eminent physicist, this volume gives an in-depth look at the physical meaning of the theory which is so important in both physical and astronomical studies. The author not only discusses the basic physics involved, but also considers these laws in light of astronomical phenomena, such as the red shift, the motion of light in the Sun's gravitational field, the motion of a planet in the Sun's gravitational field, and the curvature of space-time. The layman with some scientific background and the non-specialist will best appreciate this volume, which contains just the right mixture of descriptive text and simple mathematics.

182. Simak, Clifford D. **Wonder and Glory: The Story of the Universe**. New York, St. Martin's Press, 1969. 238p. illus. index. $7.95. LC 71-83397.

Like many similar works on the current state of astronomy and the latest, most interesting discoveries, this book is not especially unique. Nevertheless, it is a fine text that should be considered by any general reader with even a passing interest in astronomy. Written by a science fiction author/newspaper editor, the text reads smoothly, is well illustrated, and is very informative. Subjects include the standard astronomical topics like stars and galaxies, as well as "more exciting" topics like pulsars, quasars, the origin of the Universe, and life and intelligence in the universe. A prime candidate for public and home libraries.

183. Solomon, Joan. **The Structure of Space**. New York, John Wiley and Sons, 1974. 219p. illus. index. bibliog. (Physics and Humanities Series). (A Halstead Press Book). $10.95. LC 73-8543. ISBN 0-470-81221-4.

Not really an astronomy book as such, this not-so-typical volume contains discussions of many of the physical laws of the Universe, all of which are the basis of astronomical laws and theory. Among these subjects are gravity, electricity, light, magnetism, and relativity. The author tells the stories of the development of knowledge in these areas, mentioning the discoverers and their work along the way. The text is descriptive and well illustrated, and it should be of interest to the layman and beginning astronomy student. Subtitled "The Growth of Man's Ideas on the Nature of Forces, Fields and Waves," it is a fine introduction to space and the laws that govern occurrences in it, and it emphasizes an important and sometimes neglected aspect of astronomical education.

184.   Stoy, R. H., ed. **Everyman's Astronomy**. New York, St. Martin's Press, 1975, c1974. 493p. illus. index. glossary. $10.00. LC 74-81460.

   A general but serious look at astronomy, aimed at the "intelligent layman," this up-to-date work is a collection of 11 articles covering all areas of the field. Written by various astronomers, the chapters are no-nonsense treatments of history, the solar system, the stars, galaxies, the Milky Way, and telescopes. Avoiding the "gee-whiz" approach entirely, this volume is one of the best works available on a non-elementary level. Coverage of individual topics is quite complete. For example, the chapters on the planets include historical background, physical characteristics, and descriptions of recent space exploration, if any. This pattern is typical of the entire work. There are not many tables, graphs, and illustrations, but those that are included are well chosen and extremely helpful. A short but informative opening chapter on naked-eye astronomy is included for the reader so inclined; several star charts accompany this section. Although a substantial portion of the book is descriptive, the emphasis is on understanding the Universe, and the authors go into detail on stellar evolution, the workings of telescopes and other equipment, solar activity, etc. Best suited for the college graduate, the book's authors occasionally introduce some mathematics and equations, but nothing too complicated. Highly recommended for the public and home library, the work's only drawback is the lack of a bibliography.

185.   Wood, Harley. **Unveiling the Universe: The Aims and Achievements of Astronomy**. New York, American Elsevier Publishing Co., 1968. 240p. illus. index. bibliog. $11.50. LC 67-31258. ISBN 0-444-19784-2.

   An introduction to astronomy for the layman, in which the author explains basic principles, observational evidence, and the reasoning behind the subject. This is not the typical "gee-whiz" astronomy book, but it is far from being dry. The text is interesting, designed to be read straight through, and there is a section of excellent photographs in the center. The chapter on star maps is especially good. Here the author gives the reader some insight into how stars and constellations got their names, as well as explaining how to use the maps. Standard topics make up the bulk of the book: the celestial sphere, tools of the astronomer, Earth, Moon, planets, Sun, stars, the Milky Way, the Universe. Readers who become inspired by the first eleven chapters will want to delve into chapter 12, called "A Cycle of Objects." In this so-called practical chapter, the author lists tables of astronomical objects suitable for viewing with the naked eye and/or a small telescope: planets, bright stars, star clusters, double and multiple stars, variable stars, gaseous nebulae, and galaxies. For the college and public library.

For the Astronomer

186.   Doyle, Robert O., ed. **A Long-Range Program in Space Astronomy**. Washington, D.C., NASA, 1969. 305p. illus. index. refs. bibliog. $1.50pa. NASA SP-213.

110 / General Materials

This "Position Paper of the Astronomy Missions Board" describes the Board's recommendations to NASA concerning possible astronomical investigation to be carried out by the space program. Although the report itself is a bit outdated, the recommendations are still important to consider, since they represent valid research projects yet to be carried out. Among the recommended programs are x-ray and gamma-ray observations, optical studies, infrared astronomy, radio astronomy, lunar-based observations, solar astronomy, and orbiting observatories. Though it is similar to the report of the Astronomy Survey Committee of the National Research Council (see *Astronomy and Astrophysics for the 1970's*), there does not appear to be any connection. The reports make similar recommendations as to research topics, but this report concentrates on studies made with extra-terrestrial craft.

187. Gingerich, Owen, ed. **Frontiers in Astronomy: Readings from Scientific American**. San Francisco, W. H. Freeman and Co., 1970. 370p. illus. index. bibliog. LC 71-129925. ISBN 0-7167-0948-1; 0-7167-0947-3pa.

Without a doubt, some of the finest articles on astronomy are found in the pages of one of America's most respected journals, *Scientific American*. It is not surprising, then, that this volume, comprised of 39 such articles, is an excellent source of astronomical information. By documenting many of the major discoveries since 1956, the book can serve as a handy reference work, as well as provide good reading of a general nature. Among the many topics are four which the book's editor considers the most notable: quasars, pulsars, the 3° background radiation, and the intense infrared sources with the associated anomalous OH radio radiation. Each article is authored by a respected scientist and is accompanied by superb illustrations and a fine bibliography. Laymen and interested non-specialists should seriously consider purchasing this work for their home libraries, and college instructors should include it in course syllabi. For all types of libraries.

188. Gingerich, Owen, ed. **New Frontiers in Astronomy: Readings from Scientific American**. San Francisco, W. H. Freeman and Co., 1975. 369p. illus. index. bibliog. $13.00. $7.50pa. LC 75-8902. ISBN 0-7167-0520-6; 0-7167-0519-2pa.

"New" is a dangerous word to use in science; it quickly becomes "old" before one is aware of it. Nevertheless, it is an apt descriptor, for the time being, of this excellent collection of articles from *Scientific American* on the latest developments in astronomy. Like the edition published in 1970 (*Frontiers in Astronomy;* preceding entry), this work contains representative essays on a variety of fields: the planetary system, the Sun, stellar evolution, the Milky Way, galaxies, high-energy astrophysics, and cosmology. Over half the 31 articles appeared in the magazine since 1970, and there are selections from the 1960s and 1950s as well. The most profound disappointment, however, is the inclusion of 12 articles that appeared in the first edition. Though the compiler comments on the difficulty of choosing selections from the dozens of articles available, he has included more than a third of the text of the first version. There is no explanation for this in the introduction. Apparently considering this strictly a new edition of the old volume, he has dropped some articles and retained others. This is most unfortunate—the book should have included only new selections. Libraries that do not own the first version ought to order this book immediately, because it is an excellent collection that will be of interest to all types of readers. After all, it maintains the high quality of *Scientific American*. But libraries with the first edition might think twice about buying a book that will, in part, duplicate what they already have.

189. International Astronomical Union. **Highlights of Astronomy**. Dordrecht-Holland: D. Reidel Publishing Co., v.1– , 1968– . triennial.

Published after each meeting of the General Assembly of the IAU, these excellent volumes contain invited discourses, selected papers, and joint discussions from the IAU conferences. Each tome presents, as the title states, the highlights of astronomy in the current year. Thus, these volumes are a valuable record of current research and discovery, and collectively they contain hundreds of papers and thousands of references. The books belong in every observatory library and in the university astronomy library.

Volume I: 1968 (XIIIth General Assembly of the IAU, 1967). Luboš Perek, Editor and General Secretary of the Union. 548p. illus. refs. $28.30. LC 68-31894. ISBN 90-277-0137-7. Text in English, some French; 79 papers, reviews, and presentations. Joint Discussions: New Techniques in Space Astronomy; X-Ray Astronomy; The Lithium Problem; Modern Problems in Fundamental Astrometry; Extragalactic Radio Sources; Close Binaries and Stellar Evolution. Special Meetings: Lunar Probes; Coordination of Solar Observations Made at Ground-Based Observatories and with Space Vehicles.

Volume II: 1971 (XIVth General Assembly of the IAU, 1970). Cornelis De Jager, Editor and General Secretary of the Union. 793p. illus. refs. $47.50. LC 71-159657. ISBN 90-277-0189-X. 90 papers, reviews, and presentations. Special Meeting on Direct Exploration of the Moon. Joint Discussions: The Origin of the Earth and Planets; Helium in the Universe; Interstellar Molecules; Atomic Data of Importance for Ultraviolet and X-Ray Astronomy; Photo-electric Observations of Stellar Occulations; Pulsars, Cosmic Rays and Background Radiation. A Joint Meeting of five IAU Commissions: The Absolute Magnitudes of the RR Lyrae Stars.

Volume III: 1974 (XVth General Assembly and the Extra Ordinary General Assembly of the IAU, 1973). G. Contopoulos, Editor and General Secretary of the Union. 574p. illus. refs. $77.00. LC 71-159657. ISBN 90-277-0452-X. 64 papers, reviews, and presentations. Joint Discussions: Precession, Planetary Ephemerides and Time Scales; Stellar Infrared Spectroscopy; Kinematics and Ages of Stars near the Sun; Origins of the Moon and Satellites; Jovian Radio Bursts and Pulsars; The Outer Layers of Novae and Supernovae.

190. National Research Council. Astronomy Survey Committee. **Astronomy and Astrophysics for the 1970's**. Washington, D.C., National Academy of Sciences, 1973. 2v. illus. index. v.1: 136p. $4.75pa. ISBN 0-309-02029-8; v.2: 410p. $14.25pa. ISBN 0-309-02110-3. LC 72-79131.

112 / General Materials

In 1969, the Astronomy Survey Committee of the National Research Council was established at the request of the Committee on Science and Public Policy of the National Academy of Sciences. The goal of the Committee was stated as follows: ". . . to outline the present state of astronomy, to identify the most exciting problem areas in that field, and to recommend a program for the United States for the next ten years, including both major new ground-based facilities and major space-science programs." This two-volume set is the final report of the Committee's study. Volume 1, the Report of the Committee, takes a general look at the broad range of astronomy, excluding lunar and planetary considerations; and volume 2 is the Reports of the Panels, groups of individuals on the Committee who were examining specific areas of research. Among the fields examined as possible research projects were radio astronomy, optical astronomy, infrared astronomy, space astronomy, solar astronomy, theoretical astrophysics, dynamical astronomy, astrophysics and relativity. Most astronomers have probably already seen these books, but if not, they should borrow them from the library. Within these volumes are the immediate goals and plans for astronomical research in the United States, something astronomers obviously have a stake in.

AMATEUR ASTRONOMY

Despite the fact that there are many thousands of amateur astronomers in the world, little has been written about the types of books and journals aimed at this large group. While coverage in the section below is by no means complete, it does represent a healthy cross-section of the scores of materials available to non-professional astronomers, from very casual stargazers to serious amateurs. Briefly, the books listed here are those that teach, that provide tables of numerical data for quick reference or that describe telescopes and observations. Since planetary astronomy is usually of great interest among amateurs, the volumes here necessarily show a bias toward the study of the Sun, Moon, planets, comets, and meteorites. Of these books, some concentrate on one topic, like observing the Moon, while others usually try to cover all subjects of interest to the reader.

The literature for the amateur astronomer is on all levels, supporting casual activity and serious, frequent observation alike. The very beginning amateur astronomers, who may be eight or eighty but are often young, may first have their interest aroused by one of the picture books showing beautiful color photographs of the planets and stars. Or an interest in the sky might come from a friend or relative who first introduces the neophyte to the constellations one summer night. In any case, the new astronomer often first obtains a book on the constellations and sets out to make some discoveries of an elementary sort. If curiosity continues, the amateur astronomer may then purchase a pair of binoculars or a small telescope to get a better look at the heavens. A standard amateur's handbook will be close at hand. Finally, if the interest develops further, perhaps into a serious hobby, the astronomer buys or makes a larger instrument, reads an introductory text, and borrows one of the more advanced observing handbooks from the local public library. He may also join the local amateur society.

The point is that there are several levels of activity for amateurs, anything from an occasional evening sky-gazing stroll to serious, regular observing. New and experienced amateur astronomers should be able to find here any number of books, no matter how advanced or elementary, to fulfill their needs. The reader is reminded, too, to consult other sections of this guide for relevant material not classified as "amateur astronomy," like handbooks under "Reference," or books on the planets under "Solar System," or any of the books listed under "General Works." The books listed here are merely those that concentrate on non-professional observation—the serious amateur should certainly not confine his or her reading to this brief section of the guide.

A word about radio astronomy for amateurs is in order. Not surprisingly, there is very little on the subject, for "amateur" and "radio astronomy" are almost contradictory. While the basic concept of receiving radio signals from the Universe is relatively simple, the construction and use of a radio telescope is far from elementary. Further, a knowledge of and skill in electronics and radio science are necessary prerequisites for work in this field. Consequently, there are few neophytes who could even think about taking up radio astronomy as a hobby. This does not, of course, preclude the experienced backyard astronomer from learning. In any case, three of the few volumes available for the amateur are included in this section.

Amateurs who wish to find periodicals that contain articles of interest on observation and related topics should turn to the section on "General Periodicals," below.

This portion of the book is divided into four parts, covering books for the beginner, for the intermediate observer, for the advanced amateur, and books on sundials.

Card catalog subject headings: *Library of Congress*: ASTRONOMY— OBSERVERS' MANUALS; ASTRONOMY—HANDBOOKS, MANUALS, ETC.; CONSTELLATIONS. *Sears:* ASTRONOMY—HANDBOOKS, MANUALS, ETC.

Beginning

191.   Brown, Peter Lancaster. **What Star Is That**? New York, Viking Press, 1971. 224p. illus. indexes (stars, general). (A Studio Book). $15.00. LC 73-149587. ISBN 0-670-75865-5.

This excellent book is for the amateur, young or old, who is very serious about learning the constellations and identifying individual stars. Included are many illustrations, star charts, and even 15 color slides, all of which help the observer learn the stars. After some introductory material on the origin of the constellations, becoming familiar with the sky, and observing aids, the author goes into depth on the individual constellations and the names of the stars of which they are composed. There are discussions of Northern and Southern Hemisphere stars, and the skies of spring, autumn, summer and winter; there is also a chapter on planet identification. Highly recommended for the public library and its users.

192.   Cherrington, Ernest H., Jr. **Exploring the Moon Through Binoculars.** New York, McGraw-Hill Book Co., 1969. 211p. illus. index. gazetteer. $10.00. LC 68-13624. ISBN 0-07-010760-2.

Written specifically for the observer who doesn't have a telescope, this book points out that most lunar landmarks are within the range of a good pair of 7x50 binoculars. This volume contains not only many photographs for the observer's reference, but also an excellent and extensive text describing the wealth of lunar features observable with binoculars. Described, too, are each of the Moon's daily phases from new to full, and back to new. The amateur astronomer who cannot afford a telescope just yet, and who has a pair of binoculars, should get his or her hands on this excellent guide. It is sure to whet the astronomer's appetite for more.

193.   Inglis, R. M. G. **A New Popular Star Atlas (Epoch 1950).** Edinburgh, Gall and Inglis, 1974, c1949. 16 maps. index.

Useful for the beginning amateur or occasional backyard observer, this handy atlas shows all stars 5.5 visual magnitude or brighter—i.e., those visible with the naked eye. Variable stars, clusters, and nebulae are likewise shown, only if visible to the unaided eye or with binoculars. The basic format of the two polar sky maps and 14 equatorial grids is white on blue background. Stars and other objects' magnitudes are shown, along with constellation boundaries and Greek letter designation, or specific stellar name. The maps, drawn by the author, include short lists of interesting stars and non-stellar objects, in the same fashion as *Norton's Star Atlas* (16th ed., 1973) but are not nearly as lengthy. Measuring an easy-to-use 19.5x24.5cm, the book further includes an index to the constellations, a list of star and cluster names, and hints on the care and use of small telescopes. The more serious observer will prefer the maps in *Norton's*, but this little handbook will be quite sufficient for just about any situation involving naked-eye astronomy. Designed for public and school libraries, it would also be useful in certain college and university situations, possibly for beginning lab work.

194.   Joseph, Joseph Maron, and Sarah Lee Lippincott. **Point to the Stars.** 2nd ed. New York, McGraw-Hill, 1972. 96p. index. illus. glossary. $4.72. LC 71-39765. ISBN 0-07-033049-2.

A short book for the casual astronomer who wants to be able to identify the constellations and planets. Nicely illustrated with line drawings and photographs, this volume would be useful to have in secondary school and public libraries. Included are explanations of the constellations, how they got their names, and the stories behind them, as well as diagrams of the sky where they are located. A chapter on viewing artificial satellites is also included. This second edition has added short descriptions of quasars and pulsars, and has updated the planetary charts.

195.   Mayall, R. Newton, Margaret Mayall, and Jerome Wyckoff. **The Sky Observer's Guide: A Handbook for Amateur Astronomers.** New York, Golden Press; distr. New York, Western Publishing Co., Inc., 1965. 160p. illus. index. bibliog. (A Golden Handbook). $1.95. LC 65-15201. ISBN 0-307-24009-6.

This little guide is one of the better handbooks for the casual astronomer and occasional observer. Profusely illustrated with 150 paintings and photographs, the book covers a variety of subjects of interest to the layman: the planets, the Sun, the Moon, comets, meteors, stars, etc. Lists of sky objects are included for viewing with binoculars or a small telescope. Tips on observing and what to look for in the sky are important parts of this fine handbook. For public libraries and their readers, from junior high school on up, it is not as lengthy or as comprehensive as Menzel's work, *A Field Guide to the Stars and Planets*, but it is more affordable. Illustrations by John Polgreen.

196.   Menzel, Donald H. **A Field Guide to the Stars and Planets.** Boston, Houghton Mifflin Co., 1964, 1973pa. 397p. illus. index. glossary. bibliog. (The Peterson Field Guide Series). $8.95; $4.95pa. LC 63-7017. ISBN 0-395-07998-5; 0-395-19422-9pa.

In subject content, this fine handbook is no different from the multitude of guides for the amateur astronomer. It covers the standard topics—the stars, Moon, planets, comets, etc.—in an interesting but not especially unique way. However, the book does stand out in a very important area more than any other similar work—the illustrations and the star maps. Of special note are the 48 star charts, covering both Northern and Southern skies, each being produced twice, once with only the star images on a black background, and again on the opposite page with the names of stars and other celestial objects designated. These maps, which are quite good, will greatly benefit observers, whether they are occasional star-gazers or serious amateurs. In addition, the author has included a beautiful photographic atlas of the heavens, using 51 photographic plates taken at the Harvard College Observatory. Both positive and negative prints are reproduced, which is a different and worthwhile approach.

Besides the illustrations, there is an excellent text and a multitude of tables with data on the stars, planets, and other celestial bodies, all appropriately arranged for convenient use by the astronomer. The three concluding chapters on the telescope, astrophotography, and time are excellent; the reader is given clear, detailed instructions in these areas. Especially suited for the public library, the book would be a good purchase for the college library as well. Highly recommended for the amateur observer, too.

197.   Moore, Patrick. **Amateur Astronomy**. Rev. ed. New York, W. W. Norton & Co., Inc., 1968. 328p. illus. bibliog. index. (The Amateur Astronomer's Library). $6.95. LC 68-10882. ISBN 0-393-06362-3.

Instead of presenting a handbook filled with only charts and tables, the author has combined text with helpful observing techniques in a book that the amateur astronomer will want to keep for a lifetime, for casual reading and reference. Chapters include discussions of telescopes, the solar system, the Sun, the Moon, aurorae, the planets, comets, meteors, stars, nebulae, and galaxies. Supporting the lively text are many excellent diagrams, sketches, and photographs of the planets, Moon, and stars. As usual, an excellent book from Mr. Moore, the amateur astronomy authority. (Originally titled: *Amateur Astronomer.*)

198. Moore, Patrick. **Naked-Eye Astronomy**. New York, W. W. Norton & Co., Inc.; London, Lutterworth, 1966. 253p. illus. index. notes. (The Amateur Astronomer's Library). $8.95. LC 65-27466 (Norton); 67-5413 (Lutterworth). ISBN 0-393-06303-8 (Norton); 0-7188-0596-8 (Lutterworth).

A good book for the beginning amateur astronomer, this volume concentrates on what the observer can see without binoculars or telescope. Among the high points of the book are learning the constellations, finding the planets, observing the Sun and Moon, and watching the skies "change" from month to month. There are many excellent illustrations, including photographs and diagrams, which help in the learning process. Especially good for the young astronomer.

199. Muirden, James. **Astronomy with Binoculars**. London, Faber & Faber; distr. Princeton, N.J., D. Van Nostrand Co., Inc., 1963. 146p. illus. index. LC 67-3615.

Not many amateur astronomers or casual observers consider binoculars for scanning the skies; the small telescope is the usual instrument used for observation. But a great deal can be seen in the night sky with a good pair of binoculars, and this book details just what to look for. The author discusses binocular observations of the Sun, Moon, planets, comets, meteors, aurorae, and beyond, followed by interesting chapters on spring, summer, autumn, and winter stars. A good introduction to beginning stargazing as well, this work belongs in the public library. A similar book of interest is Ernest Cherrington's *Exploring the Moon Through Binoculars*.

200. Neely, Henry M. **A Primer for Star-Gazers**. New York, Harper and Row, 1970. 334p. illus. index. $8.95; $7.87 (lib bdg). LC 72-120090. ISBN 0-06-013167-5; 0-06-013168-3 (lib bdg).

Intended for those who know nothing about the stars and who don't wish to become serious students, the purpose of this book is to teach the reader how to find and recognize the constellations. It is for young people and oldsters, anyone who wants to learn a little but not be overwhelmed, because the author sees star-gazing as pure fun and little else. It includes good illustrations and dozens of star maps and charts that show the constellations. Highly recommended for the home and public and secondary school library.

201. Nourse, Alan E. **The Backyard Astronomer**. New York, Franklin Watts, Inc., 1973. 118p. illus. index. bibliog. $7.87. LC 73-4644. ISBN 0-531-02568-3.

An introduction to amateur observing for young people, this book includes illustrations in the form of diagrams that teach the observer what to look for in the sky. There is a lot of information on the constellations, as there should be, since a young person's first introduction to astronomy is often through learning to identify the constellations. Three types of observing are discussed: naked-eye, binoculars, and small telescope. The book would make a nice gift for the "scientific-minded" junior high student and would be appropriate for both public and secondary school libraries.

202. Peltier, Leslie C. **Guideposts to the Stars: Exploring the Skies throughout the Year.** New York, Macmillan Co., 1972. 176p. illus. index. bibliog. glossary. $7.95. LC 72-187797.

The majority of this volume is dedicated to teaching the young astronomer all about the stars and the constellations. The author shows that there is much that can be seen without a telescope, and he tells how the reader can look for the planets, comets, the Moon, meteors, the Milky Way, aurorae, and zodiacal light. Excellent drawings and basic stars charts are used to take the reader on a "walk" through the skies. There is a chapter on small telescopes, and useful appendices on societies and dealers, the constellations, the brightest stars, and an atlas of the stars. A good book, for junior high age on up.

203. Vehrenberg, Hans, and Dieter Blank. **Handbook of the Constellations.** 2nd ed. Dusseldorf, Treugesell-Verlag; distr. Cambridge, Mass., Sky Publishing Corp., 1973. 197p. star charts. $17.50.

This handbook/atlas is the perfect combination of astronomical charts and sky catalog. Fifty-five sections of the sky are shown, each on a separate page, in the form of a star chart or map. Facing each map is a multitude of important information useful to the observer: lists of bright stars, double stars, variable stars, galactic clusters, globular clusters, planetary nebulae, diffuse nebulae, and galaxies, all within the chart on the opposite page. Right ascension, declination, photographic and apparent magnitudes, and other useful information are included for each listed object. Every serious amateur making frequent observations will want this excellent book, as will public, special, and college libraries.

204. Whitney, Charles A. **Whitney's Star Finder: A Field Guide to the Heavens.** New York, Alfred A. Knopf, 1974. 97p. illus. index. glossary. $5.95pa. LC 73-20747. ISBN 0-394-70688-9.

Books for the casual astronomer on the constellations are pretty much alike. Most of them are good, but there is really little to differentiate among them. This short volume is an exception. Basically, it is a book for casual observers who occasionally want to know what they saw last night in the sky. The author calls it "a field guide to the heavens" and in it he explains to the reader a variety of fascinating things seen in the heavens. The big difference, however, and one that makes this item especially appealing, is the star finder, or locater wheel, that is included. With this interesting cardboard device, the backyard astronomer can identify any bright star visible from North America. By rotating the wheel, which changes the sky diagram according to the seasons and time of night, the "astronomer" has before him a representation of the sky that can be used to locate and identify the stars quickly. Also included are sections on observing the planets with binoculars, the moon, eclipses, sunset and sunrise, and some miscellaneous items of interest. Especially good for the younger astronomer, this would make a fine gift.

205. Zim, Herbert S., and Robert H. Baker. **Stars**. rev. ed. New York, Golden Press; distr. New York, Western Publishing Co., Inc., 1975. 160p. illus. index. (A Golden Nature Guide). $1.50pa. LC 61-8321. ISBN 0-307-24493-8pa (Western).

A good example of a handbook that never outlives its usefulness, this little volume was originally published in 1951 and has been in continuous use since. A general, elementary guide to the skies, *Stars* is appropriate for all types of beginning astronomers, young or old. The first section is a brief overview of amateur astronomy, the Sun, telescopes, stars, star clusters, nebulae, and galaxies. The part devoted to stars explores their numbers, distances, brightnesses, sizes, densities, motions, colors, magnitudes, classification, and much more. The middle third of the book is given to the constellations, including excellent drawings of the stars' arrangements, superimposed on a sketch of the mythical animal, person, or thing for which they are named. The final section is concerned with the most popular topic of beginner amateur astronomy, the solar system, and here the reader learns about the planets, comets, meteors, the Moon, eclipses, etc. This handbook's excellence stems from a combination of superb illustrations by James Gordon, a clearly written text, and several tables of useful statistical information. *Stars* is for the very beginner, so the enthusiastic, more serious stargazer should pass it by. It's the kind of book that should be in every family's library, for it is the very best guide, in a general sense, that is available.

Intermediate

206. Baxter, W. M. **The Sun and the Amateur Astronomer**. 3rd ed. Newton Abbot, Engl., David & Charles; New York, W. W. Norton & Co., 1973. 165p. illus. index. bibliog. £3.50 ($11.95). LC 73-159307. B 73-03070. ISBN 0-7153-5629-1 (D & C).

There are plenty of books on observing the planets and the Moon, but this is the only current volume to deal with amateur observation of the nearest star. Since it is aimed at non-professional astronomers, it is not at all technical and is easy to understand. There is one chapter about the Sun itself, its characteristics, etc., but that is not the emphasis of the book. The author concentrates on methods of observing and what to look for. Viewing the Sun with the naked eye or unfiltered telescope lens is extremely dangerous, so special equipment has been developed exclusively for solar observation. The author discusses this equipment, in particular the solar spectroscope, spectrohelioscope, telescope, and spectroheliograph, all but the latter of which can be used and constructed by amateur astronomers. A well-written, useful volume for the public library.

207. Brown, Sam. **All about Telescopes**. 2nd ed. Barrington, N.J., Edmund Scientific Co., 1975. 192p. illus. index. $4.95. LC 75-308087.

The enthusiastic amateur with designs on building or purchasing a telescope should get a copy of this book at all costs. Overwhelmingly illustrated with diagrams of equipment and the sky, this volume has everything the amateur astronomer could want in the way of help with telescope construction, auxiliary equipment, lenses, etc. Written in a non-technical way, the book goes into great detail on a variety of topics for the advanced or intermediate stargazer. The best way to describe the contents is to summarize them; the reader will immediately see why this book is so highly recommended.

1) Getting acquainted with the telescope. The author covers a variety of introductory information that will hold the interest of nearly any astronomer: history, selection (of relectors, refractors, mounts, etc.), telescope performance (pointing out that stars are not magnified by a telescope, but that their light is intensified), seeing (and what affects it), using an equatorial mount, telescope arithmetic, eyepieces, and more.

2) Observing the sky show. Brown lists the types of objects for viewing (planets, stars, galaxies, clusters, etc.), explains star maps and the importance of time in astronomy, and gives a good run-down on setting circles.

3) Photography with your telescope. One of the most comprehensive treatments anywhere is presented here, beginning with a discussion of the four telescopic photographic optical systems. The discussion about film, lenses, camera mounts, and exposure times is excellent, as is the explanation of various methods of photographing different celestial objects.

4) Mirror grinding and testing. The discussion of telescope building gets underway with this good coverage of making a reflecting telescope mirror. Topics covered include abrasives and pitch, polishing, testing equipment, conducting mirror tests, corrections, and figuring the paraboloid.

5) Telescopes you can build. A detailed discussion of reflecting and refracting instruments, complete with diagrams in the form of blueprints, makes up this fine chapter. Various sizes and types of telescopes are covered.

6) Telescope mounts. This important section introduces the reader to the types of stands on which to place the telescope tube, including the German equatorial, fork equatorial, pipe mounts, etc. A discussion of clock drives is also covered.

7) Collimation and adjustments. Here the finishing touches on your homemade instrument are presented: balancing a telescope, collimating a reflector, adjustment to the pole, and more.

8) Telescope optics. This final chapter details the different types of lenses, their uses and limitations.

The book, which is chock full of diagrams and tables of data, is just right for all types of amateur astronomers, including those of us who took the easy way out and bought a telescope. All serious amateurs should have this fine work on their bookshelves alongside *Norton's Star Atlas*.

208.   Emerson, Myron N. **Amateur Telescope Mirror Making**. New York, Carlton Press, Inc., 1969. 93p. illus. index. (A Hearthstone Book). $3.95. NUC 70-20193.

120 / General Materials

The most important part of a reflecting telescope, the favorite of many amateur astronomers, is the mirror that gathers the light from the stars and other celestial objects. Unfortunately for amateur telescope makers, the mirror is also the most difficult part of the telescope to make, and the coverage of such an important topic is frequently not adequate for the builder of the celestial instrument. Fortunately, this short but excellent book, which appeared in 1969, treats the problem with a high degree of success. In it, the author presents step-by-step instructions for making an eight-inch mirror for an amateur reflecting telescope. Not many details have been neglected in this thorough work, which covers grinding (rough, semi-fine, and fine), polishing, parabolizing, and the Foucault Test. Supplementary material includes a list of supplies and services available to the mirror maker. Anyone seriously considering building a telescope should first read this work to get an idea of the effort involved, which is considerable.

209. Glasby, John S. **The Variable Star Observer's Handbook**. London, Sidgwick & Jackson; New York, W. W. Norton and Co., 1971. 213p. illus. index. glossary. LC 72-175943. B 71-22031. ISBN 0-283-48470-5 (Sidgwick); 0-393-06377-1 (Norton).

One area in which the serious amateur may contribute significantly to astronomical research is the study of variable stars. A large telescope with auxiliary equipment is not needed—most work in this growing field can be done with a small to medium-sized telescope or even binoculars. Such research, then, is tailor-made for non-professional observers. The author covers most facets of variable star observing for both beginners and more advanced observers as he tells just how the amateur can watch the stars whose brightness varies, either rapidly or slowly. The chapters of this excellent book include Astronomical Instruments and Their Use; The Family of Variable Stars; Methods of Observation: Naked-Eye Variables; Binocular Variables; Telescopic Variables; The Light Curves; Charts and Sequences; Photographic and Photoelectric Observations; Spectroscopic Observations: The Discovery of Variable Stars; Some Recent Novae. The authority on variable stars shows his writing versatility by approaching the subject on the amateur level, and he succeeds convincingly in this highly recommended book.

210. Hartung, E. J. **Astronomical Objects for Southern Telescopes**. London, Cambridge University Press, 1968. 238p. illus. bibliog. index. LC 68-27619. ISBN 0-521-05224-6.

This handbook should be the companion of every amateur or student who observes in the Southern Hemisphere. The author begins the volume by adequately describing each type of object the observer is likely to encounter: stars, star clusters, galactic nebulae, extragalactic systems, etc. Immediately following are several pages that help the amateur astronomer prepare for actual observation, including a brief discussion of telescopes. Next, for the observer's convenience, Hartung inserts a table of 1,017 objects that can be observed between the South Pole and 50° North latitude. Arranged by right ascension and declination, each entry is briefly described as to type and the constellation in which it can be found. The majority of the book, though, is an individual detailed visual description of each object in the aforementioned table. The author's purpose here, and it is unusual, is to give the observer an idea of how the object being looked for will appear in the telescope. It is unfortunate that there is not a corresponding volume for the Northern Hemisphere. This superb volume should be in all public libraries in the Southern Hemisphere, and in many in the Northern Hemisphere.

211.   Haysham, H. **Basic Astronomy: With Projects for Amateurs and Students**. Sunderland, Engl., Thomas Reed Publications, Ltd., 1971. 234p. illus. index. bibliog. review questions. £3.25. LC 72-197336. B 71-11473. ISBN 0-901281-22-0.

Not many texts include projects for ambitious amateur astronomers, and for that reason, this particular book is especially attractive. First and foremost, this text is designed for amateurs and students who want a good explanation of and introduction to astronomy. But the author goes beyond simple explanations, by providing review questions and detailing exercises that the reader will wish to take on. This approach gives more meaning to basic astronomy, and the amateur suddenly finds the subject coming alive. Learning by doing, as it is often said, is the best way. The chapter titles indicate the topics covered: Aspects of the Sky; Stars in the Sky; Our Home in Space; Time and the Calendar; The Moon; Family of the Sun; Nature of the Stars. There are extra "time" problems as well as an astronomical quiz for self-review.

212.   Howard, N. E. **Standard Handbook for Telescope Making**. New York, Thomas Y. Crowell Co., 1959. 326p. illus. index. glossary. $9.95. LC 59-12503. ISBN 0-690-76784-6.

Although written many years ago, this volume retains its usefulness for the ambitious amateur astronomer who wishes to build a telescope. Presented in clear language for the new telescope maker, the book goes into great detail, carefully explaining, step by step, the procedures involved. Howard, who obviously knows what he's talking about, avoids oversimplification (a problem with some similar guides) and overly technical language in his text. He is quick to point out that building a telescope is not an easy, one-night project, but that it is not beyond the capabilities of nearly any eager amateur. In essence, the book describes how to build an 8-inch, f/7 Newtonian reflector, an instrument which, when completed, will provide the backyard astronomer with an excellent tool for exploring the skies. Emphasizing the cost-saving aspect, as well as the satisfaction gained from building the telescope by oneself, Howard leads the reader through mirror-making (grinding, polishing, testing, etc.), eyepieces, the tube and mounting, adjustments, and more. No details are omitted, and an ample number of photographs are included for illustration. Additional topics are telescope history, setting up a backyard observatory, celestial photography, and different types of telescopes. The author's enthusiasm pervades this excellent effort, providing a push for the wary amateur who wonders "can *I* really do it myself?" The telescope maker, old and new, needs no other volume than this to construct the instrument of his or her dreams. Highly recommended for any library, its companion is the author's fine *The Telescope Handbook and Star Atlas* (1975).

213.   Howard, Neale E. **The Telescope Handbook and Star Atlas.** 2nd ed. New York, Thomas Y. Crowell Co., 1975. 226p. illus. index. bibliog. glossary. $14.95. LC 75-6601. ISBN 0-690-00686-1.

The amateur astronomer seeking a good introduction to telescopes and practical astronomy should consider this book which is two-thirds text and illustrations and one-third atlas. The author first explains telescopes and the types of celestial objects which may be viewed; this explanation is followed by excellent star maps, a fine gazetteer, and a Messier catalog. In short, the volume is a one-stop trip for the enthusiastic beginner. In a good chapter called "The Sky," Howard gets the reader started by describing the seasonal constellations and explaining how to use star charts. The next chapter briefly introduces the reader to the basics of the celestial sphere and time, and relates them to the setting circles on a telescope. The remainder of this fine text discusses the Sun, Moon, planets, stars, etc., and tells the reader what to look for in observing each. There is even a chapter on celestial photography, detailed enough to get the amateur astronomer off and skyshooting in no time. The star atlas includes dozens of tables of both numerical and observational data. The author notes that the book is the result of 30 years' work with secondary school students. If the quality of this book is any indication, the students of Neale Howard had an excellent astronomical education.

214.   Keene, George T. **Stargazing with Telescope and Camera.** 2nd ed. New York, Amphoto; distr. Philadelphia, Chilton Book Co., 1967. 128p. illus. refs. $3.95. LC 67-25847.

An introduction to amateur astronomical photography, this short volume shows how easy it is to take good pictures in your own backyard. The author begins by discussing how to choose a telescope or pair of binoculars, telling what to look for according to the needs of the observer. He next tells how to make a reflecting telescope; the coverage here is good, but the reader should also consult other sources (e.g., Emerson's *Amateur Telescope Mirror Making*). Using the telescope and finding objects in the sky are next, followed by the meat of the text—the discussion of cameras and lenses and astrophotography. Seven helpful appendices are also included, and the references with each chapter point the way to further reading and other sources. The excellent diagrams and photographs will help the astronomer greatly, especially in the chapter on making a reflecting telescope.

215.   Mayall, R. Newton, and Margaret W. Mayall. **Skyshooting: Photography for Amateur Astronomers.** rev. ed. New York, Dover Publications Inc., 1968. 186p. illus. index. $2.75pa. LC 67-29410. ISBN 0-486-21854-6.

One of the most rewarding aspects of amateur astronomy can be the taking of pictures of the sky and individual celestial objects. It is not as difficult as it might sound, and the authors prove it here in their successful attempt to explain to amateur astronomers how to use a camera to record what they've seen. More than a "how-to" book on celestial photography, the work includes a good deal of basic astronomy, especially important for the camera buffs who are not astronomers but for whom the Mayalls' book is intended. The various chapters on the Sun, Moon, planets, stars, etc., all give advice on how to photograph these objects, best exposure times, type of film, etc. Equipment, too, is discussed, including mountings, telescopes, clock drives, and more. The information on plates and film and cameras is very good; the authors even outline topics like developing, printing, and enlarging. These latter descriptions are very brief, however, and it might be wise to get a more complete book on the subject of the darkroom or, better yet, to get help from someone who knows. A really complete guide to astronomical photography, it also includes related chapters on sidereal time, setting circles, keeping records, and advanced projects. One of the best, it should be in every serious amateur's library.

216. Norton, Arthur P. **Norton's Star Atlas and Reference Handbook (epoch 1950.0)**. 16th ed. Edinburgh, Gall & Inglis; distr. Cambridge, Mass., Sky Publishing Corp., 1973. 116p. illus. index. $12.50. LC 74-167214. GB 74-06269. ISBN 0-85248-900-5.

First published in 1910, this excellent handbook and atlas is *the* most popular and most-used volume owned by thousands of amateur astronomers. A quick examination will show why this is so: it is an excellent text for the amateur observer, combined with a better-than-average star atlas for quick reference. A wealth of useful information is included: star charts, astronomical tables, symbols, lists of objects for viewing, and more. The section on the use and care of the telescope has been especially helpful, and amateur astronomers for years have learned much from *Norton's* clear explanation of optical instruments. The extent of changes in the newest version is stated in the introduction: "The text of this sixteenth edition of the Reference Handbook has been completely re-written and brought up to date, and the lists of telescopic objects revised and enlarged." Many other new features have been added, while most of the old and familiar parts remain. A major change is the new type face in the "Reference Handbook," a welcome improvement that makes the text much easier to read. This edition was edited by Gilbert E. Satterthwaite (*Encyclopedia of Astronomy*) in consultation with Patrick Moore and Robert G. Inglis. Highly recommended for all libraries and astronomers.

217. Page, Thornton, and Lou Williams Page, eds. **Telescopes: How to Make Them and Use Them**. New York, The Macmillan Co., 1966. 338p. illus. index. glossary. (Sky and Telescope Library of Astronomy, v.4). $7.95. LC 66-22532.

What better place to obtain information on amateur telescopes than from the pages of *Sky and Telescope* magazine! One of a series of books for amateur astronomers and general readers, this work includes a collection of past articles from the journal on everything from making to using the instruments. The uniqueness and importance of this particular book lies in the fact that we are being exposed to the expertise of dozens of experienced amateurs and professionals who wrote these articles over the past years. Chapter headings include Basic Principles, Early Telescopes, Making a Reflector, The Mounting and Drive, Visual Observations of the Moon and Planets, Observing the Stars, Telescopic Accessories,

124 / General Materials

Special-purpose Telescopes, and Famous Observatories and Telescopes. An excellent practical and informative guide to the world of optical amateur telescopes, this work belongs in all public libraries and in the personal collections of astronomers everywhere.

218.　　Paul, Henry E. **Outer Space Photography for the Amateur**. 4th ed. Garden City, N.Y., AMPHOTO, American Photographic Book Publishing Co., Inc., 1976. 156p. illus. index. bibliog. $9.95. LC 67-21698. ISBN 0-8174-2407-5.

　　　　Amateur astronomers welcome the return of this popular, formerly out-of-print, handbook on how to take pictures of the heavens. The latest edition is better than previous versions, with more photographs (many by amateurs) and up-to-date information on film and the so-called "cold cameras." Encouraging the amateur astronomer/photographer to connect camera and telescope, Dr. Paul explains, step by step, what equipment is needed, how it works, and what to photograph. After an introductory chapter, the book quickly delves into a detailed description of lenses, cameras, and tripods, how they work together and separately. A good summary of the types of used or surplus lenses available for astrophotography is presented in this chapter. The following section discusses the various auxiliary devices and methods of attaching the camera to the telescope. The remainder of the volume discusses how to photograph various celestial objects from satellites to planets and stars. There are also discussions of color photography, high resolution astrophotography, low temperature astrophotography, and more. The volume is not overly detailed, however, and the amateur should also consult standard volumes like *Stargazing with Telescope and Camera* by George Keene (1967) and *Skyshooting: Photography for Amateur Astronomers* by Mayall and Mayall (1968). Paul's volume, however, is the most up to date, and it also includes a new listing of suppliers of equipment. It is the companion to the author's fine *Telescopes for Stargazing* (1970, c1966) and is appropriate for public, college, and home libraries.

219.　　Paul, Henry E. **Telescopes for Skygazing**. 2nd ed. New York, Amphoto; distr. Cambridge, Mass., Sky Publishing Corp., 1970, c1966. 160p. illus. bibliog. $6.95. NUC 73-92550.

　　　　Another of the many excellent books on instruments for the amateur astronomer, this one includes all sorts of useful information in a well-illustrated format. The types of telescopes take up a major portion of the book, along with an explanation of how they work. What to look for in the sky after purchasing an instrument is the next section of the book. There are no step-by-step instructions for building a telescope, but choosing, testing, improving, and care of equipment are all covered well. Books on constructing telescopes are listed elsewhere in this guide. Chapters on binoculars and astrophotography are also included, and the book ends with directories of equipment and suppliers.

220.　　Riemer, Marvin F. **Telescope and the World of Astronomy**. 3rd ed. Boston, Herman Publishing Co., 1967. 232p. illus. index. bibliog. glossary. $5.95. LC 67-21822. ISBN 0-89046-034-5.

A potpourri of basic astronomy and information on observing make this an ideal volume for the beginning astronomer. In this interestingly written introduction, the author combines theory and practice to prepare the reader to move on to more advanced study later if so desired. The first 150 pages are strictly text; the author talks about astronomy in general terms, introducing the stars, the solar system, comets, telescopes, photography, etc. Two excellent appendices, with valuable information for the observer, make up the next third of the book, including tables of all kinds, star charts, a glossary, bibliography, and so on. Highly recommended.

221.   Roth, Günter D. **Handbook for Planet Observers.** New York, Van Nostrand Reinhold Co., 1970. 205p. illus. index. bibliog. o.p. LC 73-126882.

This book should be read by all amateur and beginning astronomers who are interested in doing actual planetary observations. The author first discusses visual observation, covering important subjects such as telescopes, photometers, making observations, and evaluating data. Next he moves to photographic observation with a discussion of camera optics, techniques, and film processing. Finally the book talks about the individual planets, the Moon, and asteroids, explaining how to observe each and mentioning projects the observer might undertake. An excellent book, translated from the German (*Taschenbuch für der Planetenbeobachter*) by Alex Helm. Highly recommended.

222.   Thompson, Allyn J. **Making Your Own Telescope.** Cambridge, Mass., Sky Publishing Corp., 1947, 1973. 211p. illus. index. bibliog. $5.00. LC 47-5936.

After ten printings in 26 years, this well-written book still provides the amateur telescope maker with adequate detail and direction. Other than a revised one-page bibliography, the text is unchanged since its first issuance. The author begins with a 26-page section on the history of telescopes and how they work as a preparation of sorts for the uninitiated reader. Quite similar in many respects to other works on the same topic, the volume covers the following areas: grinding, polishing and testing the mirror, materials used, etc. The main theme of the book is building a six-inch f/8 Newtonian reflector, a standard size and type of instrument for the backyard astronomer. The reader will also find treatments of eyepieces, adjustments, mountings, and the diagonal mirror or prism. A chapter on optical principles, not often included in works intended for amateurs, helps astronomers understand better how the telescope works under various conditions and helps them use it to the best advantage. Diffraction images, resolving power, diffraction from obstructions within the instrument, atmospheric and thermal interference, limiting visual magnitude of a telescope, and aberration of the paraboloid are discussed in this informative section. Although the book is very "dry" in its presentation (a livelier and just as complete treatment can be found in N. E. Howard's *Standard Handbook for Telescope Making*, 1959), it is, nevertheless, an excellent, comprehensive book, worthy of any astronomy collection.

223. Webb, The Rev. T. W. **Celestial Objects for Common Telescopes.** New York, Dover Publications, 1962. v.1: 255p. illus. $3.00pa; v.2: 351p. illus. $3.00pa. LC 62-53080. ISBN 0-486-20917-2 (v.1). Gloucester, Mass., Peter Smith; $5.50ea. (hc). ISBN 0-8446-3140-X.

Successfully used by thousands of amateurs over the past century, this book provides a great deal of helpful information on the use of the astronomical telescope. The first edition appeared in 1859, and this Dover re-issue is "a revised and enlarged republication of the sixth edition . . . 1917," edited and revised by Margaret W. Mayall. Volume I, *The Solar System*, was left unchanged from the 1917 edition: " . . . the information contained in the book is of historical value in that it represents the solar system as it was known in the Rev. Webb's time, with some revisions by the Rev. Espin in 1917." Volume II, *The Stars*, has been updated by the editor. It presents stars, nebulae, clusters, and other celestial objects, arranged by the constellation area in which they are found. There are 4,000 objects in all that may be seen in a "moderate" (4 to 6-inch reflector) amateur telescope. The lengthy appendices contain Southern Hemisphere objects, precession, double stars, and other information. As important today as it was over a hundred years ago, this two-volume set should be in both college and public libraries.

Advanced

224. Beet, E. A. **Mathematical Astronomy for Amateurs.** New York, W. W. Norton and Co., 1972. 143p. illus. index. exercises. $7.95. LC 72-185014. ISBN 0-393-06388-7.

Unusual because there is little else like it, this well-written book gives meaning to astronomical laws and principles by explaining the calculations behind them. The author has given the amateur astronomer a book that has been needed for a long time, one that will increase his understanding of astronomy. As Beet points out, astronomy books often tend to have no mathematics at all or very technical equations and calculations; in either case, the serious amateur has little to go on. The volume covers (in fairly simplistic terms) the mathematics related to the study of the Earth, Moon and their orbits; time; the celestial sphere; the solar system; and stellar topics. A must for the serious amateur, it contains numerous practice exercises (pre-tested on students) which help the reader learn more. The college instructor may wish to consider this excellent work as supplementary reading or homework.

225. Heiserman, Dave. **Radio Astronomy for the Amateur.** Blue Ridge Summit, Pa., TAB Books, 1975. 251p. illus. index. $8.95. LC 74-33624. ISBN 0-8306-5714-2; 0-8306-4714-7pa.

Easily one of the clearest and most comprehensive books on a non-elementary subject, this work would be a good choice for the enthusiastic beginner. More up-to-date and less technical (without skimping detail) than

Heywood (*Radio Astronomy: And How to Build Your Own Telescope*, 1964) or Hyde (*Radio Astronomy for Amateurs*, 1962), Heiserman's volume begins with two brief chapters on historical perspective and general astronomy, laying the groundwork for the new reader by explaining optical astronomical principles and coordinate systems. The treatment of the latter essential topic is quite good, and there are numerous examples presented as illustration. Chapter five contains explanations of radio astronomy theory and technology, the author describing the three sources of extraterrestrial radio signals: thermal, non-thermal, and 21 cm hydrogen radiation. Types of radio telescopes, their configurations, design, and principles are also covered in this very important section. Following is a chapter on projects, a subject that has received minimal attention in similar books. Here we find numerous suggestions, as well as encouragement to try new ideas.

Chapters 7, 8, and 9, however, are the most important sections of the book. The first of this group on equipment outlines and summarizes antennae, receivers, record-readout devices, etc. The author makes it clear from the start that this is not an area for the beginner, and he emphasizes that a knowledge of electronics is essential. A 110 MHz radio telescope and 146 MHz interferometer systems are presented in Chapters 8 and 9, complete with details of how to build them; the instructions are clear and logically presented. The books ends with discussions of equipment and methods of observation of Jupiter and the Sun, two "nearby" and popular radio sources. A final section outlines what the author sees as immediate challenges for the amateur: "(1) develop specialized equipment and circuits that are useful, practical and easy to use; and (2) establish some standard operating procedures that promise satisfying and meaningful results." Highly recommended for public and university libraries.

226.   Heywood, John. **Radio Astronomy: And How to Build Your Own Telescope.** New York, ARC Books, Inc., 1964. 159p. illus. name index. bibliog. $.95pa. LC 62-20964.

Another of the few books for the amateur astronomer, this short, non-elementary book explains the basics for the enthusiastic and able beginner. Chapter one on radio noise tells how radio waves are propagated from a simple broadcast transmitter, giving the reader some basic information necessary for an understanding of the subject; circuits, capacitors, damped wave production, transmitters, and receivers are discussed. A second introductory chapter describes the aspects of electromagnetic radiation, its wavelengths, wavepackets, motion, etc. Following this is a brief section on the causes of extraterrestrial radio noise, including thermal emission, H_{II} regions, stars, planets, quasars, etc. Two types of simple radiotelescopes are described: the radiometer and the spaced-aerial radio interferometer. The former is usually a single aerial system feeding into a receiver, while the latter is a series of aerials connected to a receiver. The author details these two instruments, providing many schematic diagrams of circuits, etc.; various types of aerials are included in his technical discussions. The remainder of the book considers some of the various types of celestial objects observable, along with a brief summary of some past research. Like other books for the non-professional, this requires some previous or concurrent learning of radio science and electronics. Only the very serious should attempt this book. The reader interested in a general explanation of radio astronomy should read Verschuur's *The Invisible Universe* (1974), since this book only considers the technical aspects.

227.   Hyde, Frank W. **Radio Astronomy for Amateurs.** New York, W. W. Norton and Co., Inc., 1963, c1962. 236p. illus. index. glossary. bibliog. (The Amateur Astronomer's Library). $5.00. LC 63-11686. ISBN 0-393-06331-3.

For those few with the skill, money for equipment, and enthusiasm, and for those who wish an introduction to the principles behind radio astronomy, this fine book would be a good introduction to the topic. Chapters one and two approach the latter subject, discussing "What Is Radio Astronomy?" and "The History of Radio Astronomy." The reader is shown diagramatically and descriptively, for example, the difference between an optical reflecting telescope and a radio instrument. Chapter three, "Basic Astronomy," will be wasted on most readers who, one assumes, are already familiar with the heavens, though electronics hobbyists reading this text will need some sort of introduction to the stars. But the chapter is far too brief, and the interested reader should consult any of a number of books listed in this guide under "General Works" or "Amateur Astronomy."

One of the most obvious sources of electromagnetic radiation that can be observed with "simple" radio equipment is the Sun, which is described in Chapter 4. Chapters 5 and 6 cover "Basic Electronics," introducing circuit theory and components. The book goes on to cover aerial systems and radio receivers and their construction, and finally gives some guidance for constructing radio observatories. Although it is a beginning book, it is not for the beginner, so it will have a small audience. The enthusiastic radio astronomer should proceed to other texts and the journal literature, however, after this work, which is merely an introduction.

228.   Moore, Patrick, ed. **Astronomical Telescopes and Observatories for Amateurs.** Newton Abbot, Engl., David & Charles; distr. New York, W. W. Norton & Co., 1973. 265p. illus. refs. index. £4.75. $7.95. LC 73-174762. ISBN 0-7153-5615-1 (D & C); 0-393-06395-X (Norton).

An excellent introductory book for the amateur astronomer who is interested in buying or making a telescope, or in setting up a home observatory. Few books, if any, discuss the latter subject, and for that reason, the enthusiastic amateur astronomer should certainly at least browse through this work. The book's 16 chapters are divided into the two major topics mentioned above, the first discussing refracting and reflecting telescopes, mirror-making, mountings, drives, telescope adjustments, eyepieces, auxiliary equipment, celestial photography, etc. The serious reader will then wish to progress to *Amateur Astronomer's Handbook* by J. B. Sidgwick, which is more advanced. The latter section of the book examines run-off and run-off roof observatories, and observatory domes. Supplementary information along these lines can frequently be found in the pages of *Sky and Telescope* magazine. This volume should be included in the public and college library science collections.

229. Muirden, James. **The Amateur Astronomer's Handbook**. Rev. ed. New York, Thomas Y. Crowell, 1974. 404p. illus. index. bibliog. glossary. $9.95. LC 74-5411. ISBN 0-690-00505-9.

Intended as a handbook for the serious observer, this work surveys the techniques of amateur astronomy, from telescope selection to actual observation. Definitely not for the beginner, the book demands that the reader be well-versed in the basics of astronomy.

The excellent discussion of telescopes that begins the book should serve both to inform the reader and to arouse his interest as to what follows. This new edition (the first was in 1968) contains chapters on optical work for the amateur telescope maker; the detailed information in this section will surely benefit serious readers. A chapter on making observations, with an emphasis on learning to "see," follows the introductory matter. The majority of the book concentrates on how to observe the planets, moons, and other celestial bodies. The volume has four parts: Equipment, The Solar System, The Stars and Nebulae, and Optical Work for Amateurs. Several useful appendices provide statistical information for observing as well as a list of amateur astronomical societies. Not enough can be said about this excellent volume, which was also published in Great Britain as *Astronomy for Amateurs*.

230. Roth, G. D., ed. **Astronomy: A Handbook**. New York, Springer-Verlag, 1975. 567p. illus. notes. index. bibliog. $21.40. LC 74-11408. ISBN 0-387-06503-2.

The latest of a handful of volumes for the serious, advanced amateur, this fine compendium was translated and revised from the German edition (*Handbuch für Sternfreunde,* 1967) by Arthur Beer. After some very brief general comments on astronomical literature, charts, catalogs, etc., the book considers at length astronomical observing instruments, with substantial emphasis on their components. Topics discussed include lenses and mirrors, types of telescopes, eyepieces, tubes, mountings, drives, auxiliary equipment like photometers, micrometers, etc. This chapter quickly convinces the reader that the handbook is definitely not for the beginner. In fact, the entire book is on an advanced level, which raises the question of what kind of amateur astronomer the book is aimed at. For instance, many projects suggested require advanced equipment, or even access to an observatory. The book is quite good, but its interest level is not well defined. No area seems to have been overlooked in this voluminous work, which should be the companion of any advanced backyard astronomer. A few topics are radio astronomy (a good, but too brief coverage with little direction for practical projects), spherical astronomy (a fine introduction with many useful equations), applied mathematics (far briefer and more advanced than E. A. Beet's work), the Sun and Moon and their eclipses (suggestions for observation of the latter are helpful), observation of artificial satellites, the planets, comets, meteors, and aurorae, photometry (detailed, adequate explanations), and more. The user of the book should be well-prepared in astronomy, both in practice and theory, as well as geometry and trigonometry; the book is heavily mathematical. There are not many illustrations and not enough examples of equations applied to real situations. Finally, there is too much material on the solar system and not enough on the observation of stars. The reader should consult Muirden (*The Amateur Astronomer's Handbook*, 1974) and Sidgwick (*Amateur Astronomer's Handbook*, 1971) for information on advanced projects on stars and galaxies. Despite its shortcomings, there is a wealth of information here, and this book would be ideal for use in a college-level practical astronomy course.

231.   Sidgwick, J. B. **Amateur Astronomer's Handbook**. 3rd ed. London, Faber & Faber Ltd., 1971. 577p. illus. index. £5.50. LC 70-883987. B 71-19789. ISBN 0-571-0478-3.

Most amateur texts attempt to cover both observation and equipment in one volume, resulting sometimes in a neglect of one topic or the other. This particular handbook, though, concentrates only on the telescope and auxiliary equipment in its 41 chapters. The author goes into great detail, on a non-elementary level, concerning the construction, use, adjustments, and operation of telescopes, mirrors, lenses, compound optical instruments, mountings, telescope drives, micrometers, spectroscopes, clocks, and more. The volume has a superb bibliography and should be in any astronomical reference collection, in both university and public libraries. This comprehensive handbook will not be suited for casual, occasional observers or for beginning amateurs, however. It is highly recommended for very serious non-professional astronomers.

The author notes in his introduction: "*Amateur Astronomer's Handbook* deals solely with the theoretical and instrumental background to observation, together with practical implications. The actual techniques employed in the observation of the various types of celestial objects are described in *Observational Astronomy for Amateurs*, which may be regarded as a companion volume to the present work."

232.   Sidgwick, J. B. **Observational Astronomy for Amateurs**. rev. ed. London, Faber & Faber, Ltd., 1971. 376p. illus. bibliog. index. £4.00. LC 55-57280. ISBN 0-571-04738-6.

Like its companion volume, *Amateur Astronomer's Handbook*, this book is not for the casual observer; it is aimed at the amateur who is very serious about observing and learning a lot about astronomy. Explanations are detailed and non-elementary; the reader should have a good, solid background in the basics of astronomy and have some observing experience before tackling this advanced book. This volume concentrates on the various heavenly bodies, how they can be observed, and what to look for. Detail procedures are suggested, along with numerous useful data about the stars, planets, Sun, Moon, comets, meteors, etc. An excellent book with an extremely good bibliography. (The editor examined the 1961 edition.)

Sundials

233. Cousins, Frank W. **Sundials: A Simplified Approach by Means of the Equatorial Dial**. London, J. Baker; distr. New York, Pica Press, Universe Books, 1972, c1969. 247p. illus. indexes (subject and name). bibliog. $18.50. £7.25 LC 72-458111. B 70-01258. ISBN 0-212-98355-5 (Baker).

Not just an historical look at sundials, nor merely a discussion of how they work, this volume is happily both. Beautifully illustrated with photographs of old and new sundials, the book takes an in-depth look at their development and various forms. The major portion of the book, however, is devoted to explanations of the various methods of construction, along with detailed explanations of the geometry and trigonometry involved. Not for the inexperienced amateur, this is one of the best such works available on the subject.

234. Mayall, R. Newton, and Margaret W. Mayall. **Sundials: How to Know, Use and Make Them**. 2nd ed. Cambridge, Mass., Sky Publishing Corporation, 1973. 250p. illus. index. $9.50. LC 73-7642.

The second edition of this standard book for the amateur astronomer and general reader reappeared 35 years after it was first published. Only a few changes have been made in the original text, but several new chapters have been added. This work is well written, handsomely illustrated, and very comprehensive. In fact, of the four books listed here on the subject, this is the best in all respects. The authors are especially keen on showing the reader how to construct a sundial, and their explanations are quite good and easy to understand.

There are many interesting chapters, all worth mentioning: The Development of the Sundial; Why the Sundial Tells Time; How to Design and Make a Dial; Selecting the Dial to Make or Buy; Guidelines and Materials; Parts of a Dial You Should Know; Time and Standard Time Dials; How to Lay Out the Hour Lines; Dial Furniture; How to Lay Out the Lines of Declination; Portable Sundials; Variable Center Dials; The Heliochronometer; Sundial Classification; Interesting Dials of the World; Hunting Sundials.

235. Rohr, René R. J. **Sundials: History, Theory and Practice**. Toronto, University of Toronto Press, 1970. 142p. illus. refs. index. $20.00. LC 75-134636. ISBN 0-8020-1567-0.

As the title indicates, this book on sundials is a smattering of the various aspects of the ancient timepieces. The chapters on history have excellent photographs of various types of sundials through the ages and an informative text that takes the reader from Egypt to modern times. Using well-drawn diagrams and some astronomy and geometry, the author next explains how sundials work. Finally, help is given to the reader who wants to construct a sundial on his own. Translated from the French (*Les Cadrans solaires*) by Gabriel Godin, this fine work is intended for the more general reader.

236. Waugh, Albert E. **Sundials: Their Theory and Construction.** New York, Dover Publications, Inc., 1973. 228p. illus. index. bibliog. $3.50pa. LC 73-76961. ISBN 0-486-22947-5. Gloucester, Mass., Peter Smith, $6.50cl. ISBN 0-8446-4835-3.

For the historical-minded, there is very little here (although the bibliography should be helpful); for someone who wants to know how and why a sundial works, there is plenty; for the reader who wants to build a sundial, there is even more. The author covers all aspects of sundial theory and construction, and explains, step by step, how to make them. Both the serious student and interested layman can benefit from the instructions. As the author notes: the book has "simple mathematical methods ... for those who want them, but the reader who fears the numerical approach may skip them, using instead the graphical approaches which do not require ever addition or subtraction." Highly recommended.

HISTORY

A substantial portion of astronomy's literature is devoted to its history, a collection of diversified and fascinating books about the discoverers and discoveries of perhaps the oldest science. Astronomy's beginnings, of course, go back beyond written records to the dawn of civilization, when Man first looked toward the skies and considered the bright points of light above. Since then, astronomical knowledge has grown as new theory and observation have replaced the old, and the historians have fortunately recorded both the great achievements and the dismal failures in this exciting field.

Anyone who studies or reads about astronomy will quickly learn that it is impossible to do so without being exposed to the men and women behind the theories and equations. The laws of planetary motion cannot be considered without mentioning Johannes Kepler, cosmological studies cannot be complete without the story of the Greeks and their concepts of the Universe, the development of the telescope can be considered only after telling how Galileo discovered the moons of Jupiter. Astronomy has played an integral part in the history of science and civilization, and a representative sample of the large body of materials documenting this role is presented here, with an emphasis on the more recent monographs.

Two major types of works are treated in this section—biographies and general treatises, each containing both scholarly and general interest books. The former group includes volumes dealing with one particular individual, as many of the astronomical history books do, or collectively with a group of individuals. These volumes tell the stories of some extremely interesting people who happened to also be astronomers. Among the personalities who come alive in these books are geniuses like Isaac Newton, rebels like Copernicus, revolutionaries like Galileo, and "characters" like Tycho Brahe. No doubt other sciences have had interesting individuals, too, but few compare with astronomy.

The general treatises consider overviews of the astronomy as a whole, or zero in on one particular topic, highlighting, or studying in detail, its history and its place in the overall picture of the science. These works are both scholarly and general, aimed at the historian and lay reader, respectively. For a very quick overview of the history of astronomy, the reader should consult one of the general works cited elsewhere in this guide or one of the non-technical textbooks, many of which contain at least one background chapter on history.

The long list of historical works could have easily been much longer; the reader should keep in mind that this is a selective, representative group of books, and that many more are available. Scholarly historical references can also be found in the *Journal for the History of Astronomy* (see under General Periodicals), while general historical information can often be located in *Sky and Telescope* or any of the other popular magazines listed in this guide.

Card catalog subject headings: *Library of Congress:* ASTRONOMY—HISTORY; ASTRONOMY, ANCIENT; ASTRONOMY, ARABIC, [CHINESE, GREEK, etc.] ; ASTRONOMY, MEDIEVAL; names of individual astronomers. *Sears:* ASTRONOMY—HISTORY; names of individual astronomers.

Biography—General

237.   Adamczewski, Jan. **Nicolaus Copernicus and His Epoch**. Philadelphia, Copernicus Society of America; distr. New York, Charles Scribner's Sons, 1974. 162p. illus. $7.95. LC 74-174362. ISBN 0-684-13839-5 (Scribner's).

One of the best general works on Copernicus to come along in years, this book is beautifully illustrated and well written. Not only does the author describe the Polish astronomer and his work, but he also gives interesting glimpses of life in the fifteenth and sixteenth centuries. The reader will feel as though he or she is being transported back in time to witness one of the most important chapters in the history of mankind. The book is divided into four parts, each concerned with a major stage in Copernicus's life: The Country of his Childhood: 1473-1491; In the Royal Capital: 1491-1495; Under Italian Skies: 1496-1503; and In the Remotest Corner of the Earth: 1503-1543. The astronomer's view of the universe is naturally a major portion of this text, which is a translation of *Mikolaj Kopernik i jego epóka*. The book is guaranteed to appeal to the layman, student, and astronomer, so it would be suitable for public, university, and observatory libraries.

238.   Buttmann, Günter. **The Shadow of the Telescope: A Biography of John Herschel**. New York, Charles Scribner's Sons, 1970. 219p. illus. index. bibliog. refs. LC 72-85256.

Children of famous men and women frequently are overshadowed by their parents, and often find it difficult to achieve greatness themselves. Such is the case of John Herschel, son of William Herschel, discoverer of Uranus. Despite being constantly compared to his father and his achievements, the young John made a name for himself in astronomy, as well as in mathematics, chemistry, and physics. His star catalogs were, and still are, acclaimed for their accuracy and comprehensiveness. His work at the Cape of Good Hope observing the Southern skies is equally renowned. Buttmann details the scientist's exploits in mathematics and physics, too. Originally published in Germany (*John Herschel; Lebensbild eines Naturforschers*) in 1965, this first full-length biography of John Herschel was praised as a critical success. A must for the astronomical historian, the book is based on many published and unpublished works, including Herschel's personal diaries. Translated by B. E. J. Pagel and edited by David S. Evans.

239.   Crawford, Deborah. **The King's Astronomer: William Herschel.** New York, Julian Messner, 1968. 191p. index. bibliog. $3.50. LC 68-14941. ISBN 0-671-75649-4.

Young and old alike will enjoy this fascinating biography of William Herschel, famed English astronomer of the eighteenth century who discovered Uranus. In the form of a narrative, the author traces Herschel's life from young and accomplished musician in Germany to the Royal court of King George II of England. Crawford tells how the astronomer, with the aid of his sister, spent years mapping the skies; the result was a new view of the Universe for mankind. For the public and secondary school library.

240.   Heuer, Kenneth. **City of the Stargazers.** New York, Charles Scribner's Sons, 1972. 170p. illus. bibliog. glossary. index. $7.95. LC 72-1175. ISBN 0-684-12937-X.

The theme of this work is the ancient city of Alexandria, where so many early astronomers lived and carried out their observations and studies. Among these men were Ptolemy (Earth as the center of the solar system); Aristarchus (Sun as the center of the Universe); Hipparchus (precession of the equinoxes), and Eratosthenes (measurement of the Earth's circumference). The volume is descriptive and well illustrated; it should make enjoyable reading for anyone interested in ancient astronomy or science. For the public and college library.

241.   Hoffmann, Banesh, with Helen Dukas. **Albert Einstein: Creator and Rebel.** New York, Viking Press, 1972. 272p. illus. index. $10.00. LC 70-186740. ISBN 0-670-11181-3.

Not intending this to be a definitive biography, the authors "have tried, in brief compass, to give an indication of the man, letting his image come through when possible in terms of his own writings and devoting much space to his science." It is an extremely interesting narrative, well illustrated and thoroughly researched. Scientist and layman alike will enjoy this book, which not only describes what Einstein was like as a man but also relates his extraordinary achievements. A fine book, suited for both public and college libraries.

242.   Koyré, Alexandre. **The Astronomical Revolution.** Ithaca, N.Y., Cornell University Press, 1973. 531p. illus. notes. name index. $18.50. LC 72-13061. ISBN 0-8014-0776-1.

An interesting historical work in which the author examines the contributions of three astronomical revolutionaries: Copernicus, Kepler, and Borelli.

The book's purpose is stated in the foreword when the author notes that he does not intend "to review the history of astronomy in the sixteenth and seventeenth centuries from Copernicus to Newton, but only the history of the 'astronomical revolution', that is to say, the history of the evolution and transformation of the key concepts...." Included are many excerpts from the three astronomers' writings, since the author wants them "to speak for themselves." Translated from the French (*La Révolution astronomique*, 1961) by Dr. R. E. W. Maddison.

243.   Land, Barbara. **The Telescope Makers: From Galileo to the Space Age.** New York, Thomas Y. Crowell Co., 1968. 245p. illus. index. bibliog. $4.50. LC 68-21607. ISBN 0-690-80831-3.

The gains made in astronomy closely parallel the development of the telescope, from "spyglass" to Mt. Palomar to Jodrell Bank. As its power increased, so did Man's knowledge of the Universe. The author presents here ten interesting stories of some, but certainly not all, of the astronomers who contributed greatly to the development of the telescope as an astronomical tool. From the sighting of Jupiter's satellites with one of the earliest telescopes to radio studies of the Universe with a different kind of instrument, Ms. Land gives the amateur astronomer much to enjoy and ponder. The telescope makers presented in this work are Galileo, Johannes Kepler, Isaac Newton, William Herschel, Joseph von Fraunhofer, Lord Rosse, George Ellery Hale, Bernard Schmidt, Grote Reber, and Herbert Friedman.

244.   Livingston, Dorothy Michelson. **The Master of Light: A Biography of Albert A. Michelson.** New York, Charles Scribner's Sons, 1973. 376p. illus. refs. notes. chronology. bibliog. index. $12.50. LC 72-1178. ISBN 0-684-13443-8.

The improved determination of the speed of light (a much-used constant in astronomical calculations) by Ensign Albert Michelson in 1878 is only one of the highlights of this interesting volume. Although the book contains descriptions of Michelson's scientific work, it is not at all dull reading. Especially interesting is the description of the Michelson-Morely Experiment (1887) which determined that the speed of light is constant in all directions, that is, always the same, no matter what the external conditions. The description of the measurement of the diameter of Betelgeuse with the stellar interferometer is also included. The well-written and finely illustrated book contains excerpts from Michelson's letters, which adds to the value and interest of the book. Written by one of the scientist's daughters, the book will make enjoyable reading for the scientist and layman alike.

245.   Moore, Patrick. **The Astronomy of Birr Castle.** London, Mitchell Beazley; distr. George Philip and Son, Ltd., 1971. 81p. illus. refs. £1.75. LC 78-88898. B 71-26393. ISBN 0-85533-004-X.

This brief historical sketch describes the observations and work of the third Earl of Rosse at Birr Castle in Ireland over one hundred years ago. Lord Rosse built the world's largest telescope, a massive 72-inch reflector. The huge mirror could gather more light and see fainter objects than were ever imagined, and with it Rosse made discoveries of fundamental importance. His work included the observation of all bright nebulae in the Northern Hemisphere, as well as many extra-galactic nebulae, thought at that time to be part of our galaxy. Another interesting book from Patrick Moore, this work describes an astronomer whose importance has not been emphasized enough.

246.   Moore, Patrick. **Watchers of the Stars: The Scientific Revolution.** New York, G. P. Putnam's Sons, 1974, c1973. 240p. illus. index. $15.95. LC 74-78640. ISBN 0-399-11374-6.

Moore's book is the fascinating story of "the greatest scientific upheaval of all time," the removal of the Earth from its place as the center of the Universe. Not a technical work, it should be enjoyed by the student, layman, or scientist, anyone interested in the history of science. The five leading characters in this excellent book are Copernicus, Brahe, Kepler, Galileo, and Newton, the men who were responsible for the successful beginning and end of the revolution. Both the text and illustrations are excellent; college and public libraries especially will want to obtain this volume.

247.   Richardson, Robert S. **The Star Lovers.** New York, Macmillan Co., 1967. 310p. illus. index. bibliog. $7.50. LC 67-16714.

Biographical sketches of sixteen astronomers who lived between 1550 and 1960 are included in this volume aimed at the general reader. The author chose his subjects not necessarily because of their greatness, but because some aspect of their personality and work especially interested him. The reader will no doubt find several, if not all, of the sketches most enjoyable reading. The list is made up of many famous names, and several that are not so well known. In short, quite a few stellar scientists who are overlooked in most works of this type are discussed here. Each man described had different interests, backgrounds, and personalities, so each story in this book is a different one. The result is a volume of astronomical short stories which is sure to please anyone with an interest in astronomy. The roster of astronomers includes Brahe, Horrocks, Newton, Roemer, Halley, Goodricke, Schwabe, Encke, Janssen, Hall, Huggins, Lowell, Barnard, Einstein, Nicholson, and Baade.

248.   Ronan, Colin. **The Astronomers.** New York, Hill and Wang, 1964. 232p. illus. bibliog. index. o.p. LC 64-24831.

Biographical sketches of famous astronomers from antiquity to the twentieth century are presented in this volume. The theme of the book is "to sketch Man's struggle to understand the Universe in which he finds himself, by looking more closely at some of those whose work has earned them a place in the story of the advance we are still witnessing today." Excellent reading for the amateur astronomer, the book tells the stories of the Greeks, Copernicus, Tycho Brahe, Kepler, Galileo, Newton, Halley, the Herschels, and Einstein.

249. Ronan, Colin A. **Astronomers Royal**. Garden City, New York, Doubleday and Co., 1969. 224p. illus. refs. index. o.p. LC 69-10364.

A history of British astronomy, beginning in the mid-sixteenth century, during the reign of Elizabeth I. Originally published in Great Britain (*Their Majesties' Astronomers*), the book tells interesting stories of the astronomers who played important roles in the development of British and world science. Among these men were John Flamsteed, Edmond Halley, Nevil Maskelyne, John Harrison, Isaac Newton, William Herschel, Lord Rosse, John Lockyer, and others. A well-written historical account for the layman.

250. Ronan, Colin A. **Edmond Halley: Genius in Eclipse**. Garden City, N.Y., Doubleday & Co., 1969. 251p. illus. refs. index. bibliog. chronology. LC 61-12575.

The scarcity of records and details of Edmond Halley's life made this work particularly difficult to write and research. The author, however, examined the existing references and, with some "reading between the lines," has constructed a well-written, interesting, and informative biography. Halley, of course, is best known for his famous comet, last seen in 1910. Ronan, a master astronomical writer, describes the astronomer's other work, too, in this book for the general reader and student.

251. Ronan, Colin A. **Galileo**. New York, G. P. Putnam's Sons; London, Weidenfeld and Nicolson, 1974. 264p. illus. index. notes. bibliog. $14.95; £5.00. LC 74-76233. GB 74-24695. ISBN 0-399-11364-9 (Putnam); 0-297-76801-8 (Weidenfeld).

This volume on the life of Galileo is a welcome addition to the literature of the history of astronomy. The author has written an excellent book which not only tells of the man himself, but also of the times in which he lived. The book is well researched and beautifully illustrated with color and black-and-white photos. The general reader, student, and astronomer will enjoy this fine volume; it would be an important addition to any library possessing a science collection. Highly recommended!

252. Shapley, Harlow. **Through Rugged Ways to the Stars**. New York, Charles Scribner's Sons, 1969. 180p. illus. index. bibliog. chronology. (Scribner's Scientific Memoirs). LC 68-57085.

This attention-holding book contains the memoirs of Harlow Shapley, the man whose study of the location of globular clusters led to the discovery that the galaxy is much larger than previously thought. The autobiography begins with the astronomer's boyhood in Missouri, and proceeds some eighty years to his retirement. The student, astronomer, and general reader interested in science will find this story of a great man and renowned astronomer to be well worthwhile. The highlights of the book include Shapley's days at the University of Missouri, his Ph.D. study at Princeton under Henry Norris Russell, the Great Debate on the scale of the Universe with Heber D. Curtis, the many scientific awards he received, and the directorship of the Harvard College Observatory.

253. Volkoff, Ivan, Ernest Franzgrote, and A. Dean Larsen. **Johannes Hevelius and His Catalog of Stars**. Provo, Utah, Brigham Young University Press, 1971. 89p. illus. bibliog. refs. (Gift to the Friends of the Brigham Young University Library, no. 5). $8.00. LC 74-31556. ISBN 0-8425-1479-1.

A well-illustrated account of the life and work of the astronomer Johannes Hevelius, born in 1611 in Danzig, Poland. Hevelius began his observational work during the time when the telescope was revolutionizing the study of astronomy, and one of the first projects this ambitious man undertook was the drawing of a detailed lunar map, which had not previously existed. A later important work was his catalog of stars, the manuscript of which was recently acquired by Brigham Young University. This book commemorates that important acquisition.

254. Warner, Deborah Jean. **Alvan Clark & Sons: Artists in Optics**. Washington, D.C., Smithsonian Institution Press, 1968. 120p. illus. notes. (U.S. National Museum. Bulletin 274). $1.75.

This book provides some insight into the work of Alvan Clark and his two sons, the greatest telescope lensmakers in American history. The Clarks made not only telescopes but auxiliary observing equipment as well. Part I of this short volume is a biographical sketch of the Clarks; Part II lists the astronomical equipment made and remade by the family from 1844 to 1897. Most notable of their achievements was the making of objective lenses for the world's five largest (at the time) refracting telescopes, including the 40-inch at Yerkes Observatory (still the largest refractor). A good book for the astronomical historian.

255. Wright, Helen. **Explorer of the Universe: A Biography of George Ellery Hale**. New York, E. P. Dutton and Co., Inc., 1966. 480p. illus. index. bibliog. notes. LC 66-11542. ISBN 0-525-10186-1.

The fascinating story of one of America's most famous astronomers is not told here in strictly chronological order. Rather, the author has chosen to emphasize Hale's most important accomplishments, centering the book around these events (in particular, the 200-inch reflecting telescope on Mt. Palomar). Wright, who later helped edit a book of Hale's writings and other memorabilia, (*The Legacy of George Ellery Hale*) details the astronomer's life from boyhood through university life on into the most important events in his career. Among the most interesting portions of the book are the descriptions of the three large observatories to Hale's credit: Yerkes (home of the world's largest refractor); Mount Wilson (the 100-inch reflector); and Mount Palomar. This excellent volume will be enjoyed by astronomer and layman alike.

256. Wright, Helen, Joan N. Warnow, and Charles Weiner, eds. **The Legacy of George Ellery Hale**. Cambridge, Mass., MIT Press, 1972. 293p. index. illus. notes. $25.00. LC 74-148854. ISBN 0-262-23049-6.

A fascinating volume containing photographs and text on the life and achievements of one of America's great astronomers. Not strictly a biography, but a sort of "museum between covers," this book contains correspondence and writings of the astronomer as well as newspaper clippings about his work. Part 1, an essay by Helen Wright, Hale's biographer, gives an overview of his life and work. Included in this section are reproductions of photos, newsclippings, and original letters which illustrate well the text. Part 2 contains five selections of Hale's writings, while "part 3 includes essays surveying the development of large telescopes and related instrumentation, solar research, and scientific organizations—all relating Hale's original contributions to subsequent developments."

Biography—Advanced

257.   Bedini, Silvio A. **The Life of Benjamin Banneker.** New York, Charles Scribner's Sons, 1971, c1972. 434p. illus. index. bibliog. $14.95. LC 78-162755. ISBN 0-684-12574-9.

A well-researched biography of a black man, son and grandson of slaves, who achieved prominence in the late eighteenth century world of science. An unschooled tobacco planter who taught himself mathematics and astronomy late in life, Banneker calculated ephemerides published in almanacs. This biography is intended for the historian of science rather than the average astronomer, whether amateur or professional. Anyone interested in colonial science will find the volume both informative and worthwhile.

258.   Caspar, Max. **Kepler.** London, New York, Abelard-Schuman, 1959. 401p. index. (The Life of Science Library, 36). LC 59-5797.

The English translation of the classic 1948 German work by a great Keplerian scholar, this detailed book describes Johannes Kepler's life and tells how it was affected by the events of politically troubled Germany of the late fifteenth and early sixteenth centuries. The author pays special attention to Kepler's two principal works, *The New Astronomy*, and *The World Harmony*, in which his planetary laws are expounded. Any enthusiast in the history of astronomy will find the book a satisfying and complete record of the life of one of science's geniuses. Translated and edited by C. Doris Hellman.

259.   Drake, Stillman. **Galileo Studies: Personality, Tradition, and Revolution.** Ann Arbor, University of Michigan Press, 1970. 289p. bibliog. index. notes. $8.50. LC 73-124427. ISBN 0-472-08283-3.

A collection of 13 essays on Galileo, originally written by the author at various times for journals and other sources, these short treatises have been modified, combined, and edited into their present state. Each essay is basically "self-contained" and can be read and studied separately. The list of topics considered by Drake shows the breadth and depth of his expertise: Physics and Tradition before Galileo; Vincenzio Galilei and Galileo; The Scientific Personality of Galileo; The Accademia dei Lincei; The Effectiveness of Galileo's Work; Galileo, Kepler, and Their Intermediaries; Galileo and the Telescope; The Dispute over Bodies in Water; Sunspots, Sizzi, and Scheiner; Galileo's Theory of the Tides; Free Fall and Uniform Acceleration; Galileo and the Concept of Inertia; The Case against "Circular Inertia." For scholars and students, and for university libraries.

260. Flamsteed, John. **An Account of the Revd. John Flamsteed.** London, Dawsons of Pall Mall, 1966. 530p. index of names. £15.00. LC 66-70912. B 66-5567. ISBN 0-7129-0079-9.

This volume is a facsimile reprint of the first edition, printed in 1835-1837. The story of the first Astronomer-Royal, the book should be of interest to historians of science and especially of astronomy.

Title page: "An account of the Revd. John Flamsteed, the first Astronomer-Royal; compiled from his own manuscripts, and other authentic documents, never before published, and supplement to the account of the Revd. John Flamsteed [edited by Francis Baily. 1st ed. reprinted]."

261. Forbes, Eric G. **The Euler-Mayer Correspondence (1751-1755); A New Perspective on Eighteenth-Century Advances in Lunar Theory.** Amsterdam, North Holland Publishing Co.; distr. New York, American Elsevier Publishing Co., Inc., 1971. 118p. refs. notes. index. $13.75. LC 75-173862. ISBN 0-444-19580-7 (Elsevier).

The author/translator has assembled 31 letters of known correspondence between Leonhard Euler (1707-1783), a Swiss-born mathematician, and Tobias Mayer (1723-1762), a German astronomer. Both the mathematician and astronomer were interested in the theory of the Moon's motion since it then was being used "as a test case for the validity of interpreting the physical world in terms of the geometrical axioms of Newtonian mechanics. . . . " The author imparts important historical and biographical information before presenting the correspondence. A very scholarly work recommended for astronomical historians.

262. Forbes, Eric G. **Tobias Mayer's Opera Inedita.** New York, American Elsevier Publishing Co., 1971. 166p. refs. bibliog. name index. $15.95. LC 79-173863. ISBN 0-444-19578-5.

In a companion volume to *The Euler-Mayer Correspondence,* the author/translator presents the first English translation of the 1775 edition of Mayer's unpublished works. There are six treatises in this volume, four of which are concerned with his study of the Moon's motion. An introduction on Mayer's life and work precedes the *Opera Inedita*, which itself is followed by commentary on three of the treatises. The book will be of interest to historians of science.

263. Galilei, Galileo. **Dialogue Concerning the Two Chief World Systems, Ptolemaic and Copernican.** 2nd ed. Berkeley, University of California Press, 1967. 496p. illus. index. notes. $16.50; $2.95pa. LC 68-38948. ISBN 0-520-00449-3; 0-520-00450-7, CAL66pa.

An exposition and support of the Copernican theory, *Dialogue* ranks high among the classics of the history of science, and still retains its importance as a milestone in the history of the struggle for freedom of thought. Completed in 1629, it was not Galileo's greatest scientific contribution, but it was his most significant service to science itself, because it made clear the difference between authority and tradition against experiment and observation. Of significant importance, this work did more toward breaking down the religious and academic barriers against free scientific thought than anything before it. This primary source was translated by Stillman Drake, a renowned Galileo scholar, and was unfortunately unavailable to the English reader for nearly 300 years until Drake's first edition in 1953. The book will be of interest to the scholar and should be read with such critical works as Drake's *Galileo Studies* and Shapere's *Galileo: A Philosophical Study*.

264. Manuel, Frank E. **A Portrait of Isaac Newton**. Cambridge, Mass., Harvard University Press, 1968. 478p. illus. name index. notes. LC 68-140080. ISBN 0-674-69100-8.

One of the better works on the scientific genius of the seventeenth and eighteenth centuries, this biography is based on manuscripts, correspondence, and other original material. At first, one is tempted to say it must be easy to write about a man who has accomplished so much, and about whom much has already been written. But the great amount of material available does not necessarily make it an easy job. Sorting it out and condensing it is a monumental task. After all, Newton's accomplishments go beyond astronomy: he excelled in the fields of mathematics, dynamics, celestial mechanics, optics, and natural philosophy. His laws of motion are well known to all astronomy and physics students. The author expertly tells the story of Newton's life and accomplishments, concentrating on all the areas of his ability. It is a well-written, well-researched volume that should not be overlooked. For the university and public library.

265. Meadows, A. J. **Science and Controversy: A Biography of Sir Norman Lockyer**. London, Macmillan; Cambridge, Mass., MIT Press, 1972. 331p. illus. refs. $18.00. LC 72-190440. ISBN 0-333-13539-3 (U.K.); LC 72-4536. ISBN 0-262-13079-3 (U.S.).

This is the second work to deal with the well-known Victorian scientist, and the first to examine fully the controversies in which he frequently found himself involved. A scientist with no formal training, Lockyer devised an instrument for observing solar eclipses and studying the coronal spectral lines. The founder, and editor for 50 years, of *Nature* magazine, he frequently traveled around the world observing solar eclipses. But Lockyer's interest and skill went beyond astronomy, the field for which he is best known. His work included atomic structure, spectroscopy, poetry, meteorites, solar physics, and more. The controversies often arose because he was frequently the first to suggest unpopular theories, many of which turned out to be correct. A well-researched book for the student, scientst, or layman seriously interested in Victorian science.

266. Rosen, Edward, ed. **Three Copernican Treatises**. 3rd rev. ed. New York, Octagon Books, 1971. 425p. illus. index. notes. (Records of Civilization: Sources and Studies, 30). $21.00. LC 73-145545. ISBN 0-374-96913-2.

A book which could potentially be very valuable to the historian, this volume is a reprint of the 1939 edition and the 1959 revision, and it contains some new material as well. Before the author originally translated these Copernican works (the *Commentariolus* of Copernicus, the *Letter against Werner*, and the *Narratio prima* of Rheticus), no English text existed. An excellent introduction provides the uninformed reader with background information on the three treatises, and two annotated Copernican bibliographies (1939-58; 1959-70) follow the text. There is a lengthy and informative biography of the astronomer as well. The importance of this translation and the accompanying material cannot be emphasized enough.

267.   Santillana, Giorgio de. **The Crime of Galileo.** Chicago, University of Chicago Press, 1955. 339p. illus. index. $10.00; $2.25pa. LC 55-7400. ISBN 0-226-73483-8; 0-266-73484-6pa.

The author re-creates the well-known astronomer's encounter with the Inquisition in seventeenth-century Italy in this classic book. Galileo taught and defended the Copernican view of the Universe, despite strong opposition by a small faction within the Church which chose to believe otherwise. The result, of course, was persecution and a subsequent trial. Giorgio de Santillana tells well the events preceding and following Galileo's trial in Rome in this fine work, which ought to be required reading for all astronomy students. Certainly we should all reconsider the ominous meaning of Galileo's ordeal and the importance of freedom of expression.

268.   Shapere, Dudley. **Galileo: A Philosophical Study.** Chicago, University of Chicago Press, 1974. 161p. index. notes. bibliog. $9.75; $2.95pa. LC 73-92023. ISBN 0-226-75005-1; 0-226-75007-8pa.

Quite different from Colin Ronan's *Galileo*, intended for the layman, this work is more scholarly, aimed at the historian. Shapere outlines his intent this way: "The present work is the first of a projected series of detailed studies, by the present author, of important episodes in the development of science— studies which, while aiming at being as historically and scientifically responsible as the subject-matter and the capacities of the author allow, will focus on those facets of the selected episodes which are of relevance to the philosophical questions concerning the rationale of the scientific enterprise." The volume is divided into five sections: Galileo and the Interpretation of Science; The Intellectual Background; The Early Development of Galileo's Thought; Galileo and the Principle of Inertia; Reason and Experience in Galileo's Thought. Like Drake's *Galileo Studies,* this book would be best suited for the university library and its users.

269.   Shea, William R. **Galileo's Intellectual Revolution; Middle Period, 1610-1632.** London, Macmillan; distr. New York, Science History Publications, Neale Watson Academic Publications, 1972. 204p. illus. index. notes. bibliog. £4.95; $15.00. LC 72-89852. B 72-32288. ISBN 0-333-14105-9 (Macmillan); 0-88202-006-4 (Watson).

Another of the many books available on Galileo, this volume investigates the so-called "middle period" of his life, a period frequently neglected by historians. It was during this time, notes Shea, that Galileo "worked out the methodology for his intellectual revolution." The author dwells at length on Galileo's writings about hydrostatics, comets, sunspots, and the heliocentric theory. The book is not a narrative, so it is not well suited for most lay readers, although it would not be impossible to read. The text is extremely well documented, including extensive notes and a well-chosen bibliography. For the university library, the volume should be of interest to historians and students.

270. Wright, Thomas. **An Original Theory or New Hypothesis of the Universe, 1750**. Ed. by Michael A. Hoskin. London, MacDonald & Co. (Publishers) Ltd.; New York, American Elsevier Publisher Co., Inc., 1971. 178p. illus. (History of Science Library: Primary Sources). £10.00; $37.95. LC 70-139573. ISBN 0-356-0315-8 (MacDonald); 0-444-19612-9 (Elsevier).

Thomas Wright was one of the first to hypothesize that the Milky Way was a vast system of stars in a universe of countless other such systems. It was not until almost 180 years later that he was proven correct. This important volume is a facsimile reprint of Wright's famous theory of the Universe, beginning with a 37-page introduction outlining his life. The manuscript itself is preceded by text explaining its significance. The editor has done an excellent job of preparing the re-issue of this classic work, which is good source material for the astronomical and scientific historian.

General

271. Asimov, Isaac. **Eyes on the Universe: A History of the Telescope**. Boston, Houghton Mifflin, 1975. 274p. illus. index. $8.95. LC 75-15830. ISBN 0-395-20716-9.

Although this excellent book is a history of the telescope, it is also, by default, a history of astronomy, for as the telescope developed and improved, our knowledge of the skies grew in leaps and bounds. Asimov begins his narrative with a bit of pre-telescope astronomy, painting a picture of the Heavens as they were known to scientists before Galileo. Galileo did not invent the telescope, nor was he the very first to point it skyward. But he was the first to use it seriously for astronomy, and he made several significant contributions, in particular the discovery of Jupiter's four brightest satellites. The book tells this story and goes on to explain how the telescope began to change man's view of the Universe. Filled with stories of great astronomers and their discoveries, the book follows the development of the telescope from simple spyglass to large reflectors and refractors. It discusses the special telescopes of the twentieth century as well: radio, Schmidt, orbiting, infrared, etc. There is some information on how telescopes work, but this is of minor importance in this text, which concentrates on the developers of telescopes and the major and minor discoveries. Readers interested in astronomical telescopes themselves are directed to *The Astronomical Telescope* (Barlow, 1975) and *Tools of the Astronomer* (Miczaika and Sinton, 1961). Written in the author's typically clear and informative style, it comes highly recommended for the general reader and student. Asimov scores well again.

272. Baum, Richard. **The Planets: Some Myths and Realities**. New York, Halstead Press, John Wiley and Sons, 1973. 200p. illus. index. bibliog. refs. notes. $8.95. LC 73-7583. ISBN 0-470-05930-3.

Whether the reader of this book wishes to consider the events described here as history, or merely as unconfirmed reports of strange events, is unimportant. What *is* important is that the author presents, in eight chapters, fascinating descriptions of "unusual observations" which are sure to stimulate the imagination and to hold the attention of the serious amateur astronomer. No perfectly satisfactory explanations can be given for these unconfirmed phenomena, most of which were observed by famous astronomers. Among these chapters of wierd astronomical sightings are William Herschel's observations of a ring around Uranus (proven to be true in 1977), reports of a ring associated with Neptune in the mid-nineteenth century, unidentified objects near the Sun and Venus, the search for a natural lunar satellite, peaks on Venus, and more. Recommended for the public and observatory library.

273. Berendzen, Richard, ed. **International Conference on Education In and History of Modern Astronomy**. New York, New York Academy of Sciences, 1972. 275p. illus. refs. (Annals of the New York Academy of Sciences. v.198). LC 72-194959.

Astronomy professors and instructors who have not had a chance to read this volume of conference proceedings ought to do so, for the papers contained in it include some good insights into current trends in astronomical education. Sponsored by the New York Academy of Sciences and the American Astronomical Society, the meeting, held in the fall of 1971, saw astronomy educators from all over the world gather to discuss topics that are unfortunately not often considered at such conferences (or anywhere, for that matter). The text is divided into five parts: Education and Employment of Astronomers in the United States; International Issues in Astronomy Education; University Level Astronomy Education for Nonscience Concentrators; Non-collegiate Astronomy Education; and Personal Accountings of the Development of Modern Astronomy.

Especially informative is the first section, in which students and professors state their impressions of current curricula in American colleges and universities. Along these same lines, speakers in the third section discuss the programs being developed for non-science students, an important topic when one considers the number of texts available now for non-science astronomy classes (see the section on textbooks in this guide). Secondary school teachers will be interested in the section on non-collegiate astronomy education, since little seems to be available in the literature in these areas. Each of the 34 papers has references, and the brief discussions that took place after the presentations are also included.

274. Donnelly, Marian Card. **A Short History of Observatories**. Eugene, Oregon, University of Oregon Books, 1973. 164p. illus. notes. bibliog. index. $7.50. LC 73-175209.

This well-documented volume gives a brief overview of the history of the important astronomical observatories constructed since the invention of the telescope. The text is descriptive and includes names and dates connected with the observatories cited. Chapter 1, "From Galileo to Ole Rømer," describes the earliest instances in which telescopes were used in the sixteenth and seventeenth centuries, including the observatories at Leiden, Copenhagen, and Paris. The following chapter zeroes in on the use of telescopes in observatories in the form of towers, and chapter three examines the change in design from the tall structure to the low, symmetrical building in the early nineteenth century. Later in the nineteenth century, the architectural style turned to "more elaborate designs in the Classical manner." The final two sections discuss changes that occurred when the giant refracting and reflecting telescopes appeared. Heavily illustrated with pictures and floor plans, this work is intended for all interested in the history of science, and of astronomy in particular. The emphasis in this book, however, is on the history of observatory architecture rather than on the history of those institutions themselves. But the fact that both topics are included makes the work all the more interesting.

275. Grant, Robert. **History of Physical Astronomy**. New York, Johnson Reprint Corp., 1966. 637p. index. (The Sources of Science, no. 38). $28.00. LC 66-22741. ISBN 0-384-19670-5.

    A re-issue of the classic 1852 London edition, this book should be required reading for students of the history of astronomy. Based on original sources, the text follows the history of astronomy from Newton's theory of gravitation to the evolution of stellar astronomy. The volume comprehensively covers celestial mechanics of the eighteenth century, the planets, their satellites, and orbits, and Grant frequently goes into great detail on these and other topics. Comets and meteors are mentioned, too, as well as observational astronomy, eclipses, instruments, and the making of catalogs. The book's importance lies in its historical perspective, providing a valuable comparison with today's so-called "true" ideas. Anyone who has looked at old astronomy books surely has been captivated by the interesting illustrations accompanying the text; it is unfortunate that this author does not include any pictures in his book. The lack of illustrations is the only drawback to this interesting work.

276. Graubard, Mark. **Motivations, Tools and Theories of Pre-Modern Science**. Minneapolis, Burgess Publishing Co., 1967. 289p. illus. index. refs. bibliog. LC 67-13091.

    The development of science and its applications from antiquity to the late Renaissance is the subject of this work, which contains dozens of references to the history of astronomy. It is hardly surprising, though, that the author has included a great deal on early astronomy, since the science has played such an important role in the history of mankind. The first four chapters are concerned with, among other things, Babylonian and Egyptian astronomy, including the calendar and early astronomical observation. Later, Graubard deals with Greek and Roman views of the Universe, citing such familiar names as Plato, Aristotle, and Aristarchus. A quick overview of the subject matter related to the development of astronomy will find ancient astronomy, calendars, measurement of time, ancient instruments, practical astronomy, theories of the Universe, astronomy in the Middle Ages, and more. The student of astronomy, or any other science for that matter, should read this interesting work, which not only tells of the individual achievements and great minds of science, but fits them together into a well-written, overall view of the early development of science and technology. For the college and public library.

277.   Hawkins, Gerald S., and John B. White. **Stonehenge Decoded**. Garden City, New York, Doubleday and Co., 1965. 202p. illus. refs. index. bibliog. $6.95. LC 65-19933. ISBN 0-385-04127-6.

This fascinating narrative on the mysterious giant stones in England has become a classic of sorts, in both historical and astronomical literature. A popular account of the story behind Stonehenge, the book is aimed at general readers and scientists alike. The work opens with a chapter on the legends of Stonehenge, a delightful discourse on the various references to the "monument" since the Dark Ages as found in fiction and factual literature. The next several chapters delve into the people and history surrounding the structure which was built 3000 to 4000 years ago. The theories on how the stones were moved to the Stonehenge site are especially interesting—a combination of extraordinarily difficult land and sea routes. The last half of the book will no doubt be of great interest to students of astronomy, for it is here that Hawkins explains the working of the great computer that he contends Stonehenge was. In particular, he explains the eclipse predictor, the theories of which have been contested by at least one other astronomer. Finally, three of the author's articles are reprinted from *Nature* and *Science*; these explain in detail various aspects of the "observatory." Many good photos and diagrams accompany the text. This is probably the best and the best-researched volume on the subject. For all types of libraries, general and scientific.

278.   Hey, J. S. **The Evolution of Radio Astronomy**. New York, Science History Publications, Neale Watson Academic Publications, Inc., 1973. 214p. illus. refs. glossary. indexes. $10.00; $3.95pa. LC 73-80636. ISBN 0-88202-027-7; 0-88202-030-7pa.

This is the story of radio astronomy from its beginnings in the 1930s through 1970. The first half of the volume looks at the early discoveries and the beginning of various research programs, including Karl Jansky's discovery of radio waves emanating from the center of the Milky Way, and Hey's own discovery of solar radio emissions. The book also includes descriptions of some of the world's great radio telescopes, including the pioneer of the big ones, the instrument at Jodrell Bank (see *The Story of Jodrell Bank* by Sir Bernard Lovell). As the text progresses, the author looks at the achievements that have been made in radio astronomy and analyzes their significance. Among these events are the discovery of the 21cm hydrogen line, radio waves from other galaxies, radio studies of the planets, quasars, radio stars, and much more. A good overview for anyone interested, the book contains an excellent list of references on the radio astronomy literature.

279. Jones, Bessie Zaban, and Lyle Gifford Boyd. **The Harvard College Observatory: The First Four Directorships, 1839-1919.** Cambridge, Mass., Belknap Press of Harvard University Press, 1971. 495p. illus. index. notes. $16.50. LC 73-143228. ISBN 0-674-37460-6.

The history of the Harvard College Observatory is, in many ways, the history of astronomy in the United States. Not only is it one of the oldest American observatories, but it has been the home of some of the most famous astronomers in the world. This excellent book tells the exciting history of the first 80 years of the observatory's existence under the directorships of William Cranch Bond (1839-1859), George Bond (1859-1866), Joseph Winlock (1866-1875), and Edward Charles Pickering (1876-1919). The text actually begins with the period before the observatory was established, 1642-1839, when students were taught astronomy as part of the regular curriculum and when progress was being made toward the building of a telescope structure. The authors have written an interesting text, spiced liberally with quotations from correspondence and with old photographs, and the directors and the great astronomers come alive in Jones and Boyd's well-documented volume. This is the fourth major account of HCO's past; the third appeared in 1931.

280. Jones, Bessie Zaban. **Lighthouse of the Skies: The Smithsonian Astrophysical Observatory: Background and History, 1847-1955.** Washington, D.C., Smithsonian Institution, 1965. 339p. index. notes. (Smithsonian Publications. 4612.). $5.00. LC 66-60190.

Although the Smithsonian Astrophysical Observatory was not formally established until 1890, it had its beginnings in 1846 when the Institution began. At that time, President John Quincy Adams felt that an observatory would be an excellent use of James Smithson's bequest to the people of the United States. The author traces the history of the SAO from its beginnings in 1846 to 1955, when it became associated with Harvard after its move to Cambridge. Relying mainly on Smithsonian sources such as the *Annual Reports*, she tells the history of this great observatory and the development of astrophysics there.

281. Koestler, Arthur. **The Sleepwalkers: A History of Man's Changing Vision of the Universe.** New York, Grosset & Dunlap, 1963, c1959. 624p. illus. index. notes. bibliog. (The Universal Library, UL159). $3.45pa. LC 63-6841/CD. ISBN 0-448-00159-4.

Of the many excellent "serious" histories of astronomy, this fine volume ranks at the top, along with Pannekoek's *A History of Astronomy*. It is a critical narrative in which the characters in the play come alive, along with in-depth discussions of their theories and their implications. The author would contend that this book is not a history of astronomy, but rather a history of cosmology, a survey of how mankind has viewed the Universe and his place in it. And since the emphasis in the book is not on events and discoveries, although they are mentioned, the author's contention is correct. In any event, it is a superb work, sure to withstand the test of time. Koestler investigates the intertwining threads of science and religion as they weave through history, and he explains and discusses their importance to astronomy. The psychological process of discovery, as the author calls it, is considered as the book examines the lives of Copernicus, Kepler, Brahe, and Galileo. An excellent account for historian, student, teacher, and layman, it should be in every astronomy collection.

282. Kopal, Zdeněk. **Widening Horizons: Man's Quest to Understand the Structure of the Universe.** New York, Taplinger Publishing Co., Inc., 1971, c1970. 176p. illus. index. bibliog. glossary. $6.95. LC 73-99307. ISBN 0-8008-8320-9.

This introductory volume for the general reader and student presents a history of how mankind has perceived his Universe since antiquity. A more than usual amount of space is devoted to the ancient astronomers: Aristotle, who asserted that the Earth is shaped like a ball; Eratosthenes, who measured the Earth's size to a remarkable degree of accuracy; Aristarchus, who made some of the earliest suggestions on the size of the solar system; and so on. Kopal's descriptions of the period are especially good. The Renaissance men (Copernicus and Tycho Brahe, for example) are considered next, and the author shows how they substantially changed some of the then still-accepted ideas of the ancient astronomers. Then comes an era of "new astronomy" led by the founding fathers— men like Kepler, Galileo and Newton—who established many of the laws upon which astronomy is based. All in all, it is an excellent beginning book for the newcomer who wants a brief overview of astronomy. The author, who has written so many excellent books on the Moon, has succeeded again in this thought-provoking treatise on the development of astronomical ideas.

283. Ley, Willy. **Watchers of the Skies: An Informal History of Astronomy from Babylon to the Space Age.** New York, Viking Press, 1969. 529p. illus. index. bibliog. (A Viking Compass Book). $8.50; $2.95pa. LC 78-4129. ISBN 0-670-75043-3; 0-670-00254-2pa.

A well-written, often-cited account of the history of astronomy, this work is both comprehensive and well researched. Divided into three major sections, it is one of the best such works available: 1) a chronological arrangement of history from antiquity to the eldest Herschel, 2) the solar system, and 3) histories of special astronomical problems. The second division is the largest; here Ley discusses each planet individually, describing its study from ancient times to the present, as well as including various facts and figures on its physical make-up. The first section, however, is the most important. Of the major characters in astronomical history, Ley discusses not only their scientific contributions but their life histories as well. Familiar faces appear as expected: Galileo, Copernicus, Kepler, Ptolemy, Brahe, Newton, Halley, and many more. The very first chapter deals with Egyptian and Babylonian astronomy, providing some interesting insight into the early periods of astronomical history. The third part of the book is more "current," delving into more recent developments on stars, galaxies, the Sun, distances, motions, etc. Intended for all readers, it will be best appreciated by students and laymen who want a serious but not overly scholarly treatment. Ideal for the public library, it has an excellent bibliography.

284. Lovell, Sir Bernard. **The Story of Jodrell Bank**. New York, Harper and Row, 1968. 265p. illus. index. LC 68-17043.

The re-telling of an historical event is frequently more interesting when done by someone who participated personally; such is the case of this excellent volume. Professor Lovell began his work at Jodrell Bank using a trailer filled with "ex-Army radar equipment," back in the early days of radio astronomy. He eventually directed the construction of one of the greatest radio telescopes in the world, the main subject of this book. The author details the many difficult years of the project, complete with construction delays, severe and repeated financial troubles, and opposition from fellow scientists. Nevertheless, the project was completed in 1957, and the telescope and Sir Bernard got its first good publicity when it was used to detect the first Russian Sputnik in October of that year. It was the telescope's role in the early years of the Space Age which assured its success and its future.

285. Moore, Patrick. **The Picture History of Astronomy**. 4th ed. New York, Grosset & Dunlap, London, MacDonald, 1972. 253p. illus. index. chronology. $7.95. LC 72-171341. ISBN 0-448-01548-X (G & D); 0-356-04182-4 (MacDonald).

Beautifully illustrated with hundreds of photos, diagrams, and drawings, the fourth edition of this popular book presents many aspects of astronomy in a manner most interesting to the person pursuing the science as a hobby. Aimed mainly at the teenage reader, the book should also be immensely enjoyed by adults. The text mentions not only the great men of astronomy but also the great telescopes and observatories of the world. Although much of the volume concentrates on pre-twentieth century events, there is also much on the various recent developments like radio astronomy, quasars, life in the Universe, and the space age. An excellent two-page section that highlights "landmarks in the story of astronomy" is especially good. A history of astronomy written for amateurs, by an amateur, says Moore in his foreword of this highly recommended work.

286. Pannekoek, A. **A History of Astronomy**. London, George Allen and Unwin; distr. Totowa, N.J., Rowman and Littlefield, Inc., 1961. 521p. illus. index. notes. $12.50. LC 61-66763. ISBN 0-87471-365-X.

Translated from the 1951 Dutch edition (*De Groei Van Ons Wereldbeeld*), this well-written famous text examines the history of astronomy with regard to the growth of man's concept of the world (i.e., the Universe). One of the first sciences, the author points out, astronomy played an important part in early history. For example, the heavens were the source of the gods and religious practices. But astronomy evolved from a religious act to a science as man progressed culturally. Unlike many of the traditional sciences, the author notes, it has contributed little to changing our world. Rather, it has evolved into a pursuit of the physical knowledge of our universe. Pannekoek considers all these things in his excellent account of astronomy's history and growth.

287. Ronan, Colin A. **Astronomy**. Newton Abbot, Engl., David & Charles; distr. New York, Barnes and Noble Books, 1973. 112p. illus. (Illustrated Sources in History). £3.95; $10.50. LC 73-159129. ISBN 0-06-495970-8 (B & N); 0-7153-5593-7 (D & C).

The author created this fascinating "scrapbook" of astronomical history by drawing upon the writings of famous astronomers down through the ages. Extracting pages and paragraphs of their works and adding appropriate illustrations (many of which accompanied the original texts), Ronan gives the reader an excellent look at the development of various astronomical theories. Each excerpt is annotated and footnoted, giving the reader the proper background on each piece. The first chapter starts the book off with a bang, investigating theories of the planetary system, allowing us to read the works of the "big three," Aristotle, Ptolemy, and Copernicus, followed by the thoughts of Kepler, Galileo, Newton, etc. The other five chapters are just as good, as Ronan prepares overviews of the study of the stars, the nebulae, cosmology and more. The book's strong points are the excellent, unusual arrangement of the text and the superb illustrations. This work should not be missed by any astronomy student, professor, or librarian.

288. Ronan, Colin A. **Discovering the Universe: A History of Astronomy**. New York, Basic Books, Inc., 1971. 248p. illus. index. (Science and Discovery Books). $6.95. LC 72-135556. ISBN 0-465-01670-7.

This story of the development of astronomy begins over six thousand years ago when primitive man first wondered about the heavens, and thought about what caused day and night. From antiquity, the book progresses to the present, mentioning discoveries and breakthroughs as they occurred. Ronan does not overlook the main cast; he discusses the exploits of the ancient astronomers, and of Copernicus, Galileo, Kepler, Lockyer, Herschel, Hale, and many more, all of whom played an integral part in the discovering of the Universe. An excellent volume by a fine astronomical writer, this book will be enjoyed by the layman or student; it teaches the reader many astronomical concepts as well.

289. Shapley, Harlow, and Helen E. Howarth, eds. **A Source Book in Astronomy**. New York, McGraw-Hill Book Co., Inc., 1929. 412p. illus. index. (Source Books in the History of the Sciences.) LC 29-1109.

A different type of historical work, this book contains reprints of original statements, papers, or reports of scientific discoveries in astronomy. An ideal volume for the astronomical historian and student, its contents include the Copernican theory, Kepler's planetary laws, Galileo's telescope, Newton's theories of physics, Halley's comet, Adams and Leverrier on Neptune's discovery, and many more. In all, there are 63 entries describing the major breakthroughs in astronomy up until the twentieth century. The arrangement allows the reader to examine, first-hand, the ideas and equations of the famous astronomers, and to interpret them as he or she sees fit. A remarkable book that ought to be reprinted.

290. Shapley, Harlow, ed. **Source Book in Astronomy: 1900-1950.** Cambridge, Mass., Harvard University Press, 1960. 423p. illus. index. (Source Books in the History of the Sciences). $20.00. LC 60-13294. ISBN 0-674-82185-8.

A follow-up to Shapley and Howarth's 1929 work, this volume begins at 1900 and provides excellent glimpses of astronomers and their discoveries during the first half of the twentieth century. Like its predecessor volume, this collection of contributions to astronomy vividly illustrates what Shapley calls "the vigorous march of astronomical science." Mathematical formulation and technical language are avoided as much as possible, in order to best serve the general reader, but such information is not excluded when it represents important breakthroughs in the field. Sixty-nine contributions make up this superb book.

291. Smart, William M. **The Riddle of the Universe.** New York, John Wiley and Sons, Inc., 1970, c1968. 228p. $5.50. LC 68-25830. ISBN 0-471-79914-9.

Not only an historical work, this book is also a kind of introduction to astronomy. The author describes how man's concept of the Universe has changed over the centuries, while at the same time explaining various astronomical ideas pertinent to the subject matter. The three stages of our exploration of the Universe discussed by the book are 1) up to medieval times (Earth, Moon, Sun and five known planets were the Universe); 2) nineteenth and early twentieth centuries (the Milky Way became the Universe); 3) very recently (when other galaxies and systems of galaxies were discovered—the Universe exploded in size).

292. Struve, Otto, and Velta Zebergs. **Astronomy of the 20th Century.** New York, Macmillan, 1962. 544p. illus. index. notes. glossary. bibliog. $12.50. LC 62-21206.

Not meant to be comprehensive and detailed in its account of astronomy, this book describes, by the use of specific examples, some of the important gains made by astronomers since 1900. The coverage of topics is understandably uneven, but, then, certain discoveries were more important than others. Some of the research discussed includes photography of the Milky Way, radial velocities, photometry, radio astronomy, spectral classification, stellar structure, unusual stars, and galaxies. The book brings up to date (to the early 1960s) many historical works that stop in the late nineteenth century, but an updated edition of this fine volume is needed, since the 1960s were a period of significant and frequent discovery. The reader with some background in astronomy will profit most from this book, which is aimed at the student and lay reader.

293. Whitney, Charles A. **The Discovery of Our Galaxy.** New York, Alfred A. Knopf, 1971. 308p. illus. index. glossary. $10.00. LC 76-154942. ISBN 0-394-46068-5.

The story of how astronomers determined the shape and size of the Milky Way is the subject of this interesting book. The author begins with the ancients' concept of the heavens (the Earth as the center of the Universe, which then was not considered to be large) and works up to the "final" picture of our galaxy as seen in the 1950s (a spiral-shaped system of billions of stars in an almost limitless Universe of billions of galaxies). Among the astronomers discussed in this volume are Brahe, Kepler, Galileo, Newton, Herschel, Lord Rosse, and Harlow Shapley, all of whom made observations of the skies, put forth theories about the "nebulae," and laid the groundwork for the ultimate "discovery" of the Milky Way's appearance. An excellent book for all readers, from high school age on up, it is appropriate for the public and college library.

294.   Woodbury, David O. **The Glass Giant of Palomar**. rev. ed. New York, Dodd, Mead & Co., 1970. 390p. illus. LC 77-135210. ISBN 396-01919-6.

A significant footnote to the history of astronomy, this book tells the story of the 12-year project to build what was then the world's largest optical telescope, the 200-inch reflector. The account begins with the history of earlier telescopes at Mount Wilson Observatory, but the majority of the book deals with how George Ellery Hale planned and built the great telescope on Palomar Mountain. The difficult construction of the mirror for the reflecting instrument (at Corning Glass in New York) is detailed as well, since it gives the reader insights into the problems that accompany the building of a "super-telescope." One chapter that did not appear in earlier editions, "Toward the Future–1970," recounts briefly the recent discoveries made at Palomar and takes a look at what is ahead for the observatory. A fine book for public, college, and observatory collections, it should be read in conjunction with Helen Wright's *Explorer of the Universe* and *The Legacy of George Ellery Hale.* A new edition, or even a reprint, would be a valuable addition to the literature.

Scholarly Works

295.   Cotter, Charles H. **A History of Nautical Astronomy**. New York, American Elsevier Publishing Co., 1968. 387p. illus. index. bibliog. $13.75. LC 68-12049. ISBN 0-444-19959-4.

The author uses descriptive text, good illustrations, and trigonometry in this work, which covers the history and theory of finding a ship's position at sea by astronomical methods. The book examines the many diverse problems of astronomical navigation and explores the methods of solution. The following chapters are included in this fine volume: The Development of Nautical Astronomy (a general historical overview from the Babylonians and Phoenecians to the early twentieth century); Astronomical Methods of Time-Measuring at Sea (including discussions of units of time, the calendar, mechanical clocks, marine chronometers, etc.); The Altitude-Measuring Instruments of Navigation (involving descriptions of the seaman's quadrant, the astrolabe, the cross-staff, the back-staff, and more); The Altitude Corrections (here the author examines mathematically such areas as refraction, depression, parallax, and so on); Methods of Finding Latitude (using the Pole Star, solar meridian atltitude, meridian altitude of a star, etc.); Methods of Finding Longitude (from eclipse observations, observations of Jupiter's satellites, lunar occultations and transits, etc.); Position-Line Navigation; and

Navigational Tables. Two brief appendices on spherical astronomy and spherical trigonometry follow. Intended for the student and astronomer with interests along these lines, the text does require some background in trigonometry.

296. Daumas, Maurice. **Scientific Instruments of the Seventeenth and Eighteenth Centuries.** New York, Praeger Publishers, 1972. 361p. illus. index. bibliog. refs. $38.50. LC 77-112019.

Translated from the 1953 French work (*Les Instruments scientifiques aux XVII et XVIII siécles*), this fascinating volume examines the development and use of the various instruments used in scientific research. The book was well researched (there are several hundred references) and will provide scientific historians with a remarkable look at the tools of early researchers. Two sections of the volume are concerned with the instrument-making industry, and another examines factors that influenced the evolution of the industry. There are numerous references throughout the book to astronomical devices, including the telescope, astrolabe, quadrant, etc., but there is no particular emphasis placed on these instruments. This remarkable and expensive book, translated by Mary Holbrook, includes 142 photographs or sketches of the instruments described.

297. Dicks, D. R. **Early Greek Astronomy to Aristotle.** Ithaca, N.Y., Cornell University Press, 1970. 272p. illus. index. notes. (Aspects of Greek and Roman Life). $12.50. LC 76-109335. ISBN 0-8014-0561-0.

The only book since Sir Thomas Heath's two editions (*Aristarchus of Samos*, 1913, and *Greek Astronomy*, 1932) to deal exclusively with Greek astronomy, this work is concerned with the development of astronomy as a science, rather than with cosmology or cosmogony. Therefore, there is little mention of what Dicks calls the "cosmological fantasies of the pre-Socratics." The book concentrates on Plato (the author says his importance in the development of Greek astronomy is underestimated), whose theory of the Universe said that the world must be a perfect sphere and that all motion (especially that of the heavenly bodies) must be in perfect circles at uniform speed. Eudoxus is also paid special tribute by this work, describing his "spheres within spheres" theory, which tried to account for the apparent irregularities of the motion of the planets in Plato's perfect Universe. The book also covers the origin of the constellations and the influence of Babylonian astronomy on the Greeks. It concludes with Aristotle, so it is shorter than Heath's second book.

298. Hellman, C. Doris. **The Comet of 1577: Its Place in the History of Astronomy.** New York, AMS Press, 1971, c1944. 488p. indexes. refs. (Columbia University Studies in the Social Sciences, no. 510). $18.50. LC 72-110569. ISBN 0-404-51510-X.

For centuries, comets were considered to be omens of terrible events, spirits or bodies which moved high in the Earth's atmosphere. They were certainly not thought to be astronomical in nature. The great comet of 1577 changed all that thinking, however, and the author, in the reprint of her dissertation, explains just how that particular comet brought about the change. While the book may not be of great interest to laymen, it should spark some curiosity among scientific historians. For the university library.

299.   Jaki, Stanley L. **The Milky Way: An Elusive Road for Science.** New York, Science History Publications, Neale Watson Academic Publications, Inc., 1975. 352p. refs. name index. $6.95pa. LC 72-87334. ISBN 0-88202-022-6.

This well-researched volume tells the story of the history of the investigation of the Milky Way from the Greeks to the twentieth century. The author dug through dusty manuscripts, rare books, and journals to piece together a picture of how our galaxy has been perceived by the great (and not-so-great) astronomers for the past 2000 years. Among the men discussed are Aristotle, Roger Bacon, Galileo, Isaac Newton, Thomas Wright, William Herschel, and Edwin Hubble. Unlike Whitney's *The Discovery of Our Galaxy,* this book is a scholarly, dry presentation not intended for the layman. The scholar/historian will best appreciate the heavily footnoted essay in this work.

300.   Lovell, Sir Bernard. **Out of the Zenith: Jodrell Bank, 1957-1970.** New York, Harper and Row, 1974, c1973. 255p. illus. index. notes. $12.50. LC 73-14269. ISBN 0-06-012719-8.

Radio astronomers and students should be interested in this volume, an in-depth description of the 250-foot radio telescope (Mark I) at Jodrell Bank, England. During its 14 years of operation, the Mark I was used to make many important "radio" discoveries, including identifying quasars and pulsars. The volume is not aimed at amateur astronomers or lay readers, as it contains many technical passages describing the work done with the telescope. The reader interested in a more general treatment of the telescope and its history should read Lovell's *The Story of Jodrell Bank.* The Mark I became the Mark IA in 1970 after a complete overhaul and many improvements. It remains one of the greatest radio instruments in the relatively short history of radio astronomy.

301.   McGucken, William. **Nineteenth-Century Spectroscopy; Development of the Understanding of Spectra: 1802-1897.**Baltimore, Johns Hopkins Press, 1969. 233p. illus. index. bibliog. $12.50. LC 74-94886. ISBN 0-8018-1059-0.

The importance of spectroscopy to astrophysics is well known to the astronomer. Nearly all that is known about a particular star is gained from the study of its spectral lines: its mass, temperature, rotation, type, etc. This well-researched volume, which provides some interesting insights into the discoveries made in the development of spectroscopy, may be of interest to student and astronomer alike. In it, the author tells the important stories behind the research from William Hyde Wollaston's observation of dark solar lines in 1802 to J. J. Thomson's discovery of the electron in 1897. For the university and observatory library, there is no comparable work.

302. Meadows, A. J. **Early Solar Physics.** Oxford, Elmsford, N.Y., Pergamon Press, Inc., 1970. 312p. index. refs. (The Commonwealth and International Library. Selected Readings in Physics). $8.50; $6.00pa. LC 74-103021. ISBN 0-08-006653-4; 0-08-006654-2pa.

Probably the only book to deal with the history of solar studies, this work is divided into two distinct sections. The first is an overview of the period in question (1850-1900) and the development of knowledge about the Sun during those years. The second part is a selection of original articles from books and journals of the period on solar physics, written by the scientists who made the discoveries and carried out the research. This format works quite well and should be considered by other authors writing on the history of astronomy. For the university and observatory library.

303. Meadows, A. J. **The High Firmament: A Survey of Astronomy in English Literature.** Leicester, Engl., Leicester University Press; distr. New York, Humanities Press, Inc., 1969. 207p. index. refs. $7.50. LC 75-406782. B 69-02159. ISBN 0-7185-1082-8.

This unusual volume takes a look at references to astronomy that appeared in English literature between 1400 and 1900. As such, then, it is a book not on astronomy but, rather about literary analysis. Nevertheless, its place in this guide is justified because of its potential use to scientific historians. The author states the reasons for writing his book: "There are, perhaps, two main reasons for studying how scientific trends are reflected in non-scientific literature. In the first place, the scientific outlook has changed so vastly over the past few centuries that unless a special analysis is made many literary references to a science become imcomprehensible to a modern reader. On the other hand, the world view of even educated non-scientists has sometimes differed appreciably from that of their scientific contemporaries. As a result, a study of non-scientific literature can be of value in the history of science since it provides an insight into the diffusion of scientific ideas throughout society as a whole." If you're an astronomer with humanistic leanings, read this one-of-a-kind volume.

304. Nakayama, Shigeru. **A History of Japanese Astronomy: Chinese Background and Western Impact.** Cambridge, Mass., Harvard University Press, 1969. 329p. illus. index. bibliog. (Harvard-Yenching Institute. Monographic Series. v.18). $11.50. LC 68-21980. ISBN 0-674-39725-8.

One of the few works dealing with Japanese astronomy in any Western language, this scholarly book describes how the Japanese adapted Chinese astronomical theory and how they accepted various Western ideas beginning in the late sixteenth century and continuing up to the 1880s. The author's main contention is that there was little that was original in Japanese astronomy, but from historical perspectives it is interesting to consider how the Chinese and Europeans affected the development of the science there. In examining the Chinese influence, Nakayama considers early Chinese cosmology, astrology and the occult, and calendrical science. The impact of the West includes the Jesuit influence, Aristotelian cosmology, and the Copernican and Newtonian theories. The reaction to and the extent to which these ideas were gradually accepted is the author's main thesis. He also discusses at length the reform of the calendar in Japan. For the university library.

305. Newton, Robert R. **Ancient Astronomical Observations and the Accelerations of the Earth and Moon.** Baltimore, Johns Hopkins Press, 1970. 309p. refs. index. $15.00. LC 70-122011. ISBN 0-8018-1180-5.

This well-researched book is divided into two parts, the first concerned with ancient observations of the Sun and planets, and the second involved with the variations in motion of the Earth and Moon. The first section is mainly concerned with eclipses, since these astronomical events were usually recorded by the ancients. Eclipse reports from all parts of the ancient world are included, as well as equinox observations by specific individuals such as Ptolemy and Hipparchus. In the last portion of the volume analyses are made of the various observations, along with a discussion of Earth-Moon accelerations. The author has done an excellent job of researching old records to present his information in a logical and interesting format.

306. Nietro, Michael Martin. **The Titius-Bode Law of Planetary Distances: Its History and Theory.** Oxford, Elmsford, N.Y., Pergamon Press, 1972. 161p. illus. refs. indexes (subject, author-name). (International Series of Monographs in Natural Philosophy, v.47). $11.00. LC 78-178682. ISBN 0-08-016784-5.

Astronomers and science historians will both be interested in this book, which explains and discusses at length one of the most interesting footnotes to the history of astronomy. The law known as Bode's Law was, as the author points out, originated by Johann Daniel Titius von Wittenberg, and not Johann Elert Bode. The latter received credit for the law, however, after using it and expounding its theories. The introduction and historical background lay the ground work for the theory behind the Law, which is rather technical and is not suited for the lay reader. The general reader can obtain a simplified explanation from any one of a number of introductory texts. For the university library.

307. Thom, A. **Megalithic Lunar Observatories.** Oxford, Engl., Oxford University Press, 1971. 127p. illus. bibliog. indexes (author and subject). £3.00. $10.25. LC 73-565933. B 71-03143. ISBN 0-19-858132-7.

In this well-researched book, the author describes in great detail how Megalithic man studied the Moon. The follow-up to *Megalithic Sites in Britain* (1967), it will mainly interest the historian of astronomy, but it can be read by the general reader as well. Included are many diagrams showing the observatory sites and the methods of lunar and solar observation. Chapter headings include Introduction; Astronomical Background; Refraction and Parallax; Solstitial Sites; An Analysis of the Observed Declinations; Extrapolation Theory; The Stone Rows and Their Use; The Work at the Observatories; and Conclusion.

TEXTBOOKS

It must be difficult for an instructor to choose a textbook nowadays—the number of such materials has skyrocketed in the recent past. The librarian who must select these books for the collection will also find it impossible to choose and/or afford the great many texts available. It is hoped that the annotations in this section will point the way to several excellent textbooks either for class use or for the library, thus making the job of selection a little less painful. By the time this guide is published, though, it is certain that another dozen or so new texts will have appeared; it is a losing battle.

Most texts listed here are introductory college-level books, for the astronomy or physics major or for the non-science student. Currently, the latter category by far outnumbers the works for the astronomy student. Some advanced texts are also included, as examples, for the upper level undergraduate and beginning graduate student. The emphasis, though, is on the beginning texts.

Like the "gee-whiz" general works mentioned elsewhere in this book, the astronomy text for the non-science student tends to emphasize the "wonder and glory" of the Universe. The books are basically descriptive in nature: there are a multitude of photographs, diagrams, and charts illustrating the planets, stars, Sun, Moon, etc., and there is an obvious lack of mathematics. Further, these books often concentrate on the concepts and theories of astronomy and do not dwell at length on the details and mathematics involved. Historical matter is often an important part of this type book.

Non-technical texts have an important place in the literature of astronomy, since they try to reach those who have no scientific background; they should not be overlooked as part of the astronomy collection. Several of these texts could be used at the senior high school level, and they are suitable for the general reader as well. Many could have easily been included in the General Works section of this guide.

Less frequent are the texts for the science major, the student who is prepared in mathematics and science. These books also have a great many good illustrations, which is important, since they are also introductory volumes. The text in this type of book is less descriptive, as might be expected, and mathematics is highly emphasized, ranging from algebra and trigonometry to elementary calculus. The diagrams are more detailed and illustrate the higher level of material being presented. Problem sets are more difficult and provide an opportunity for the student to test skills at solving various astronomical problems.

Topics glossed over in the non-science texts become more important in this type of work. Stellar evolution, for example, is often mentioned only briefly or eliminated from the more general text, but it is very basic to the text for the astronomy major. In general, the technical texts cover more topics in greater detail and on a higher level. Historical and descriptive matter is often de-emphasized. The more advanced texts show a broad, detailed coverage of particular topics, like celestial navigation, spherical astronomy, statistical astronomy, and astrometry.

Small colleges will mainly wish to collect the more general text for non-science students, along with a few of the more advanced works for science students, while larger colleges and universities will want as many as possible in both categories, according to its needs and budget.

158 / General Materials

Also included here is a handful of materials meant to supplement the basic introductory texts: various sets of laboratory exercises for science and non-science classes. This type of material is just as important as the text, and the instructor frequently chooses such a work to use as a guide in preparing the laboratory work. The librarian, too, will want to acquire a few such aids for the use of students and instructors.

Card catalog subject headings: *Library of Congress:* ASTRONOMY; ASTROPHYSICS; ASTROMETRY; STATISTICAL ASTRONOMY; ASTRONOMY, SPHERICAL AND PRACTICAL. *Sears:* ASTRONOMY; ASTROPHYSICS.

For the Liberal Arts Student

308.   Berman, Louis. **Exploring the Cosmos**. Boston, Little, Brown and Co., 1973. 477p. illus. glossary. index. refs. questions. $12.95. LC 72-11454. ISBN 0-316-09178-2.

Written with the general reader in mind, this book is lavishly illustrated in an appealing format. It covers the standard topics mentioned in similar texts, but it also touches on some related subjects that are currently quite popular. Among these are cosmology, space astronomy, exobiology (life on other worlds), and interstellar spacecraft and communication. Excluded entirely are discussions of coordinate systems and time—two fairly important, but not vital, topics. Their absence does not detract from this text. The author has de-emphasized the mathematical side of astronomy, and what is included along these lines is printed in blue, so that the reader can skip these sections, if desired, without loss of continuity.

The book could also be used for a college liberal arts course (an instructor's manual comes with the book). General review questions accompany each chapter, along with bibliographies of selected reading. A high point of this work is the 16 appendices, which cover a variety of topics from scientific number notation to a glossary.

309.   Brandt, John C., and Stephen P. Maran. **New Horizons in Astronomy**. San Francisco, W. H. Freeman and Co., 1972. 496p. illus. index. (A Series of Books in Astronomy and Astrophysics). $14.00. LC 74-178298. ISBN 0-7167-0338-6.

Although there are many texts for the non-science college student, this is easily one of the better ones available. Beginning with the Earth and a discussion of geological and biological development, the authors prepared the reader for the main theme, that of astronomical evolution. This approach is a welcome one— by relating astronomy to such topics as biology, geology, and chemistry, the authors make the text much more relevant than most. The standard topics follow, with a heavy emphasis on the solar system, and the student will receive a good overview in this section. The last portion of the book explores some of the current topics which are so frequently included in this sort of text. Among the latest discoveries mentioned are pulsars, quasars, etc. The overall approach is non-mathematical—which is sure to be appreciated by the liberal arts student—and the multitude of diagrams and photographs add a final touch to this excellent text.

310. Cole, Franklyn W. **Fundamental Astronomy: Solar System and Beyond**. New York, John Wiley and Sons, Inc., 1974. 476p. illus. index. glossary. bibliog. questions. $16.50; $10.95pa. LC 73-12146. ISBN 0-471-16472-0; 0-471-16473-9pa.

About one-half of this introductory college text is concerned with the solar system and its population, the Earth, Moon, Sun, planets, comets, and meteors. Consequently the book would be especially suitable for secondary and elementary school teachers who wish a basic overview of astronomy, with emphasis on the solar system, a topic usually studied in elementary school science. The book is fairly typical in its arrangement—a brief historical introduction followed by discussions of the solar system, stars, and the Universe. The text is not overly crammed with illustrations, and those that appear are usually good. Each chapter contains review questions, and there are several useful appendices and a glossary. It is neither a gee-whiz text nor a technical introduction— it is simply a straightforward introduction to astronomy for the non-science student or the serious layman.

311. Frederick, Laurence W., and Robert H. Baker. **An Introduction to Astronomy**. 8th ed. New York, D. Van Nostrand Co., 1974. 453p. illus. index. refs. glossary. questions. $10.95pa. LC 73-18624. ISBN 0-442-22436-2.

One of the best texts for the liberal arts student, this standard volume has been used by uncountable numbers of students since the first edition. Emphasizing the observational aspects, the authors present astronomy as it should be—as an exciting, always-changing science with its roots in antiquity. Like any good textbook for the non-science student, this one stresses the descriptive aspects of the subject. Heavily illustrated, the book takes the reader on an interesting, educational tour of the Universe, from the Earth out to the edge of infinity. The subject coverage is not unusual; the standard topics of telescopes, Moon, Sun, planets, stars, comets, galaxies, cosmology, etc., are listed in a logical fashion. The descriptions and explanations are better than most, however, accounting for this volume's longevity and popularity. Like other recent books, this includes lively discussions of astronomy's latest discoveries: pulsars, black holes, and so on. Added features are narrow columns with wide margins for notes, numerous tables and graphs of numerical data, and excellent review questions. Highly recommended to the astronomy instructor and college library.

312. Gainer, Michael K. **Astronomy: Observational Activities and Experiments**. Boston, Allyn and Bacon, Inc., 1974. 131p. illus. bibliog. $6.25. LC 74-4094.

Most books of astronomy laboratory exercises are aimed at the science student who will most likely major in astronomy or physics. Happily, this book is an exception to the rule and is intended for the non-science student. Many "non-science" college courses in astronomy do not have labs, but some do, and the instructor should consider this excellent effort for class use.

About half the "experiments" do not require observation and could be done at home. These cover valuable topics like the celestial sphere, evaluation of experimental data, using *The American Ephemeris and Nautical Almanac*, using celestial photographs, and so on. The remainder of the 22 exercises involve observation, and a few require the use of a 35-mm camera. Each exercise includes detailed procedure and a section guiding the student in the analysis of results. The public library might also wish to purchase this excellent work.

313.   Hodge, Paul W. **Concepts of Contemporary Astronomy**. New York, McGraw-Hill Book Company, 1974. 547p. illus. index. bibliog. problems. questions. observations. $10.95pa. LC 73-20283. ISBN 0-07-029125-X. Instructor's Manual: $2.95.

The arrangement of this excellent text is atypical—the author has selected four major topics and has expanded around them. Aimed at the college liberal arts student, the book examines 1) the astronomer, his research, his tools, and his goals; 2) how the solar system began; 3) stellar evolution; and 4) the structure of the Universe. As the title indicates, the emphasis is on concepts and ideas that are the basis of astronomy. Instead of long, drawn-out discussions of any one particular topic, there are brief, clearly written discussions of many different topics, presented in a manner that should hold the student's attention even through long reading assignments. Illustrations are numerous and quite good; the format is similar to that of a popular magazine. Also included are discussion-type questions for classroom work, experiments for home or lab, references for further reading, and lists of visual aids. Designed for a one-semester or one-quarter course.

314.   Hynek, J. Allen, and Necia H. Apfel. **Astronomy One**. rev. ed. Menlo Park, Calif., W. A. Benjamin, Inc., 1972. 402p. illus. index. $11.95. LC 75-186625. ISBN 0-8053-4749-6.

Another of the many introductory texts for the non-science college student, this book is intended especially for one-semester survey courses in community colleges. Though it is not particularly outstanding overall, it is more than adequate in its lively descriptive explanations of astronomy. Its arrangement consists of four parts: The Architecture of the Universe; Stars; The Solar System; The Universe of Galaxies. The first section provides an overall picture of the Universe, preparing the student for the remainder of the material; less useful texts attack topics one at a time and reserve the overall picture for the end of the book. The first section of *Astronomy One* reviews the important astronomical discoveries, in chronological order, illustrating the problems faced by astronomers, and showing what text lies ahead. The volume is well illustrated; only the simplest math (algebra and geometry) is used.

315.   Inglis, Stuart J. **Planets, Stars, and Galaxies; An Introduction to Astronomy**. 3rd ed. New York, John Wiley and Sons, Inc., 1972. 498p. illus. index. bibliog. questions. problems. $13.25. LC 79-169164. ISBN 0-471-42740-3.

The most refreshing thing about this introductory text is its non-standard arrangement of topics. After a brief lead-in chapter called "Man and the Heavens," the author talks about light, telescopes, and the atom before turning to a discussion of the physical characteristics of the planets. Particularly welcome is the chapter on the age and origin of the solar system, a topic usually neglected or glossed over in beginning texts. The remainder of this volume for the liberal arts student covers the standard topics in a non-mathematical mode—that is, more description than equation. High points of the book are the good illustrations, thoughtful questions, and excellent bibliographies. Suited for a one-semester course.

316.   Jastrow, Robert, and Malcolm H. Thompson. **Astronomy: Fundamentals and Frontiers.** 2nd ed. New York, John Wiley & Sons, 1974. 519p. illus. index. questions. $14.50. LC 78-174770. ISBN 0-471-44078-7. Study Guide: $4.95. ISBN 0-471-10170-2.

In order to present astronomy to the non-science college student in an attractive, interesting manner the authors have included a multitude of beautiful illustrations and photographs of the universe, and have avoided, as much as possible, complicated equations and technical terminology. Their approach succeeds quite well. A welcome change from similar works is that the authors deviate from the usual order of subject presentation, discussing the solar system after stars and galaxies.

In general, this is a good text, but it lacks important supplementary material, such as appendices, a glossary, and bibliographies. Why was no effort expended along these lines? The book's strongest points are its superb illustrations and the chapters on stellar evolution.

An unusual addition to the textual matter is Chapter 17, which explores the possibility of "Life in the Cosmos." While interesting, this chapter is far too brief, and its value here is questionable.

A *Student Study Guide* by Michel Breger and William Jeffreys was published in 1975 and is intended to be used along with the text. It is quite good and will give the student aid in both learning the material and preparing for examinations. It includes suggestions for studying the material in each chapter and short tests (with answers). The instructor choosing this text will want to consider this guide as a supplement to course work.

317.   King, Ivan R. **The Universe Unfolding.** San Francisco, W. H. Freeman and Co., 1976. 504p. illus. index. glossary. questions. $14.95. LC 75-33369. ISBN 0-7167-0521-4.

One of the better new texts for the non-science college student and layman, this volume is extremely well written. The presentation is clear and non-mathematical (as far as possible), and it makes downright enjoyable reading. In the deluge of elementary textbooks, both good and bad, it is a rare and pleasant situation to find one that reads so well and is interesting! King skillfully combines history and basic astronomical theory with the latest discoveries in a book that it is hoped will become a standard source. Though the topics covered are fairly typical, the arrangement is not; but it is not at all illogical. The author first presents telescopes, then a look at the home planet; this is followed by the important section on time and coordinate systems. These three chapters prepare the reader for the rest of the volume by providing a good overview. Remaining chapters cover the Moon (incorporating manned exploration), orbital motion, the Copernican revolution, the solar system, life in the Universe (currently a hot topic), stars of all types, starlight and how we interpret it, the Sun, interstellar gas and dust, the Milky Way, galaxies, radio galaxies and quasars, and cosmology. The book is not overly detailed, and some supplementary reading is suggested. The volume is an excellent introduction and should be seriously considered by the instructor. Appropriate for any library with astronomy materials, it is also very well illustrated.

318.   Menzel, Donald H., Fred L. Whipple, and Gerard de Vaucouleurs. **Survey of the Universe.** Englewood Cliffs, N.J., Prentice-Hall, 1970. 860p. indexes (subject and name). illus. bibliog. questions. $16.95. LC 79-149820. ISBN 0-13-879163-5.

This extensive introduction to astronomy can be tackled by anyone who has a secondary school education and modest training in mathematics—specifically, algebra and geometry. It is fairly comprehensive, covering both historical and current topics in its 35 chapters. It is well illustrated with many black-and-white photographs and diagrams, but its main purpose does not appear to be impressing the reader with pictures of the Universe. Rather, the authors successfully attempt to give the uninitiated reader a thorough look at astronomy's various aspects, while teaching the basic concepts.

The book contains a good bibliography and an extensive set of textbook questions if the reader wishes to test his learning. It could be used as a text, but it is really not suited for this purpose. It is too long for a one-semester non-science course, but not technical enough for an astronomy course for science majors.

319.   Oster, Ludwig. **Modern Astronomy**. San Francisco, Holden-Day, Inc., 1973. 448p. illus. index. glossary. refs. questions. experiments. exercises. $14.95. LC 72-83247. ISBN 0-8162-6523-2.

The level of this general college text falls somewhere between the liberal arts course and the course for science students, but it could be used by the former group. Like most similar works, this volume gives a general overview of as many topics as possible in order to introduce the student to the broad spectrum of astronomical subjects. The one or two-semester text is divided into three major parts, the solar system, the stars, and the star systems. Although overall it is not a particularly unique work, it does have some good points worth mentioning. The questions and exercises are quite good—they make the student think, rather than just regurgitate facts. Selected words within the text itself are printed in boldface type to emphasize important points for the student. The thorough chapters on stars discuss stellar evolution to a greater extent than do most non-science texts. The brief section on the history of astronomy could be longer, though, especially since the text seems to be aimed at non-science students.

The nine excellent appendices include tabular data as well as textual information on astronomical instruments, celestial objects, a brief look at basic physics, simple mathematics, and other topics. In short, this fine volume is worth considering.

320.   Pananides, Nicholas A. **Introductory Astronomy**. Reading, Mass., Addison-Wesley Publishing Co., 1973. 344p. illus. index. glossary. review questions. $12.95. LC 72-1942. ISBN 0-201-05675-5.

There is really nothing unique about this text for non-science college students. The subject material is standard, logically arranged, and fairly comprehensive. There is an abundance of illustrations, another characteristic of the text for the non-science course, and each chapter has review questions which usually require short answers. One suspects that these questions were formulated using lecture notes, because the questions (and answers) sound like an outline for a lecture. The high point of the book is the inclusion of Johnny Hart's "B. C." comic strips, which are used by the author at strategic points in the text. The numerous appendices contain tabular data on stars and other celestial objects, as well as other useful information.

321.   Payne-Gaposchkin, Cecilia, and Katherine Haramundanis. **Introduction to Astronomy**. 2nd ed. Englewood Cliffs, N.J., Prentice-Hall, 1970. 610p. index. illus. bibliog. $15.50. LC 70-95752. ISBN 0-13-478107-4.

An introductory text for students or general readers with some basic mathematics and physics background, this book could be used by a liberal arts student fulfilling a science requirement. The level of the material falls somewhere between the science and non-science text, though, so it is difficult to decide which group it would best suit. It is intended for a full year course, but could be adapted easily for a shorter period—the authors have marked chapters and paragraphs that can be skimmed if necessary. One minor drawback to this well-illustrated volume is that there are no questions or exercises. The authors have briefly annotated their bibliographies, something which, unfortunately, is rarely done.

322.   Swihart, Thomas L. **Astrophysics and Stellar Astronomy**. New York, John Wiley and Sons, 1969, c1968. 299p. illus. index. $11.50. LC 73-603. ISBN 0-471-83990-6.

This text does not fall cleanly into the two general astronomy text categories for science or for non-science students. It is aimed at the former, to be sure, but on the sophomore or junior level. It is definitely not a survey text in any case—it concentrates on astrophysics, stars, and galaxies and cosmology. The student should have had sophomore physics and calculus before attempting this volume, which appears to be a follow-up to any general astronomy text aimed at the science major. Previous astronomical knowledge is not required, though.

The problems are generally non-trivial exercises that will keep students busy. Happily, the author has seen fit to work out the questions at the end of the book, which is rarely done in any text. Each section has an excellent, extensive bibliography, with many references to technical journals and books. A good text for serious students.

323. Woods, John A., and Duncan R. Hazard, eds. **The Science of Astronomy.** New York, Harper and Row, 1974. 466p. illus. glossary. index. $9.50pa. LC 73-10684. ISBN 0-06-041446-4.

A joint effort by several authors, this up-to-date text is designed for an introductory, one-semester college course for non-science students. It is almost purely descriptive, requiring little or no mathematical background. The book includes descriptions of most of the latest developments and discoveries in the field, so it will hold the student's attention. Other areas include the standard topics: history, the solar system, the Moon, the stars, the Sun, the Milky Way, stellar evolution, cosmology, etc.

There are good questions with each chapter, good illustrations, and supplemental material (e.g., star charts)—in short, the material expected in such a text. It is a good text, but, compared to other similar efforts, it has little that is unique.

For the Science Student

324. Abell, George O. **Exploration of the Universe.** 3rd ed. New York, Holt, Rinehart, and Winston, 1975 738p. illus. index. bibliog. glossary. exercises. $15.00. LC 74-20790. ISBN 0-03-089665-7.

One of my college professors once told an anecdote about which two books he would take to the hereafter if it were possible. In my case, one of the two books would be this text. A complete revision of the second edition (1969), this version is only slightly larger. The major differences are the deletion of outdated material, the rearrangement of the retained information, and the addition of totally new material. In particular, the new information includes data on the planets gained by spacecraft in recent years, an update of the chapter on comets, the latest theories on quasars, pulsars, and neutron stars, current ideas on the interstellar medium, and more. New chapters or sections are astrology, the atomic nucleus, relativity, and life in the Universe.

The basic format has been retained in this printing—a double-column page, with most of the old, and many new, illustrations. More exercises have been added to many of the chapters, and some problems include answers. The more advanced material is printed in smaller type so that it can be skipped if necessary. On the whole, though, the book is aimed at the average college student, taking a one-year or even a one-semester course. The appendices, as usual, are excellent and numerous. The strong point of this text, which makes it stand out among the flood of such books is the clear, well-written material, which is both logical and interesting. This latest edition is much better than the previous volumes, both of which were superb.

325. Abell, George. **Exploration of the Universe: Updated Brief Version.** New York, Holt, Rinehart and Winston, Inc., 1973. 483p. illus. index. bibliog. glossary. exercises. LC 73-173505. ISBN 0-03-075950-1.

This short version of Prof. Abell's standard text is intended for a one-semester astronomy course for liberal arts students. Among its strong points are excellent illustrations, easy-to-understand text, useful appendices, and its chapters on the history of astronomy. Mathematics are de-emphasized, which is important for the non-science student, and diagrams and graphs are instead used freely to illustrate theory and fact. Chapters include exercises that the instructor may find useful for both homework and in-class assignments. The book is arranged like most beginning texts, except that the historical matter precedes the usual material on the planets, stars, and Universe. Highly recommended.

326. Frederick, Laurence W., and Robert H. Baker. **Astronomy**. 10th ed. New York, D. Van Nostrand Co., 1976. 559p. illus. refs. glossary. indexes (subject and name). problems. $16.95. LC 75-31297. ISBN 0-442-22444-3.

First published in 1930, this standard text has undoubtedly been used by more students and instructors than any other similar work. An introductory college textbook, it is primarily aimed at the scientifically oriented student who may or may not be planning to major in astronomy. A not unusual arrangement of topics is employed, starting at the Earth and moving out into the Universe, touching on the Sun, planets, Moon, stars, galaxies, and so on. Features of note are the wide margins for note-taking, excellent and numerous illustrations, review questions, topical bibliographies, and comprehensive appendices. The book is not as mathematics-oriented as Motz and Duveen's *Essentials of Astronomy* (1971, c1966), but it is advanced enough for any science student. The fact that there are not as many equations and numbers as in Motz and Duveen makes it more appealing to a wider audience. This latest edition, like the previous versions, was published to update "old" information in the predecessors. Changes of note in the 10th edition include new "photographs never before reproduced," a list of 34 interstellar molecules, new material on the origin of meteorites, and more. Highly recommended for university classes, it is comparable to Abell's *Exploration of the Universe* in completeness of information and clarity of text.

327. Harwit, Martin. **Astrophysical Concepts**. New York, John Wiley and Sons, Inc., 1973. 561p. illus. index. refs. $17.95. LC 73-3135. ISBN 0-471-35820-7.

Rather than including detailed discussions of astronomical bodies, the author of this book emphasizes the concepts of astrophysics; the stars, galaxies, and other celestial objects are mentioned when they pertain to physical principles discussed in the text. Numerous topics are covered in this comprehensive and well-written advanced text for college seniors and beginning graduate students. The first chapter takes an overall look at astrophysics, a non-mathematical approach to just where and how it fits into the scheme of the Universe. After that, the book gets down to business, and the author proceeds to discuss the cosmic distance scale, dynamics and masses of astronomical bodies, random processes, photons and fast particles, electromagnetic processes in space, quantum processes in astrophysics, stars, cosmic dust and gas, and the structure of the Universe. Problems are scattered throughout the text, at appropriate points within the subject matter. Up to now there have been no other comparable texts on this level for the college student. For this reason, this book should be part of the college astronomy library.

328. Kelsey, Linda, Darrel Hoff, and John Neff. **Astronomy: Activities and Experiments.** Dubuque, Ia., Kendall/Hunt Publishing Co., 1974. 183pp. illus. bibliog. $6.95pa.

The authors have expanded their excellent first volume of 30 exercises to 40 suitable for use in an introductory college astronomy lab course. Aimed at the science student, the book contains a variety of problems that cover most of the basics of astronomy. Of particular note are the exercises on astronomical coordinate systems, determining the Earth's mass, plotting the Moon's orbit, an eclipsing binary star demonstration, determining lunar mountain heights, and four exercises involving the Pleiades star cluster. Each problem includes a statement of purpose, references, introduction, procedure, and discussion questions. The material is clearly presented, well planned, and logically organized for class use. Supplementing the text is an excellent set of appendices containing suggestions for writing lab reports, suggestions for special projects, a bibliography, and tables. The astronomy instructor should consider this work as a possible companion to the introductory text and as a guide to preparing laboratory sessions.

329. Minnaert, M. G. J. **Practical Work in Elementary Astronomy.** Dordrecht-Holland, D. Reidel Publishing Co.; distr. New York, Springer-Verlag, 1969. 247p. illus. refs. subject index. $9.50. LC 71-76256. ISBN 90-277-0133-4 (Reidel); 0-387-91027-1 (S-V).

Practical work for astronomy undergraduates is often de-emphasized because of lack of time or equipment, and the lab work that *is* done is frequently trivial or abbreviated. While practical work can be by-passed in liberal arts astronomy courses, it should be an important part of beginning astronomy courses for science students. This excellent volume should be considered by the instructor in such a course as a guide for good, worthwhile lab exercises.

The problems in this book are of two types: actual observation and laboratory work. Observational exercises require small telescopes and one larger instrument, while the laboratory work needs long tables, preferably in the library where the proper astronomical reference tools are at hand. Although the topics covered span a wide variety of areas, the instructor should not attempt to use all or even most of the problems. Certain exercises that are applicable to the course outline can be chosen at his discretion. Each problem contains an explanation and set of instructions as well as references. This fine book includes a total of 40 problems in the following areas: space and time, instruments, motions of celestial bodies, planets and satellites, the Sun, and the stars.

330. Motz, Lloyd, and Anneta Duveen. **Essentials of Astronomy.** New York, Columbia University Press, 1971, c1966. 711p. illus. indexes (subject and name). bibliog. $15.00. NUC 73-3644. LC 65-10031. ISBN 0-231-03632-9.

Designed for astronomy students and other college science majors, this fine text should be seriously considered by the instructor. The arrangement of the subject material in this full-year text is not unusual. The four major sections cover the solar system, stellar properties and the structure of the stars, stellar systems and the structure of the Milky Way, and galaxies and cosmology. Profusely illustrated, the text relies heavily on discussions of physical principles, resulting in an excellent treatment of the relationships between physics and astronomy. The authors emphasize mathematics (trigonometry and logarithms) in their discussions, but this is to be expected in a text for science students. While not stressing the descriptive aspects of astronomy, the authors do not ignore them entirely, either.

The exercises are good, involving both knowledge of concepts and manipulation of formulae. The division of chapters into many subtopics (some more advanced than others) allows the instructor to include as much or as little of each area as time allows. The two appendices on optical astronomical instruments and radio astronomy are excellent (especially the former, which goes into detail on light and lenses).

331. Smith, Elske V. P., and Kenneth C. Jacobs. **Introductory Astronomy and Astrophysics**. Philadelphia, W. B. Saunders Co., 1973. 564p. illus. bibliog. index. problems. $15.95. LC 72-88853. ISBN 0-7216-8387-8.

An astronomy text for the "serious scientific student," the undergraduate (sophomore or junior) with some trigonometry, calculus, and physics preparation, this volume is structured for a one-year course but can be adapted for a one-semester survey. The usual arrangement of material is followed—the Earth first, then the solar system, stars, galaxies, and the Universe. Physics are appropriately heavily emphasized, illustrating to the student the intimate relationship between astronomy and physics. The chapter on celestial mechanics, an important topic for the astronomy major, is developed well, beginning with historical background, followed by discussions of Newton's and Kepler's laws. The many problems that accompany each chapter are good exercises which teach the student to apply the many laws and formulae presented. Like most good texts, this work has extensive reading lists and useful appendices, including one on mathematics (a brief review of trigonometry and calculus).

332. Wyatt, Stanley P. **Principles of Astronomy**. 2nd ed. Boston, Allyn & Bacon, Inc., 1971. 686p. illus. index. bibliog. problems. $14.95. LC 70-131204.

Written for college students majoring in the sciences, this introductory text presents a comprehensive view of contemporary astronomy. The subject arrangement is standard: Earth, planets, stars, galaxies, and the Universe, in that order. Like most current texts, this volume is lavishly illustrated, and many photographs are in color. Along the same line, the author frequently uses diagrams to describe the physical state of the Universe, an approach that succeeds quite well. Each chapter includes problems, many of which require trigonometry or logarithms. These exercises are very helpful, and they illustrate well the theories presented in each area. The good bibliography contains references to many of the classic works, as well as to contemporary books. Suitable for a two-semester course, this second edition reflects discoveries made in recent years and incorporates the metric system of units and measurements.

168 / General Materials

333.   Wyatt, Stanley P., and James B. Kaler. **Principles of Astronomy: A Short Version.** Boston, Allyn & Bacon, Inc., 1974. 487p. illus. index. bibliog. problems. $10.95pa. LC 73-84910.

Basically the same text as the longer version, this book is designed to be used in a one-semester, science-oriented, college-level astronomy course. A page-by-page comparison of the two editions shows that certain illustrations and secondary topics have been eliminated to produce the compact version. The most notable omissions are the entire chapters on "Atoms, Radiation, and Spectra" and "Star Clusters and Associations." The omissions, though, do not hinder the book's usefulness, and the volume should seriously be considered by the college instructor.

Advanced or Supplementary Material

334.   Atanasijević, I. **Selected Exercises in Galactic Astronomy.** Dordrecht-Holland, D. Reidel Publishing Co.; distr. New York, Springer-Verlag, 1971. 144p. illus. refs. indexes (subject and name). (Astrophysics and Space Science Library, v.26). $12.80. LC 73-159652. ISBN 90-277-0198-9 (Reidel); 0-387-91087-5 (Springer).

Most volumes on practical work in astronomy are aimed at the beginning astronomy student and cover problems running the gamut of topics. This excellent book, however, is a set of eight not-so-brief exercises for the advanced undergraduate or beginning graduate student, and its subject coverage is limited to the Milky Way. "Limited" here, however, does not mean that all the problems look at the same area; quite the contrary. The student begins by determining the position of the galactic equator, and then analyzes the distribution of globular clusters on the sky. The following problem deals with determining the distance of an open star cluster and its speed with respect to the Sun. Noncluster star fields are the subject of problems 4 and 5. Galactic rotation is the theme of the next two sections, and the orbit of a star in the Milky Way is considered in the last exercise. All but one of the problems can be solved using a hand calculator; the other requires the use of a computer. An important addition to the astronomical teaching literature.

335.   McNally, D. **Positional Astronomy.** New York, John Wiley and Sons, 1974. 375p. illus. index. (A Halstead Press Book). $15.95. LC 74-4819. ISBN 0-470-58980-9.

Written for first-year college astronomy students, this recent text explains the fundamental concepts behind astrometry, the study of the angular separations of celestial bodies on the sky. Similar books on a more advanced level already exist (Woolard and Clemence, *Spherical Astronomy*; Van de Kamp, *Principles of Astrometry*), but up until now there was little for the beginner. The arrangement of topics is standard: definitions and terminology are laid out first, followed by discussions of the various astronomical coordinate systems; from Horizon to Galactic and several in between. Time and its place in astrometry are next considered, and the author then explores various aspects of positional astronomy such as correction factors (parallax, refraction, aberration, precession, nutation, proper motion), eclipses and occultations, orbital motion, and more. Mathematics is a very important part of astrometry, and equations abound here. Fortunately for the reader, this is restricted to trigonometry, algebra, and geometry, nothing unusual for a high school graduate. Further, important equations throughout are summarized in the appendices.

The book is well written and concepts are clearly presented, but it is doubtful that first-year astronomy students are its major audience. It would probably be more suitable for second-year students who have had one year of general astronomy. While astrometry may be basic, the in-depth treatment presented in this text is probably a bit too advanced for the beginning student. Finally, the fact that there are no problems of examples is distressing; it would have been helpful if the abstract equations had been related to real celestial situations.

336.   Mueller, Ivan I. **Spherical and Practical Astronomy As Applied to Geodesy.** New York, Frederick Ungar Publishing Co., Inc., 1969. 615p. illus. refs. index. $18.50. LC 68-31453. ISBN 0-8044-4667-9.

This voluminous book has much to offer to the student and specialist interested in determining the exact locations of terrestrial points from measurements on natural and celestial bodies. First of all, it is quite comprehensive— it covers a variety of topics of interest to geodesists, thus making a good handbook for those interested. A brief overview of the subject matter is appropriate. After covering in detail, with lots of equations, the basics like the celestial sphere and its coordinate systems, variations in the celestial coordinates, and time systems, the author covers several subjects that are not always covered adequately in other sources. Especially good is the section on star catalogs, which defines, lists, compares, and relates most of the major works, new and old. The chapters on equipment and timekeeping are excellent, too, the latter describing the various time services like stations WWV and WWVH. Also of interest is the chapter on eclipses and occultations, in which prediction of these events is described. Finally, each chapter contains an excellent list of references.

337.   Podobed, V. V. **Fundamental Astrometry: Determination of Stellar Coordinates.** Chicago, University of Chicago Press, 1965. 236p. illus. indexes (subject and name). $11.00. LC 64-15810. ISBN 0-226-67149-6.

While this volume may be "fundamental," it is not elementary; it should be considered only by the graduate student or astronomer who plans to study or specialize in determining stellar coordinates. The emphasis is on the equipment used in astrometry, as well as the importance of precision measurements with that equipment. Like many Russian texts, this one presents the material in a dry, straightforward fashion. A listing of the topics covered will give the reader a good indication of the work's coverage: fundamental astrometry, instruments used, the horizontal axis, the instrument tube, the graduated circle, the eyepiece micrometer, relative or differential determination of coordinates, determination of right ascensions and declinations, photographic method for determining relative coordinates and proper motions, and catalogs of positions and proper motions of stars. Math is used heavily in this volume, which is sure to be of use to the serious student of spherical astronomy. This English edition was edited by A. N. Vyssotsky and translated by Scripta Technica, Inc.

338. Trumpler, Robert J., and Harold F. Weaver. **Statistical Astronomy.** Berkeley, University of California Press, 1953. 644p. illus. indexes (subject and name). bibliog. LC 53-7419.

The applications of statistics to astronomy cover a multitude of areas, but in particular they occur in the study of the structure, constitution, and dynamics of the Milky Way. This classic book devotes most of its text to these problems related to our galaxy by introducing the student to the principal statistical problems in astronomy, their mathematical formulation, and methods of solution. Further, considerable attention has been paid to numerical methods, since theory alone cannot prepare the student for actual calculations. Among problems considered in this standard work are stellar distribution (spectral and space), stellar motions near the sun, and galactic rotation. This very technical work is old but not outdated; it is still very useful.

339. van de Kamp, Peter. **Principles of Astrometry; With Special Emphasis on Long Focus Photographic Astrometry.** San Francisco, W. H. Freeman, 1967. 227p. illus. index. bibliog. notes. (A Series of Books in Mathematics). $7.00. LC 66-22077. ISBN 0-7167-0318-1.

The author presents in this book a good summary of the basics of astrometry ("the space-time relations of stellar positions on the celestial sphere"), and, in particular, the concepts and methods of long-focus astrometry. In his explanations of the study of stellar position, van de Kamp has divided his book into three major sections: I, Spherical Astrometry (includes spherical trigonometry, the celestial sphere, stellar positions, and proper motions); II, Plane Astrometry (includes relation between plane and sphere, long focus photographic astrometry, stellar paths, parallax, orbital motion, etc.); III, Analysis of Observations (includes theory of errors and least squares method). For the reader with an elementary background in solid geometry, trigonometry, calculus, and astronomy, this work is recommended for college and public libraries.

340. Voigt, Hans-Heinrich. **Outline of Astronomy.** Leyden, The Netherlands, Noordhoff International Publishing, 1974. 2v. 556p. illus. index. refs. Dfl. 60.00 (about $26.00). LC 72-97241. ISBN 90-01-71270-3. (Available from Academic Book Services. Holland, P.O. Box 66, Groingen, The Netherlands.)

Intended for upper-level undergraduates and graduate astronomy students, this excellent book is not a textbook, but rather a summary of knowledge already learned. It is based on the author's prepared lecture notes, which have been revised and expanded into a comprehensive work suitable for the student (and astronomer) who wishes to review certain points of astronomical theory. The book would be ideal as an extra aid in studying for graduate qualifying exams. Definitions, equations and their derivations, summaries and explanations of dozens of topics abound in this concisely written, well-organized manual. The book is not a collection of astronomical data, although many numbers appear, but rather a concise outline of astronomical concepts and methods. Readers seeking numerical data should consult Allen's *Astrophysical Quantities* or Robinson's *Astronomical Data Book*.

Approximately half the first volume is devoted to spherical astronomy, the solar system, electromagnetic radiation, etc., and the remainder deals with solar and stellar atmospheres. Volume two is concerned with completing the discussion of stars (structure, evolution, variables, binaries, etc.) and the Milky Way and other galaxies. As implied earlier in the statement about the intended audience, the user of this work should be familiar wtih astronomy on an intermediate or advanced level and should know at least calculus to best appreciate and use this volume. Keep in mind that this book is not intended to teach astronomy; it is meant only as a review. It would be a good choice for the university and observatory reference collection. This superb book was translated (*Abriss der Astronomie*) by L. Plaut and Nancy Houk.

341.   Woolard, Edgar W., and Gerald M. Clemence. **Spherical Astronomy**. New York, Academic Press, 1966. 453p. illus. index. $28.50. LC 65-26416. ISBN 0-12-762750-2.

One of the basic, but often neglected, subjects of astronomy is the study of the apparent positions and motions of celestial bodies on the sky. Spherical astronomy is usually introduced to the beginning astronomy student during discussion of celestial coordinates and time. After this modest beginning, however, the topic is seldom mentioned throughout the student's career as an undergraduate and graduate. This text solves that problem by exploring spherical astronomy in depth. Suitable for an upper level undergraduate course, this book develops in detail the principles used to determine the apparent positions of objects on the celestial sphere; this process is known as astrometry, the essence of this volume.

Prepared by astronomers at the U.S. Naval Observatory, the book includes recent developments in spherical astronomy, as well as basic material like the celestial sphere and astronomical coordinate systems. The authors go on to discuss parallax, refraction, and aberration, followed by reference systems and their variation. Other topics are precession and mutation, the ecliptic and equatorial systems, time systems, star catalogs, star systems, and some mention of the historical development of the terminology of spherical astronomy. The student should be familiar with trigonometry and differential calculus before attempting this volume, one of the few available on this topic.

342.   Wright, Frances W. **Celestial Navigation**. Cambridge, Md., Cornell Maritime Press, Inc., 1969. 137p. illus. index. refs. $6.00. LC 70-76202. ISBN 0-87033-000-4.

The college student and amateur astronomer wishing to learn how to navigate by the stars should consider this fine volume. The material is not overly

172 / General Materials

technical, and the author proceeds slowly in introducing the basics of this fascinating subject. A chapter called "descriptive astronomy" describes for the uninitiated reader the basic background material on the celestial sphere, explaining and defining terms used frequently in navigation. This good beginning section includes several well-planned diagrams that illustrate latitude, longitude, zenith, horizon, celestial equator, etc. The second chapter appropriately teaches the reader how to find his or her north (or south) celestial poles, which is the very first step in learning to navigate. From that point, the author moves to the sextant, going into great detail on its use. This is followed by a practice "cruise" on a mythical ship in which the reader puts to use the skills learned previously. The use of tables (in almanacs) and the determinate of various times are an important part of this useful section. Techniques for the determination of latitude at solar meridian passage are stressed. Students who master this text will be on their way to becoming competent celestial navigators.

GENERAL PERIODICALS

Amateur astronomers and general readers have several journal titles to choose from, but it should be remembered that the amateur's main source of information is usually books and other monographic materials. Nevertheless, the backyard observer needs and wants to keep up with current events in the field, and one periodical in particular suits those needs—*Sky and Telescope*. The most popular and important magazine available in astronomy, it has articles on all sorts of subjects for all levels of amateurs (and for many professional astronomers as well).

Astronomy, a new journal that is also mainly for the amateur astronomer and general reader, has become quite popular, too. Its emphasis is on color photography and feature articles. Several other lesser-known but important titles are also listed below.

The public library, depending on its size, will want a few amateur or "general" astronomy journals, as well as a handful of the most important technical publications. The size, education, and interest of the clientele must be considered, as well as the overall budget. At the very least, the public library will, or should, have *Sky and Telescope* and *Astronomy*. School libraries (and most observatory libraries) will also want to stock one or the other or both.

Entries here follow the same format as those items listed under Professional Journals in Chapter 1 including subscription information and abstract.

343.   **Astronomy: The World's Most Beautiful Astronomy Magazine.** Milwaukee: AstroMedia Corp. v.1– , 1973– . Monthly. illus. LC 73-645009. ISSN 0091-6358.

Color, color, and more color! The subtitle of this recent addition to the astronomical serial literature is by no means an overstatement. It is indeed a superbly illustrated journal, with dozens of full-color photographs and paintings depicting the beauty of the Universe. The illustrations are of such high quality that most are suitable for framing. Aimed at both amateur astronomers and general readers, this periodical's articles consist of non-technical features on current topics like space exploration, creation of the Universe, cosmology, etc. There are many pieces to aid the beginning or infrequent backyard observer, complete with color star-charts and diagrams and photos. They are occasionally too elementary for the advanced amateur, but they are not intended for that audience.

The regular departments in each issue include Astro-News (the latest from the world of astronomical research), Astro-Mart (free classified advertisement), Star Dome (the monthly sky with star maps and special events), Astronomy Reviews (excellent book critiques), and Planet Finder (a two-page color chart showing where each wandering member of the solar system can be found). Admittedly, the gee-whiz approach is taken here, but this is a refreshing and exciting alternative to the usual drabness of astronomical reporting. Astronomy, after all, is amenable to such an approach, unlike some of the other sciences. This journal should be held by any library that can afford it (and that should cover 99 percent of the libraries); this fine publication is a wonder to behold.

Write: ASTRONOMY, New Subscriber Services, P.O. Box 186, Westchester, Ill., 60153. Cost: $12.00/year.

344. **Griffith Observer.** Los Angeles: Griffith Observatory. v.1– , 1937– . Monthly. illus. LC 42-12723.

Articles and essays of general astronomical interest are included in this 24-page publication intended for layman, amateur, student, and astronomer. The format resembles a newsletter on glossy paper, but the content is of high quality. Astronomers and other scientists are the primary authors of articles on history, research, observations, and speculation. The journal is not at all technical, so would be suitable for all types of libraries, from high school to observatory. It would be an excellent purchase for the doctor's office waiting room.

Black-and-white photographs abound, many from amateur astronomers; all of the photographs are quite good; there is even a two-page centerfold of a heavenly body or star field. Besides regular features such as a sky calendar and simple star charts, there are anecdotes, poetry (!), and articles for the active amateur astronomer. Finally, there is no advertising; financial support appears to come solely from subscriptions, which are inexpensive and well worth the price.

Write: Griffith Observatory, Box 277, Los Angeles, California, 90027. Cost: $5.00/year; $9.00/2 years.

345. **Irish Astronomical Journal.** N. Ireland: Armagh Observatory. v.1– , 1950– . Quarterly. illus. refs. summaries. vol. indexes. LC 54-37143. ISSN 0021-1052.

Both general readers and astronomers will find articles of interest in this publication, which is issued under the auspices of the Armagh and Dunsink Observatories. It is not, however, an observatory publication and is not available on exchange. Topics include historical considerations and current research. The articles for the specialist are only somewhat technical and can probably be read by advanced amateurs. Besides articles on astronomical subjects, there are regular features such as editorials, book reviews, obituaries, and news. Though it is not a primary source, most observatory libraries will nevertheless wish to subscribe.

Write: Managing Editor, Irish Astronomical Journal, Armagh Observatory, Northern Ireland. Cost: $5.00 (£2.00) for individuals; $7.50 (£3.00) for institutions.

346. **Journal for the History of Astronomy**. Cambridge, Engl.: Science History Publications, Ltd. v.1– , 1970– . 3 issues/year. illus. refs. volume indexes. LC 73-618135. ISSN 0021-8286.

Unique among scientific journals, this is the only publication to deal entirely with the history of a science field. Written for the historian and interested astronomer, the articles are scholarly essays, heavily footnoted and thoroughly researched. Biographies of astronomers, historical notes on observatories, overviews of the development of astronomical theory, and historical analysis are just a few of the areas that appear in this periodical. Besides the standard papers, there are both essay reviews and book reviews, the former being longer, more in-depth versions of the latter. Further, there is a section called "Notices of Books," a selected, annotated list of new publications of interest to science and astronomy historians. Other occasional features include lists of dissertations on the history of astronomy, descriptions of manuscript collections at various observatories, and historical notes (brief papers). The text is mostly in English, but some articles appear in French or German.

Write: Science History Publications, 156– Fifth Avenue, Room 502. New York, New York, 10010. In Europe, write: Science History Publications, Ltd., Churchill College, Cambridge CB3 0DS, England. Cost: $37.17/year.

347. **Mercury**. San Francisco: Astronomical Society of the Pacific. v.1– , 1972– . Bi-monthly (odd months). illus. notes. refs. volume index. LC 72-624567. ISSN 0047-6773.

General articles aimed at laymen and amateur astronomers are the basis of this publication, which provides a refreshing departure from the standard professional and amateur astronomy periodicals. Reporting original research is not the intention here; rather, the authors present overviews, opinions, and theories about some of currently "hot" topics of astronomy. Articles are presented in a gee-whiz style with bold headlines, striking photographs, and a newspaper-type three-column format. Regular features include Book Trek, a well-written column with selected reading lists on a different topic each issue, aids and articles for the astronomy instructor, interviews with famous astronomers, and notes about articles of interest in other journals. Highly recommended for both public and observatory library, this periodical also occasionally carries information about *ASP* activities.

For subscription information see under *Astronomical Society of the Pacific.Publications.*

348.   **Observatory: A Review of Astronomy.** Sussex, Engl.: Royal Greenwich Observatory. v.1– , 1877– . Bi-monthly. illus. refs. volume, name, and subject indexes. ISSN 0029-7704.

The editors of this publication expertly blend a variety of materials aimed at the astronomer and layman alike into a most enjoyable review of the field. Leafing through the pages one finds non-technical articles on current topics, short reports of research (with equations, diagrams, and footnotes), reprinted lectures, the minutes of the Royal Astronomical Society meetings, and book reviews. These reviews are among the most opinionated, critical, and enjoyable pieces anywhere. The librarian should subscribe, if only in order to read these book critiques, penned by astronomers. Included, too, are reports from observatories, correspondence, and brief news notes. Recommended for all astronomy libraries, it has a cumulative index covering the first 75 volumes.

Write: Royal Greenwich Observatory, Herstmonceux Castle, Hailsham, East Sussex BN27 1RP, England. Cost: $12.00/year: £5.00/year.

349.   **Physics Today.** New York: American Institute of Physics. v.1– , 1948– . Monthly. illus. annual author, subject, and book review indexes. ISSN 0031-9228.

Although this popular journal concentrates on physics, it includes a fair number of articles on astronomy. It is basically a news magazine for the physicist, containing a variety of features: articles, current events, letters, book reviews, calendar, editorials, information exchange, and advertising. The general interest articles on current physics/astronomy topics resemble the pieces that appear in *Scientific American*, except that they are not usually as long and involved. Often an entire issue will be devoted to one subject, such as Soviet physics. For all types of libraries, this journal is a must for the astronomer and astrophysicist.

Write: American Institute of Physics, 335 E. 45th St., New York, N.Y. 10017. Cost: $24.00/year to non-AIP members. Microfilm available.

350.   **Sky and Telescope.** Cambridge, Mass.: Sky Publishing Corp. v.1– , 1941– . Monthly. illus. semi-annual indexes. LC 44-30805. ISSN 0037-6604.

Amateur and professional astronomers alike read this excellent journal to keep up with current events and research in the field. Aimed primarily at the amateur observer and student, *Sky and Telescope* is a popular-type periodical with a professional touch. The magazine abounds with photographs (many in color), but, unlike *Astronomy* (another popular journal), it avoids the gee-whiz approach, presenting a straightforward, no-nonsense look at the Universe.

Articles (written by amateur and trained astronomers, as well as by *S & T* staff) include news stories, features, and regular columns. There is a heavy emphasis on astronomical research carried out by NASA, and a multitude of photographs and pictures accompany these very frequent reports. Special events, like the opening of new observatories, and lunar and solar eclipses are frequently spotlighted on the cover and inside. There are scientific articles that review various astronomical topics, as well, and these papers are written for the advanced amateur and astronomer.

Regular features intended for the amateur include "Amateur Astronomers" (reports of activities of groups), "Books and the Sky" (reviews), "Celestial Calendar" (monthly and annual celestial events of note), "Gleanings for ATM's" (spotlights on home-made telescopes and observatories), "News Notes," "Observer's Page" (reports of amateur observations, projects, etc.), and a pull-out monthly star-chart. The January issue contains a "Graphic Time Table of the Heavens," a diagramatic, two-page almanac showing the rising and setting times of the Sun, Moon, and planets, the duration of twilight, and more. In short, any amateur astronomer who does not subscribe to this gold-mine of information is missing out. The book reviews are among the best anywhere, and cover both advanced and elementary texts. Librarians and astronomers should regularly peruse this section, which also lists and briefly annotates new books received. The advertising is one of the highlights of this publication, in addition to classified ads, it includes attractive layouts of backyard and observatory gear for sale, books, and trips. There is no other astronomical periodical to compare with this.

Write: Sky Publishing Corp., 49-50-51 Bay State Road, Cambridge, Mass. 02138. Cost: $10.00/year; $18.50/2years; and $27.00/3years. Microfilm edition available from University Microfilms, 300 N. Zeeb Road, Ann Arbor, Michigan 48106.

3

DESCRIPTIVE ASTRONOMY

SOLAR SYSTEM

Not surprisingly, there are more books on the solar system than on any other area of astronomy. The planets have been extensively explored by telescope and spacecraft, the Moon has been studied by astronomers and astronauts, and the Sun has been observed in detail to tell the scientist more about the other stars. The solar system is often the first subject mentioned in an astronomy text, and it is a very popular topic with amateur astronomers. Much of our science fiction has a scenario on or near one of the planets. In short, this broad area of planets, Moon, Sun, comets, meteors, etc., is a very popular section of astronomy, so it constitutes a large segment of the literature.

This lengthy section is divided into eight separate parts: 1) general works; 2) the Sun; 3) the Moon; 4) the planets; 5) comets and meteors; 6) atmospheres; 7) gravity; and 8) celestial mechanics. Some sections have fairly equal representations of works for general and advanced readers, while others display a definite bias toward one type or the other. Furthermore, laymen who wish to find a detailed book on a particular topic may not find a non-technical book because there is none, or none current. They will find it necessary to consult a general text.

General works in the solar system literature include books for both amateur astronomers and general readers, as well as for astronomers, although the former category is usually larger. The selection listed here is only a small portion of the publications available.

There are quite a few good books on the Sun, too, including both general and advanced works. Mostly aimed at the astronomer and graduate student, these books reflect the current interest in the solar spectra, structure, and energy output and production, along with the solar wind.

Interest in the Moon has skyrocketed in the last 15 years, and there are great numbers of books on our natural satellite, particularly technical volumes and conference proceedings. Included in this category are atlases, catalogs, geologic works, and more, but not all are listed in this section. Atlases, for example, are found in their appropriate category under Reference Works.

Like the Moon, the planets have received and will continue to receive a lot of attention from astronomers. The present interest in the planets is partially due to the so-called Grand Tour of the outer planets now in progress and the successful Mariner and Viking missions to Mars, as well as recent successful spacecraft studies of Venus and Mercury. A large majority of the books in this category are intended for the general reader, but there are also a substantial number for the student and astronomer.

The wanderers of the solar system, the comets and meteors, have a section of the literature all their own, consisting of many popular and technical publications, with an emphasis on the former. Comet hunting, always a desirable topic among amateurs, is even more popular today, resulting in an unusually large collection of comet books.

The small section on atmospheres, mainly technical in nature, includes works on aurorae and airglow, as well as studies of planetary atmospheres. Current interest in these topics is the result of solar wind and Earth-Sun relationship research, and exploration of the planets. The few proceedings volumes listed here are a very small segment of a growing body of literature in this increasingly studied field. (Information on stellar atmospheres can be found under Stars—Structure and Evolution.)

There are also three selections on gravity, a topic of interest to physicists and astronomers alike, in light of relativity studies and research on strange objects like black holes and collapsed stars.

Finally, a section on celestial mechanics is included to illustrate one of the more theoretical and rigorous portions of solar system studies. The branch of astronomy that deals with the motions and gravitational attractions of the bodies of the solar system, it is a subject typically studied by the advanced student and astronomer. Relying heavily on classical mechanics and mathematics, it is not a subject for the beginner. (Practical celestial mechanics, astrodynamics, is listed under Space Science.)

Card catalog subject headings: *Library of Congress:* SOLAR SYSTEM; SUN; MOON; PLANETS; VENUS (PLANET), etc.; EARTH; METEORS; COMETS; SOLAR SYSTEM—PICTORIAL WORKS. *Sears:* Same headings.

Library of Congress: UPPER ATMOSPHERE; AURORAS; AIGLOW; COSMIC PHYSICS; PLANETS—ATMOSPHERES; VENUS (PLANET)—ATMOSPHERE. *Sears:* ATMOSPHERE, UPPER; AURORAS.

Library of Congress: GRAVITY. *Sears:* GRAVITATION.

Library of Congress: MECHANICS, CELESTIAL; ORBITS; PERTURBATION; PROBLEM OF THREE BODIES; PROBLEM OF MANY BODIES. *Sears:* DYNAMICS.

General Works

351. Berlage, H. P. **The Origin of the Solar System.** Oxford, Elmsford, N.Y., Pergamon Press, Inc., 1968. 130p. illus. author index. bibliog. (Pergamon International Science Series). $3.50; $2.50pa. LC 67-31500. ISBN 0-08-012742-8; 0-08-012741-Xpa.

Although this book is aimed at the layman, it is not written in the popular style that often oversimplifies ideas and includes no technical language. Divided into 75 very short chapters discussing the various hypotheses on the origin of the planetary system, this volume provides an excellent chance for the reader to contrast the range of ideas put forth. The emphasis is mainly on the classical cosmogony theories, however, slighting some of the more recent ideas, such as clues from meteorites on planetary origin. Selected topics include Laplace's theory, the tidal theory, planetary evolution, the primeval nebula, the chemical composition of the planets, and many others. Beginning astronomy students interested in the evolution of the solar system should consider this interesting volume, but they should also read the book by the same title from the *Sky and Telescope* Library of Astronomy to get a more complete picture.

352. Brandt, John C., and Paul W. Hodge. **Solar System Astrophysics.** New York, McGraw-Hill Book Co., 1964. 457p. illus. notes. refs. LC 64-19502.

A variety of topics are introduced in this text for advanced students, ranging from celestial mechanics to the solar interior to planetary atmospheres. Briefly, several subjects which are often treated alone in separate volumes are presented here in an overview of the workings of the solar system. In many cases, the treatments of individual subjects are, in themselves, too brief, but these introductions are good starting points for the student. Further, the extensive lists of references will lead the reader to more complete and often more advanced works. The prerequisites for this text are knowledge of calculus, and familiarity with astronomical terminology, usually gained from an elementary course. The range of subject matter is broad, but there is an emphasis on the Sun (five chapters) including structure, energy processes, magnetic field, and spectrum. Comets, meteors, meteorites, and asteroids, usually grouped together in solar system books, have separate chapters. Individual planets are not spotlighted; rather, the authors discuss general concepts like atmospheres, interiors, surfaces. This otherwise excellent book has no problem sets, an unfortunate omission.

353. Cousins, Frank W. **The Solar System.** New York, Pica Press; distr. New York, Universe Books, 1972. 300p. illus. index. notes. refs. $22.50. LC 70-175855. ISBN 0-87663-708-X (Pica).

From early theories to current observations, this book will provide any reader with an excellent introduction to the Sun and its planets. Beginning with two well-researched chapters on classical and Renaissance views of the solar system, the author sets the stage for this, one of the better books on the subject. In these two sections we are briefly introduced to some of the great astronomers of the past and the roles they played in current views of the planetary system. Next is a unique chapter called "Models of the Solar System," a look at some early orreys, mechanical devices that show the relative positions and motions of the bodies of the solar system. There are about two dozen excellent photographs showing these fascinating instruments, as well as celestial globes and astronomical clocks. The following chapter on the planets and their satellites is not very long, but this is not a major drawback. Readers wanting more descriptions of the individual planets, however, may wish to consult other books in this section. Next are two general parts on planetary atmospheres and interiors in which the author discusses the makeup of the various planets. Following this are sections on the Earth-Moon system; comets, meteors, etc.; the Sun; and interplanetary dust and gas. The latter part of the book discusses the stability and origin of the system, emphasizing historical developments. The final section is concerned with life in the solar system and beyond, and Cousins raises some interesting questions here. For the general reader, serious amateur and student, this volume would be an excellent addition to any library.

354.   Hartmann, William K. **Moons and Planets**. Tarrytown-on-Hudson, New York, Bogden & Quigley, Inc.; distr. Belmont, Calif., Wadsworth Publishing Co., Inc., 1972. 404p. illus. index. refs. $13.95. LC 70-170777. ISBN 0-8005-0032-6 (B & Q); 0-534-00321-4 (Wadsworth).

An overall descriptive view of the solar system is the theme of this recent book for the layman or student. After some introductory information and photographs on the individual planets, the author discusses celestial mechanics, the science concerned with the motions of bodies in space. Here the reader is presented with the basics of this important topic: Kepler's laws of planetary motion, Newton's law of gravitation, circular velocity, escape velocity, tidal forces, and more. The author skillfully combines history and an explanation of the physical laws with a minimum of mathematics, giving a clear presentation of each concept. This well-written text continues on to a historical review of the origin of the solar system, including ancient and contemporary theory. An interesting chapter on the formation of stars follows, which may or may not belong in this book; the author justifies it as a parallel comparison of the origin of planets and stars. The origin of the solar system discussion continues in "The Growth of the Planets," and then Hartmann covers the sometimes slighted comets, asteroids, and meteorites. Chapters 10-12 cover planetary interiors, atmospheres, and surfaces, speaking in both general and specific terms, and relying on Earth situations for comparisons. The book ends with a short chapter on life in the Universe. An attractive layout combined with a good arrangement and coverage of topics make this a good choice for the public or college library.

355.   Kaula, William M. **An Introduction to Planetary Physics; The Terrestrial Planets**. New York, John Wiley and Sons, Inc., 1968. 490p. illus. bibliog. indexes (author and subject). (Space Science Text Series). $20.75. LC 68-27148. ISBN 0-471-46070-2.

Intended as an introductory text, this technical volume explores the make-up of Mercury, Venus, Earth, Mars, and the asteroids and meteorites. Grouped together because of their similarities in composition and structure, the terrestrial planets are also the closest and are more easily studied. Still, until we can actually stand on their surfaces and analyze the atmosphere and soil, many of the conclusions drawn are made by extrapolating information from what we know about the Earth. Therefore, the book spends the first three chapters on the Earth's interior, mechanical and thermal aspects of planetary structure, and the electromagnetism of a planetary interior, drawing parallels and applying known quantities to get answers. The second section of this often mathematical book concerns solar system dynamics, including a look at celestial mechanics, Earth-Moon dynamics, etc. Next comes planetary observations; and geology of the Moon and Mars, while the fourth looks at meteorites. The book ends with a general discussion of the composition and origin of the terrestrial worlds. Parts of the information here are now outdated, but the book retains much value for the advanced student. The text includes problems. The 586 references are arranged at the end of the book but also separately with each chapter in paragraph form; the author emphasizes the particular usefulness of each.

356.   Kopal, Zdeněk. **The Solar System**. London, Oxford University Press, 1972. 152p. illus. index. glossary. $6.00; $1.95pa. £2.50; £1.25pa. LC 73-173797. ISBN 0-19-885061-1; 0-19-88061-8pa.

Written for the uninitiated reader, this short book summarizes our present knowledge of the solar system, emphasizing recent research results. Not a "gee-whiz" book, it is a down-to-Earth look at a variety of topics written in an easy-to-understand, straightforward manner. The author, an eminent lunar scholar, covers the planets, moons, asteroids, comets, and meteors, and more, hitting the high points of each. Particularly suited for non-astronomer scientists and students, the book is a capsule summary of the physical characteristics of the planets, their motives, and their importance. The final few chapters are the most interesting—discussions of the origin of our solar system and the possibility of other planetary systems in the universe. More comprehensive and detailed treatments than this work can be found easily, but for its purpose of summarizing succinctly the current state of knowledge, this work is one of the best.

357.   Link, F. **Eclipse Phenomena in Astronomy**. New York, Springer-Verlag, 1969. 271p. illus. refs. subject index. $28.20. LC 68-56208. ISBN 0-387-04646-1.

A unique volume in the literature of astronomy, this book is the only work to deal exclusively with eclipses and their implications. Written for the advanced reader, this mathematical tome discusses eclipse phenomena from three viewpoints: theoretical, experimental, and historical. One notices immediately that among the book's seven chapters there is no mention of solar eclipses— certainly the most important aspect of the phenomena. The author justifies this by saying that the emphasis here is on the eclipsing body itself. Nevertheless, this omission is most disturbing. Almost half the text is devoted to lunar eclipses, including an introduction, photometrical theory of the umbra, photometry, lunar luminescence, shadow increase, and thermal phenomena observable during eclipses. The following chapter discusses the eclipses of artificial Earth satellites

182 / Descriptive Astronomy

and what we can learn from them. Occultations and transits are also discussed, as well as eclipse phenomena and radio astronomy, and Einstein's deflection of light theory. Certainly not a must item for every astronomy library, but some institutions may wish to acquire it.

358.   Moore, Patrick. **The New Guide to the Planets.** 3rd ed. New York, W. W. Norton & Co., 1972, c1971. 224p. illus. index. $7.95. LC 72-39165. ISBN 0-393-06319-4.

A good, general survey of the solar system, this volume is not a handbook for the observer. (The author refers the reader to his *The Amateur Astronomer*, 7th ed., published by Lutterworth Press in 1971.) The major intent of the book is to describe the planets, their physical characteristics and discoveries, in a non-technical, understandable manner. Successful, as usual, Moore begins with some general comments on the "wandering stars," giving the new reader some perspective on how our solar system is arranged. He next describes briefly the theories advanced on the formation of the planets, followed by a discussion of their motions and relative orbits. Giving the book an up-to-date touch, he concludes this introductory material with a look at possible spacecraft exploration of Mars, Jupiter, etc. Then, one by one, the author describes each planet from Mercury to Pluto, including the asteroids and the Moon. Highlights include the rings of Saturn, the Martian polar caps, and the clouds of Venus. A good choice as a first book on the solar system, this should not be considered as anything beyond that. Public libraries will find it ideal for beginning amateur astronomers.

359.   Page, Thornton, and Lou Williams Page, eds. **Neighbors of the Earth.** New York, The Macmillan Co., 1965. 336p. illus. index. glossary. bibliog. (Sky and Telescope Library of Astronomy, v.2). $8.95. LC 65-21994.

Dwelling on the physical characteristics of the planets, comets, meteors, asteroids, and interplanetary junk (dust and gas), this book shows how our knowledge about these subjects has changed during the middle third of this century. Consisting of articles from past issues of *Sky and Telescope*, the work begins with Mercury and Venus, the two planets closest to the Sun; of particular interest here are the discussions of the temperatures and atmospheres of the two celestial bodies. This material, like a great deal of the rest of the book, is somewhat dated now, but that does not detract from its value since the theme is somewhat historical anyway. Mars, probably the most popular planet for astronomers and sci-fi writers, is discussed with a fair amount of detail: size, surface, the vegetation question, geology, atmosphere, etc. The remaining planets are covered in the following chapter, which emphasizes Jupiter, the largest body in the solar system (excluding the Sun, of course). The rings of Saturn are discussed also. The asteroids, or minor planets, are covered next in a short section with much historical background. Comets and meteors are the subjects of chapters 5 and 6, respectively, highlighted by articles on Halley and his famous comet, and large meteorite craters like the Barringer in Arizona. The atmospheres of several of the planets, as well as aurorae, are taken up in the next section, followed by a description of the interplanetary debris in the form of dust, solar particles, and tiny meteors. The volume does not include information on the Sun or the Earth (see *The Origin of the Solar System,* 1966, v.3 of this series). Not to be considered a comprehensive treatment, this is instead supplementary reading.

360. Page, Thornton, and Lou Williams Page. **The Origin of the Solar System.** New York, The Macmillan Co., 1966. 336p. illus. index. glossary. bibliog. (Sky and Telescope Library of Astronomy, v.3). $7.95. LC 66-15028.

Not entirely concerned with cosmogony, this collective work contains many pieces on the solar system not found in related books in the *Sky and Telescope* series. In particular, there are two chapters on the Sun, one concerned with its production of energy and why we depend on it, and the other a description of the Sun's outer layers. The latter is primarily concerned with physical characteristics and processes, like sunspots, flares, granulation, chemical composition, spectra, radio emission, and more. The following two chapters describe the Earth as a planet and its atmosphere, respectively. Selected topics in the former are rotation, age, internal structure, shape, etc., while the latter concerns the various atmospheric layers and their exploration. The book finally turns to the main theme, solar system origin, by discussing the various evidence pointing to a common beginning for the planets and Sun. The trends in theories on cosmogony from the 1930s to the 1960s are approached next, giving the reader an idea of current thoughts. For classical theory, the reader will find it necessary to consult other sources like Berlage's *Origin of the Solar System* (1968). The possibility of life on other planets rounds out this solar system scrapbook. Editorial comment accompanies most articles, reprints from past issues of *Sky and Telescope.* An excellent source for amateur and student, it is subtitled "genesis of the sun and planets, and life on other worlds."

361. **The Solar System.** San Francisco, W. H. Freeman and Co., 1975. 145p. illus. index. bibliog. (A Scientific American Book). $8.50; $4.50pa. LC 75-28113. ISBN 0-7167-0551-6; 0-7167-0550-8pa.

One of the most informative and up-to-date books on the Sun and planets, this work is a reprint of twelve articles that appeared in the September 1975 issue of *Scientific American.* As expected, each piece is a well-written, beautifully illustrated essay by an expert in the field. The emphasis in this series of articles is on the latest data gathered using spacecraft; the majority of the excellent photographs in the book were taken with cameras on such interplanetary vehicles. The first chapter, by Carl Sagan, presents the solar system in general, describing each of the planets briefly and overviewing the last 18 or so years of exploration by spacecraft. Two excellent charts are included here, one on the physical characteristics of the planets and one on the various space missions to the planets; each contains a multitude of facts for learning and settling arguments. The following article, by A. G. W. Cameron of Harvard on the origin and evolution of the planetary system, is far more theoretical and is extremely thought-provoking. The remainder of the book spotlights the Sun, Mercury, Venus, the Earth, the

184 / Descriptive Astronomy

Moon, Mars, Jupiter, and the outer planets, followed by the meteorites, comets, asteroids, and satellites, and interplanetary particles and fields. Photographs of the planets' surfaces (from Mercury to Jupiter) are especially informative about the geologic and weather conditions there. Informed and neophyte readers alike will enjoy this fine compilation.

362.   Strong, James. **Search the Solar System: The Role of Unmanned Interplanetary Probes.** Newton Abbot, Engl., David & Charles, Ltd.; New York, Crane, Russak and Co., 1973. 160p. illus. index. bibliog. £3.25. $8.50. LC 73-80427 (C & R); 73-178708 (D & C). ISBN 0-8448-0213-1 (C & R); 0-7153-6031-0 (D & C).

The story of how manned and unmanned spacecraft have explored and will explore the planets and many of their satellites is the theme of this fascinating volume. Written for the general reader, the book describes in vivid detail the various space probes and their journeys, as well as the planets. Thus, the text of this work, in addition to describing a portion of our space effort, is also a good source of information on our neighbors in space, from Mercury to Pluto. Since each planet has varying surface and atmospheric conditions, each presents different problems to the designers of the probes. The author discusses these aspects in enough detail to provide the reader with basic information.

363.   Whipple, Fred L. **Earth, Moon and Planets.** 3rd ed. Cambridge, Mass., Harvard University Press; distr. Cambridge, Mass., Sky Publishing Corp., 1968. 297p. illus. index. bibliog. (The Harvard Books on Astronomy). $8.75; $2.75pa. LC 68-21987. ISBN 0-674-22400-0; 0-674-22401-9pa.

A semi-popular work for high school and college students concerning our solar system, this book is well illustrated and clearly written. The standard topics are covered—the terrestrial planets, the giant planets, Mars, origin of the solar system, etc. The arrangement of chapters is a bit curious: after a general introduction and a chapter on gravity, Whipple discusses the discoveries of Neptune and Pluto, followed by a section on the distances and masses of the planets. Five chapters on the Earth and Moon are followed by chapters on the other planets. Like the other Harvard Books on Astronomy, this volume, now in its third edition, would be an excellent choice for the public and college library. It is devoid of mathematics and technical descriptions, emphasizing instead the descriptive aspects of the subject. Five excellent appendices supplement the work and include hints for observing the planets with an amateur telescope.

Sun

364.   Brandt, John C. **Introduction to the Solar Wind.** San Francisco, W. H. Freeman and Co., 1970. 199p. illus. index. refs. (A Series of Books in Astronomy and Astrophysics). $10.00. LC 75-89919. ISBN 0-7167-0328-9.

Aimed at the advanced student, this volume is a brief overview of the corpuscular radiation flowing from the Sun. In the first chapter, the author gives some historical background, outlining the various attempts at proving the existence of the phenomena. Next, he summarizes solar physics for the uninformed reader, pausing to explain solar structure, the chromosphere and the corona. An understanding of these processes and physical conditions is a necessary prelude to comprehending the solar wind. The theoretical basis for the phenomenon, beginning with Parker's hydrodynamic theory (1958), is detailed in the mathematical chapter 3. Ground-based observation methods (including ionic comet tail observations, radar and radio observations, and geomagnetic observations) are considered next, followed by a chapter on space observations. The effects of the solar wind on the solar system are reviewed in the following important chapter. Here are explanations of interactions with comets (producing tails), the Earth's atmosphere (geomagnetic activity and aurorae), the Moon, Mars and Venus, Jupiter, etc. The final section, on the impact of the phenomena on astrophysics, is a summing-up of current thoughts. Overall, a good introduction suitable for the astronomy collection.

365.   Bray, R. J., and R. E. Loughhead. **The Solar Chromosphere**. London, Chapman and Hall; distr. New York, Halstead Press, John Wiley and Sons, 1974. 384p. illus. refs. indexes (subject and name). (The International Astrophysics Series). $33.75. LC 73-8101. ISBN 0-412-10730-9 (Chapman); 0-470-09807-4 (Wiley).

A comprehensive and up-to-date treatment of the chromosphere is the result of this technical volume aimed at the solar physicist and astrophysicist interested in wave phenomena in stellar atmospheres. This region of the solar atmosphere between the photosphere and corona is one of the most interesting to scientists because of its inhomogeneity and variety of physical activity. This volume, then, summarizing the important work on this area, is a significant contribution to the literature. The contents are almost entirely devoted to the "quiet" chromosphere, however, since an understanding of it is essential in solving the problems of varied and complex phenomena associated with the active regions of the Sun. This excellent work embodies the quality of the authors' previous books on solar granulation (1967) and sunspots (1964).

CONTENTS: 1. Historical Introduction. 2. Spicules and Other Fine Structures at the Solar Limb. 3. The Morphology and Dynamics of the Quiet Chromosphere Observed on the Disk. 4. Physical Conditions in the Quiet Chromosphere. 5. The Fine Structure of the Active Chromosphere. 6. The Propagation and Dissipation of Waves in a Compressible Gravitationally-Stratified Atmosphere. 7. Theories of the Heating of the Chromosphere and of the Origin of Spicules.

366.   Bray, R. J., and R. E. Loughhead. **Sunspots**. London, Chapman and Hall, Ltd.; New York, John Wiley and Sons, Inc., 1964. 303p. illus. refs. indexes (name and subject). (The International Astrophysics Series, v.7). $16.00. LC 65-7091 (Wiley); NUC 65-106910 (Chapman). ISBN 0-470-09955-0 (Wiley); 0-412-07600-4 (Chapman).

The solar scientist, student, and interested layman will all find useful information in this excellent and comprehensive work. From general historical information on sunspots to technical descriptions of their nature, the authors leave no stone unturned. The historical introduction is quite good; we learn about the earliest known reference to a sunspot in the fourth century B.C., the first telescopic observations by Fabricius, Galileo, Scheiner, and Harriot, and the development of advanced observing techniques. The structure and composition of individual sunspots is discussed in detail, accompanied by dozens of excellent photographs; the pores, "young sunspots," are described as well. Physical conditions in the umbra and penumbra are presented, along with sunspot spectra. Magnetic fields surrounding these entities are an important part of their study, and the authors discuss field strength, variations, measurement methods, etc. Sunspot groups and their characteristics are the subject of chapter 6, a detailed look at how the groups form, their lifetime, motion, and so on. Statistical considerations on sunspot group occurrence and distribution are covered as well. The book concludes with sunspots and activity centers, magnetohydrodynamic theories of the origin of sunspots, and the solar cycle. A high point of this book is the extensive lists of references which accompany each chapter.

367. De Jager, C., ed. **The Solar Spectrum.** Dordrecht-Holland, D. Reidel Publishing Co., 1965. 417p. illus. refs. (Astrophysics and Space Science Library, v.1). $26.00. LC 65-10715. ISBN 90-277-0119-9.

Fourteen contributed papers and several recorded discussion sessions comprise this conference volume, which summarizes the importance of the solar spectra. "Forty years of Solar Spectroscopy," a review given by M. G. J. Minneart (the symposium honoree) begins the book, summarizing the last four decades of research, highlighting the photosphere, sunspots, the chromosphere, the corona, and radio-emission. Following this informative paper, the book is divided into five major sections: The Photosphere; The Quiet Chromosphere; The Active Photosphere and Chromosphere, Spots, Plages, Flares and Prominences; The Corona; Particle and Radio Emission from the Sun. The specialist working solar research will appreciate this advanced text, based on a symposium held at the University of Utrecht, The Netherlands, August 26-31, 1963.

368. Ellison, M. A. **The Sun and Its Influence.** 3rd ed. New York, American Elsevier Publishing Co., 1968. 240p. illus. index. bibliog. $7.50. LC 68-31848. ISBN 0-444-19943-8.

Although the primary emphasis of this long-standing volume is on solar-terrestrial relations, there is also an excellent introductory description here of the Sun itself, a good preface to the main body of the book. In this vein, the first few chapters provide an excellent overview of the Sun's radiation, atmosphere, and physical properties, with an emphasis on those features which affect the Earth: sunspots, solar flares, and prominences are all discussed at length. The late Professor Ellison's text goes on to explain how the Sun's radio and cosmic waves, magnetic fields, and flares influence the Earth and its upper atmosphere. The chapter on the ionosphere is very good, and the author highlights the commonly known phenomena of aurora, or the Northern Lights. There is not much new material in this edition, except for the short passage on the solar wind and the Van Allen Belts. A newer edition to include the many recent advances in Sun-Earth studies would be highly desirable, since there is little for the layman on this subject. Not too simple or too technical, this fine book is appropriate for college and public libraries.

369.   Gibson, Edward G. **The Quiet Sun.** Washington, D.C., NASA, 1973. 330p. illus. index. refs. $6.20. LC 72-600092. NASA SP-303.

Written for the upper level undergraduate and beginning graduate student, this text is one of the most up-to-date and well-presented available. Basically a survey of the structure and physical processes in our star, the book is well-illustrated with photographs and diagrams of the Sun's surface and interior. After a brief introductory section, the author presents a description of the solar core, convective zone, photosphere, chromosphere, corona, etc.; in short, the reader is shown the interior of the Sun from the inside out. The next four chapters discuss, in depth, the interior (including a discussion of the energy-producing processes), the photosphere (the layer of the Sun which emits energy into space), the chromosphere (the outer layer of the Sun), and the corona (the halo which is responsible for the solar wind). The text is a good combination of description and mathematics, with a substantial list of references with each chapter, making it an ideal starting point for the reader interested in solar studies.

370.   Hundhausen, A. J. **Coronal Expansion and Solar Wind.** Berlin, New York, Springer-Verlag, 1972. 238p. illus. subject index. refs. (Physics and Chemistry in Space, v.5). $21.60. LC 72-85398. ISBN 3-540-05875-3 (Berlin); 0-387-05875-3 (New York).

A text suitable for the graduate student or scientist unfamiliar with solar studies, this volume brings up to date and consolidates much of our current knowledge on the radiation streaming out from the Sun. A highly mathematical book, it covers the following topics: history and background; the identification and classification of some important solar wind phenomena; the dynamics of a structureless coronal expansion; chemical composition of the expanding coronal and interplanetary plasma; high-speed plasma streams and magnetic sectors; and flare-produced interplanetary shock waves. Both theoretical and observational aspects are covered in this comprehensive review, which emphasizes the relation between the corona and the solar wind phenomena. Effects on the Earth, comets, etc. are mentioned, but there is little detail here. Well-researched, the book contains a bibliography with 337 entries.

371.   Kiepenheuer, Karl. **The Sun.** Ann Arbor, The University of Michigan Press, 1959. 160p. illus. index. $5.00. LC 59-7294. ISBN 0-472-00110-8.

An excellent, compact treatment of the nearest star is the outcome of this book for the student or well-prepared layman. Written in descriptive, general terms, this volume begins with an overview of the subject in which the author spews out dozens of pertinent facts on the Sun's size, temperature, energy output, and composition. The chapter ends with a three-page chronology of solar research since 450 B.C. The purpose of the next chapter, "The Solar System," is questionable. The author describes each planet briefly in an attempt to illustrate the solar system family, but such information is easily obtained elsewhere. Chapter three, on the Sun's surface, tells the reader of the so-called "granules," cell-shaped areas that appear and disappear due to violent convective currents, and chapter four discusses sunlight, how it is observed, and what it tells us about the Sun. The longest and most interesting chapter, "The Changing Face of the Sun," explores in great detail sunspots, faculae, flares, prominences, filaments, and storms, all fascinating, spectacular events in the Sun's outer layers. Several other topics are mentioned, but the above are the most important. The volume, very well illustrated with detailed closeup photos of the Sun, would be a good book for all types of libraries—but especially for public libraries.

372.   McIntosh, Patrick S., and Murray Dryer, eds. **Solar Activity Observations and Predictions.** Cambridge, Mass., The MIT Press, 1972. 444p. illus. refs. index to contributions. (Progress in Astronautics and Aeronautics vol. 30). $20.00. LC 72-5953. ISBN 0-262-13086-6.

Comprised of just the right combination of review papers and reports of research results, this is an excellent, up-to-date overview of the Sun and its effects on the immediate space environment. The volume begins with a good review of solar activity, highlighting the solar atmosphere, the "quiet Sun," solar magnetic fields, solar flares, etc. The remainder of the book's first section on "Solar Observations and Theories" contains papers reporting on particular research of the aforementioned topics, including a piece on solar radio astronomy. The effects on the space environment are discussed in the second section, "Interplanetary Medium," whose primary topic is the solar wind. The third section explores solar-terrestrial interaction, with papers on ionospheric effects and geomagnetic response, to name a few. The substantial final section considers the prediction of types of solar activity, a difficult and not-yet-perfected process. Of interest here is the paper surveying the current solar forecast centers. Written for scientists and graduate students, this excellent survey would be a good acquisition for astronomy libraries.

373.   Macris, Constantin J., ed. **Physics of the Solar Corona.** Dordrecht-Holland, D. Reidel Publishing Co.; distr. New York, Springer-Verlag, 1971. 345p. illus. refs. name index. (Astrophysics and Space Science Library, v. 27). $29.60. LC 76-154741. ISBN 0-387-91091-3 (Springer); 0-90-277-0204-7 (Reidel).

In part updating works like Shklovskii's *Physics of the Solar Corona* (1966), and presenting new data from recent observations, this volume consists of lectures given at a NATO Advanced Study Institute (Athens, 6-17 September 1970). The 23 papers in this book cover a good portion of current knowledge in the field, including atomic processes in the solar corona, magnetic fields, coronal monochromatic emissions, effects of flare-associated waves, models of the active and quiet solar atmosphere, radio emissions, and others. The volume's usefulness is unfortunately seriously hindered by the lack of a subject index;

finding information on the solar wind, for example, is most difficult, especially since none of the papers listed in the table of contents is specifically on that topic. The advanced student and astronomer will find this volume useful as a good overview of the subject.

374. Menzel, Donald H. **Our Sun**. rev. ed. Cambridge, Mass., Harvard University Press, 1959. 350p. illus. index. (The Harvard Books on Astronomy). $9.50. LC 59-12975. ISBN 0-674-64750-5.

One of the best and most comprehensive general works on the Sun, this standard volume is a good blend of current knowledge and historical perspective. The introductory chapter in any book is the most important, and in *Our Sun* this is especially true. Menzel grabs our attention immediately by describing what would happen to the Earth should our star burn out suddenly, illustrating well the Sun's importance to our survival. From that point the introduction discusses early solar science and the solar distance and mass, overwhelming us with large figures. Chapter two tells the reader briefly how to observe the Sun and highlights some of the conspicuous features observable including sunspots and prominences (special equipment required). A later chapter on sunlight describes how the light is gathered and analyzed and what it means. Further on, Menzel discusses in detail sunspots, granulation, prominences, and the corona. Excellent photos accompany these chapters, and the uninitiated reader will be able to see, for the first time, detailed illustrations of the solar surface. Eclipses are next, followed by a chapter on the Sun's place in the galaxy and universe; here Menzel discusses briefly stellar evolution and how the Sun compares to other stars as well as the origin of the solar system. There is even a chapter on solar energy, a bit outdated now, but a good introduction. Intended for the general reader and beginning college student, the book is an excellent introduction to solar studies. A more up-to-date volume is desirable, but this version can stand on its own for quite a while. For the college and public library.

375. Moore, Patrick. **The Sun**. New York, W. W. Norton and Co., Inc. 1968. 128p. illus. index. bibliog. $4.95. LC 68-27145. ISBN 0-393-06276-7.

There are few books for the amateur or layman on the Sun. Fortunately this, one of those few, was written by one of the best and most prolific general astronomy authors. Moore's text begins with an overview of the Universe, a brief survey which results in the revelation, not surprising to some, that the Sun is insignificant in the overall picture. Having set this perspective, the author then proceeds to explain how important the Sun is to planet Earth. The following chapter, "The Sun as a Star," is an introduction of sorts to the Sun in the history of astronomy. From this point on, however, the physical characteristics of our star are the main theme of the book. Topics include the distance of the Sun, its surface, spectrum, and interior. Moore wisely cautions the reader never to look at the Sun directly with the naked eye or telescope, but rather only by projecting its image onto a piece of cardboard. Other subjects of interest are eclipses, radio emissions, and the Sun's life cycle, and supplementary information on great solar astronomers, observing the Sun, and numerical data. An excellent treatment of the topic, this work is highly recommended for the home and public library.

376. Shklovskii, I. S. **Physics of the Solar Corona.** 2nd ed. Oxford, Pergamon Press, 1966. 475p. illus. refs. indexes (subject and author). (International Series of Monographs in Natural Philosophy, v.6). $8.75pa. LC 65-9284. ISBN 0-08-013752-0.

This slightly dated volume is one of the few texts on the coronal layer, so it is important to the literature of this subject. Chapter one, covering the information gathered from observation of the corona, is fairly long and includes photometric observations, physical processes, polarization, the continuous coronal spectrum, emission and absorption lines, and non-eclipse observations using the coronograph. The introductory chapter is a good beginning for this advanced text, which then goes on to consider the solar coronal spectrum, a quantitative chemical analysis of the corona, and ionization. These chapters make up one of the better sources on the intrinsic properties and processes of the corona. The remainder of the volume concentrates on the effects of the corona on the Earth and interplanetary space, a frequent theme in the literature. Discussions consider the emission of ultraviolet and soft x-ray radiation, and radio wave emission. While more recent information on these subjects can be found in the serial literature, the student and astronomer specializing in solar studies should read this book as background information. Translated from the Russian (*Fizika Solnecho: Korony*) by Louis Anderson Fenn, the book was edited by A. Beer; additional material was translated and edited by A. J. Meadows.

377. Sonnett, C. P., P. J. Coleman, Jr., and J. M. Wilcox, eds. **Solar Wind.** Washington, D.C., NASA, 1972. 717p. illus. refs. $6.00. LC 72-600187. NASA SP-308.

This massive volume of conference proceedings for the specialist contains invited reviews, contributed papers, and transcripts of a round-table discussion. A combination of current theory and reports of observations, the book discusses in depth the solar wind phenomena, an area of growing interest over the last 10 to 15 years. Each of the 53 papers includes an abstract, dozens of references, and many illustrations. Appropriate for the astronomy or aerospace library, it contains the following sections: I, Photospheric and Coronal Magnetic Fields; II, The Interplanetary Magnetic Field; III, Large-Scale Features of the Solar Wind Plasma; IV, Solar and Stellar Spin Down; Angular Momentum of the Solar Wind; V, Microstructure, "Turbulence," and Hydromagnetic Waves in the Solar Wind; VI, Shock Waves and the Structure of Large Scale Disturbances; VII, Radio Observations of the Solar Wind; VIII, Observational Evidence and Theory of the Ionic Composition of the Solar Wind and Fractionation Effects at the Coronal Base and in the Solar Wind; IX, Interaction of the Heliosphere and the Galactic Medium: The Distant Solar Wind. From a NASA-sponsored conference, Pacific Grove, California, March 21-26, 1971, this volume is one of the most complete available.

378. Strickland, A. C., general ed. **Annals of the IQSY (International Years of the Quiet Sun).** Cambridge, Mass., The MIT Press, 1968-70. 7v. illus. bibliog. indexes. LC 67-11390.

In 1957-1958, the period of greatest sunspot activity ever recorded, a worldwide venture known as the International Geophysical Year (IGY) collected an enormous amount of data on the solar events taking place. It became evident that the data could be fully appreciated only if it were compared to information gathered during a period of minimum solar activity (few sunspots). In 1962 the International Year of the Quiet Sun (IQSY) was organized by a Special Committee of the International Council of Scientific Unions (ICSU) as the project for minimal solar activity study during 1964-65, a predicted period of solar inactivity. The results of the study are presented in these seven volumes, which are certain to be of interest to astronomers and other scientists as well.

Volume 1: *Geophysical Measurements: Techniques, Observational Schedules, and Treatment of Data.* C. M. Minnis, ed. 1968. 398p. $25.00. ISBN 0-262-09005-8.

Volume 2: *Solar and Geophysical Events 1960-65 (Calendar Record).* J. V. Lincoln, comp. 1968. 297p. $20.00. ISBN 0-262-09006-6.

Volume 3: *The Proton Flare Project (The July 1966 Event).* 1969. 511p. $25.50. ISBN 0-262-09007-4.

Volume 4: *Solar-Terrestrial Physics: Solar Aspects (Proceedings of the Joint IQSY/COSPAR Symposium, London, 1967, Part I).* 1969. 414p. $25.00. ISBN 0-262-09008-2.

Volume 5: *Solar-Terrestrial Physics: Terrestrial Aspects (Proceedings of the Joint IQSY/COSPAR Symposium, London, 1967, Part II).* 1969. 460p. $25.00. ISBN 0-262-09009-0.

Volume 6: *Survey of IQSY Observations and Bibliography.* 1970. 589p. $30.00. ISBN 0-262-09010-4.

Volume 7: *Sources and Availability of IQSY Data.* 1970. 345p. $25.00. ISBN 0-262-09011-2.

379.   Zirin, Harold. **The Solar Atmosphere.** Waltham, Mass., Blaisdell Publishing Co., 1966. 402p. illus. diagr. list of symbols. index. refs. (A Blaisdell Book in the Pure and Applied Sciences). LC 65-21458.

Despite the fact that this volume is now 11 years old, it is one of the better texts available on the advanced level. Beginning with a brief descriptive chapter on the Sun's physical characteristics, the author displays his clear, logical writing style. The following chapter is fairly unique among works of this type, consisting of an excellent discussion and well-chosen illustrations of solar "gadgetry," the equipment used to observe the Sun. The various solar telescopes and spectrographs are explained graphically, and this section gives added meaning to the more "heavy" topics which follow. Three more preliminary sections precede a discussion of the solar atmosphere, however: plasmas, interpretation of spectra, and a review of atomic spectra. An understanding of these topics necessarily must be acquired before proceeding. The outermost portion of the Sun, the corona, is the center of attention in Chapters 6 through 8, including descriptive matter and a discussion of the physical processes there. The chromosphere and the photosphere are covered next, followed by chapters on prominences, solar activity (sunspots, etc.), and flares, all important phenomena in discussions of the solar atmosphere. Zirin closes with a short chapter speculating on future research on these topics. An excellent source of information, this book should be in every astronomy library.

Moon

380.   Alter, Dinsmore. **Pictorial Guide to the Moon**. 3rd ed. New York, Thomas Y. Crowell Co., 1973. 216p. illus. index. glossary. $8.95. LC 73-9869. ISBN 0-690-00096-0.

If you're a lunar "freak" and enjoy pictures of the Moon, this is just the book for you. Written by a lunar and planetary specialist, this new edition is better than ever. Beginning with a description of early lunar observations, the author dwells at length on some of the more important lunar astronomers and includes a selection of early Moon maps. The book then turns to a physical description of the Moon, beginning with a discussion of surface conditions. Next is a superb chapter on the identification of lunar features, including a list of over 240 prominent topographic aspects like craters, mountains, rilles, valleys, seas, etc. Twenty-seven photographs accompany this section, showing each of the aforementioned features, properly labeled. This unique attraction is one of many high spots in this fine book. The Moon's phases, too, are dealt with thoroughly, accompanied by some excellent photography of each stage from crescent to full. Following is a descriptive survey of various lunar features including craters, mountains, valleys, etc. Supplementary topics are discussions of the tides and the evolution of the Moon. Overall, this text is one of the best, especially considering the hundreds of photographs and the clearly written, comprehensive text. Revised by Joseph Jackson, this volume is not only an appropriate coffee-table book but also an excellent candidate for the public library.

381.   Kopal, Zdeněk. **The Moon**. Dordrecht-Holland, D. Reidel Publishing Co., 1969. 525p. illus. index. bibliog. $34.00. LC 77-504388. ISBN 90-277-0123-7.

One of the few texts available for a graduate or advanced undergraduate course on the Moon, this excellent volume requires prior knowledge of astronomy, physics, chemistry, and mathematics. Divided into four major parts, the book begins with a basic treatment of the Moon's motion and the dynamics of the Earth-Moon system. The author begins by describing the lunar distance, size, and mass, and then discusses the Moon's movement under the influence of the Sun, Earth, and other planets. Part two discusses the structure of the Moon's interior in six chapters that are no doubt now somewhat outdated. In any case, the author considers such things as thermal and stress history, chemical composition, and hydrostatic equilibrium. Part three is concerned with lunar topography, mentioning relative coordinates, morphology of the surface, origin of lunar formations, etc. The chapter on the mapping of the Moon adds some historical perspective and has several excellent illustrations of old and new lunar maps. Finally, the volume ends with six chapters discussing the Moon's radiation; included are explanations of the photometry of Moonlight, thermal emission, luminescence of the lunar surface, and structure of the surface. A comprehensive text, substantially updating the author's *An Introduction to the Study of the Moon* (1966), it should be in every astronomy library.

382. Kopal, Zdeněk. **The Moon in the Post-Apollo Era.** Dordrecht-Holland, D. Reidel Publishing Co., 1974. 223p. illus. name index. refs. (Geophysics and Astrophysics Monographs, v.7). $25.00; $16.50pa. LC 74-26877. ISBN 90-277-0277-2; 90-277-0278-0pa.

In his latest book, the eminent astronomer summarizes the results of manned and unmanned lunar exploration since 1959. The information here updates substantially his earlier work, *The Moon* (1969), which only included minor information on the Apollo 11 mission. A significant contribution to the literature, this book summarizes and highlights the incredible amount of data collected as a result of the U.S. and Soviet space flights—an extremely difficult task, but one handled smoothly and expertly by Professor Kopal. This excellent volume, which belongs in all university and observatory libraries, contains the following sections: 1, Exploration of the Moon by Spacecraft; 2, Manned Exploration: Apollo (1969-1972); 3, Basic Facts: Distance, Size and Mass; 4, Shape and Gravitational Field of the Moon; 5, Internal Structure of the Moon; 6, Morphology of Lunar Formations; 7, Surface Structure and Chemical Composition; 8, Stratigraphy and Chronology of the Lunar Surface; 9, Lunar Exosphere; 10, Origin and Evolution of the Moon.

383. Kopal, Zdeněk, ed. **Physics and Astronomy of the Moon.** 2nd ed. New York, Academic Press, 1971. 303p. illus. refs. indexes (subject and author). $21.00. LC 75-107571. ISBN 0-12-419340-4.

The first edition of this non-elementary volume appeared in 1962 at a time, comparatively speaking, when our knowledge of the Moon was meager. Relying entirely on observations from Earth, astronomers theorized as best they could about the lunar surface, interior, and atmosphere. As spacecraft orbited, crashed, and soft-landed on the Moon, new data quickly began to replace old, and the need for a new edition of this excellent book became a necessity. Edited by a well-known lunar specialist, an updated volume appeared soon after the first lunar landings, containing almost entirely new material. Six lunar scientists contributed the following essays: The Motions of the Moon in Space; Librations of the Lunar Globe; Dynamics of the Earth-Moon System; Geometrical and Dynamical Properties of the Moon; Optical Properties of the Lunar Surface; and Origin and History of the Moon. For professional astronomers and graduate students, this work skillfully updates one of the better available texts.

384. Levinson, Alfred A., and S. Ross Taylor. **Moon Rocks and Minerals.** New York, Pergamon Press, 1971. 222p. illus. subject index. glossary. LC 70-140580. ISBN 0-08-016669-5.

One of the earliest volumes on lunar samples brought back by the astronauts, this book condenses and highlights the scientific reports from the proceedings of the Apollo 11 Lunar Science Conference (Jan. 5-8, 1970, Houston). Aimed at the non-geologist graduate student and scientist, this technical book considers four major areas of study: 1, Mineralogy and Petrology; 2, Chemical and Isotope Analysis; 3, Physical Properties; 4, Bioscience and Organic Geochemistry. The most emphasis in the book is on the mineralogy and chemical analysis of the samples, and the former section is highlighted by several cross-section photographs of the rocks, along with a detailed description of the types of minerals found. Both color and balck-and-white photos abound in the volume, giving the uninitiated reader an excellent look at the strange rocks from the Moon. A brief chapter at the end discusses various theories of lunar origin and the significance of the samples along these lines. This volume's importance to the scientific community lies in its clear presentation and its summary of an overwhelming amount of data.

385.　　Lucas, John W., ed. **Thermal Characteristics of the Moon**. Cambridge, Mass., The MIT Press, 1972. 340p. illus. refs. index to contributors. (Progress in Astronautics and Aeronautics, v.28). $20.00. LC 79-39803. ISBN 0-262-12058-5.

Literally hundreds of technical papers have been published since the gathering of new lunar data, and this book is presented as a good example of the kinds of material being studied. Drawing upon Earth-based observations with telescopes and actual experiments on the Moon by astronauts, the contributors to this work describe well some facts and theories about thermal studies. Topics on Earth-based measurements include infrared measurements of the lunar surface, lunar microwave emissions, and radar mapping of the Moon's surface. A second section of the book describes on-site thermal experiments, including the Apollo 15 Lunar Heat Flow Measurement, and the third part of the volume concentrates on thermal characteristics of lunar-type materials. Ending with a thermal history of the Moon, the book includes hundreds of references to related technical literature.

386.　　Levinson, A. A., ed. **Apollo 11 Lunar Science Conference. Proceedings**. Elmsford, N.Y., Pergamon Press, 1970. 3v. 2492p. illus. index. refs. (*Geochimica et Cosmochimica Acta.* Supplement 1). LC 72-119485. ISBN 0-08-016392-0.

387.　　Levinson, A. A., ed. **Lunar Science Conference. Second. Proceedings**. Cambridge, Mass., The MIT Press, 1971. 3v. 2818p. illus. indexes (subject and author). refs. (*Geochimica et Cosmochimica Acta.* Supplement 2). $25.00/vol ($70.00/set). LC 78-165075. ISBN 0262-12051-8(v.1); 0-262-12054-2(v.2); 0-262-12057-7(v.3).

388.　　King, Jr., Elbert A., Dieter Heymann, and David R. Criswell, eds. **Lunar Science Conference. Third. Proceedings**. Cambridge, Mass., The MIT Press, 1972. 3v. 3263p. illus. refs. indexes (author and subject). (*Geochimica et Cosmochimica Acta.* Supplement 3). $32.00/vol ($90.00/set). LC 72-5496. ISBN 0-262-12064-X.

389.　　**Lunar Science Conference. Fourth.Proceedings**. Elmsford, N.Y., Pergamon Press, 1973. 3v. 3290p. illus. refs. indexes (author and lunar sample). (*Geochimica et Cosmochimica Acta.* Supplement 4). LC 73-15974. ISBN 0-08-017909-6; 0-08-17910-Xpa.

390. **Lunar Science Conference. Fifth. Proceedings.** Elmsford, N.Y., Pergamon Press, 1974. 3v. 3172p. illus. index. refs. (*Geochimica et Cosmochimica Acta.* Supplement 5). $100.00/set. LC 74-165220. ISBN 0-08-018318-2.

Easily one of the most monumental sources of technical information on the Moon, the proceedings of this annual conference have yielded (as of 1975) over 1,000 technical papers dealing with the study of lunar samples brought back by the U.S. astronauts. In general, the three volumes of each set have been concerned with I, Mineralogy and Petrology; II, Chemical and Isotope Analyses, Organic Chemistry; and III, Physical Properties. There have been slight variations in a couple of volumes and some additional information in still another, but the three major topics have been followed fairly closely. Included in each volume are photographs of samples of lunar material and their cross-sections, as well as hundreds of charts and graphs. The amount of material is overwhelming, and it is hoped that, when the last conference is held, a cumulative subject and author index will be compiled. At least one attempt has been made to condense and render understandable the multitude of advanced papers in these volumes; that was Levinson and Taylor's book called *Moon Rocks and Minerals* (1971), which compacted and highlighted the first conference listed above. A similar book by the same Stuart Taylor in 1975 (*Lunar Science: A Post-Apollo View*) not only condensed the findings of the conferences on mineralogy, etc., but added information on overall lunar geology and topography. Not essential for every astronomy library, it should at least be in those collections with a strong lunar and planetary emphasis. The volumes would be good acquisitions for geology libraries as well.

391. Mason, Brian, and William G. Melson. **The Lunar Rocks.** New York, John Wiley & Sons, Inc., 1970. 179p. illus. index. refs. $12.00. LC 73-129659. ISBN 0-471-57530-5.

Although this is not a "popular" book by any means, it is less technical than similar works (e.g., *Moon Rocks and Minerals*, Levinson and Taylor, 1971), and therefore could be handled by the knowledgeable layman as well as the non-specialist scientist and student. Chapter one, a brief overview of lunar geologic knowledge before manned landings, is followed by a short description of the Apollo 11 and 12 landing sites and activities of the astronauts. Better, and more detailed, descriptions of the former can be obtained from any of a number of pre-Apollo (before 1969) works on the Moon, but the discussion will suffice for the purposes of this text. The most important chapters of the book, however, are those on mineralogy (a discussion of the types of individual minerals found in the samples, and their percentages) and petrology (an examination of the three basic types of rock: igneous, microbreccias, and fines). Chapter six is one of the most interesting sections, for here the lunar rocks are compared with terrestrial rocks, an important area not to be overlooked. Here the authors discuss basic similarities (composition) and differences (age) of several samples. Lunar geochemistry, a run-down of the elements found in the rocks, round out this fine book, appropriate for the college, special, and public library. Its only drawback is that there are not enough photographs and illustrations of the samples.

392. Moore, Patrick, and Peter J. Cattermole. **The Craters of the Moon; An Observational Approach**. London, Lutterworth Press; distr. N.Y., W. W. Norton & Co., 1967. 160p. illus. index. refs. $5.95. LC 67-91375. B 67-11987. ISBN 0-393-06355-0 (Norton).

    Much of the information in this book has been updated by the manned lunar flights in recent years, yet this work remains, and probably will for some time, a good starting point for the amateur astronomer who wants an informative introduction to the Moon, in particular the lunar surface. Approximately two-thirds of this non-technical yet not elementary work is devoted to theories of the formations of craters and other lunar features. In particular, the meteoric and volcanic points of view are discussed, and terrestrial geologic activity is compared and extended to the Moon. There are also lengthy discussions of maria (so-called dark areas, or "seas"), mountains, various walled formations, linear elements (faults, rilles), and lunar rays. Well illustrated with photos and diagrams, the book would be of great use for background information for the amateur lunar observer. A volume that emphasizes the historical development of lunar studies, it contains a large number of references to the literature.

393. Taylor, Stuart Ross. **Lunar Science: A Post-Apollo View**. Elmsford, N.Y., Pergamon Press, Inc., 1975. 372p. illus. indexes (subject and author). glossary. refs. notes $16.50; $9.50pa. LC 74-17227. ISBN 0-08-018274-7; 0-08-018273-9pa.

    Whereas most volumes on the Moon since 1969 have concentrated mainly on the mineralogy and geochemistry of the lunar samples (including Taylor's *Moon Rocks and Minerals*, with A. A. Levinson, 1971), this excellent book combines those topics with an overall view of lunar geology. Written in review style, with hundreds of references to previous work, the volume looks in retrospect at the thousands of pages of scientific papers produced since the beginning of the Moon landings. The first major chapter, on lunar geology, takes a descriptive look at the physical characteristics of the surface (like craters, peaks, ringed basins, lava flows, rilles, ridges, etc.). Taylor also discusses the various theories advanced in explaining the formation of the geological features. The nature of the surface material itself is considered at length in the following chapter, examining the regolith (the layer of debris over the lunar bedrock), the soil, glasses, craters, meteorite flux, and more. The types of rocks and their mineralogy are detailed in the chapters on the maria (the "seas," or lowlands) and the highlands; followed by a description of the lunar interior and its physical characteristics. The drastic changes in our theories of the Moon's origin and evolution complete this book, a capsule summary of this century's greatest scientific achievement. For the graduate student, geologist, and astronomer.

Planets

394.   Alexander, A. F. O'D. **The Planet Saturn; A History of Observation, Theory and Discovery.** New York, MacMillan, 1962. 474p. illus. index. refs. LC 62-52625.

This volume remains the standard work on the ringed planet, since no general treatises have appeared since. Because it is slightly dated, there are no up-to-the-minute reports of the latest findings and no mention of the new satellite discovered in the 1960s. Nevertheless, the book is an excellent piece of writing and a comprehensive look at the history of man's interaction with Saturn. Observations by the ancients, by Galileo, by Cassini, and by Herschel, to name a few, are the highlights of this fine book. As might be expected, a great deal of space is devoted to the magnificent rings that circle the second-largest planet, and Alexander tells of the various theories related to their composition, rotation, and size. Other topics not to be overlooked are studies of the belts or zones on the planet's surface, the moons of Saturn (including Titan, its largest satellite), and various observations. Although frequent references are incorporated into the text, a separate bibliography would have been more desirable. Any library would want this book, which is now unfortunately out of print.

395.   Alexander, A. F. O'D. **The Planet Uranus.** London, Faber & Faber; New York, American Elsevier, 1965. 316p. illus. index. bibliog. LC 65-22522.

In a follow-up to his book on Saturn, the author presents a comprehensive look at the third largest planet in our solar system. The first and only volume devoted entirely to Uranus, it begins with the official discovery by William Herschel in 1781. Like *The Planet Saturn* (1962), this work is extremely well researched and given to great detail as the author tells the history of the "green gas giant," not to be confused with the jolly green giant. One of the highlights of the book is the chapter on how Uranus was named, in which Alexander tells of the many proposed names like Herschel, Astraea, Georgium Sidus, and several other alternatives. J. E. Bode (of Bode's law of planetary distances) was the eventual winner in this sweepstakes, which was not decided overnight. The orbit, satellites, size, cloud bands, polar flattening, and various observations are a few of the subjects discussed by the author. Despite being over ten years old, the book is still fairly up to date, although a new edition describing the discovery of rings around the planet (1977) would be desirable. Like the companion book on Saturn, though, its emphasis is historical and it will retain its value for a long time to come.

396.   Asimov, Isaac. **Jupiter: The Largest Planet.** New York, Lothrop, Lee & Shepard Co., 1973. 224p. illus. index. glossary. $6.95. LC 72-9359. ISBN 0-688-40044-2.

Hundreds of facts and figures, historical references, and speculation comprise this excellent book for the layman, from teenage reader on up. Well illustrated and well written (as all the author's books are), this fascinating volume is sure to arouse the interest of young and old alike. Beginning with a general overview of the planets, Asimov gives some background on their discoveries and discusses their relative orbits and brightnesses. The following chapter begins with a slew of superlatives on "the largest planet," which is the subject of the remainder of the book. Jupiter's various physical characteristics are compared with those of the other planets, and the reader is shown that it is the largest, most massive planet, possessing the most satellites and emitting the most radio wave radiation of all the bodies in the solar system. A great deal of space is devoted to the 12 satellites, including their orbits, densities, masses, magnitudes, and so on. Other major topics are the Jovian gaseous atmosphere and features, the giant Red Spot, and what we could see from the surface of its moons. The book ends with a discussion of some unanswered questions about its inner structure, surface, and radio emission. Recommended for school and public libraries.

397. Bronshten, V. A., ed. **The Planet Jupiter**. Jerusalem, Israel Program for Scientific Translations, 1969. 89p. illus. bibliog. $3.00pa. NASA TT F-563 LC 73-606057.

Various ground-based mid-1960s observations of the largest planet comprise this text of 11 papers written by Russian astronomers connected with the USSR Academy of Sciences. Written for scientists, this collection is one of the few advanced-level works available on Jupiter. There is no overall theme (except maybe "Jupiter in 1963-64") however, and the usefulness of the work is questionable. Planetary and Jovian specialists will benefit most from this volume, which discusses the Jovian atmosphere, cloud belts, rotation, and various observations. "The Features of Jupiter" is the most descriptive (and least technical) piece in the book. It presents, in minute detail, a description of the various cloud bands and the Red Spot in 1963/64, complete with several sketches. This typical report will give the unfamiliar reader a good overview of the planet's appearance, taken from a specific set of observations.

398. Gehrels, Tom, ed. **Jupiter: Studies of the Interior, Atmosphere, Magnetosphere and Satellites**. Tucson, University of Arizona Press, 1976. 1254p. illus. index. refs. glossary. (I.A.U. Colloquium no. 30). $38.50. LC 75-36124. ISBN 0-8165-0530-6.

Far overshadowing its predecessors in depth of coverage and comprehensiveness of subject matter, this huge volume brings up to date the "current" status of knowledge on the "great gas giant planet." Containing 44 review papers averaging 28 pages each, this conference volume relies heavily on information gained from the Pioneer spacecraft flybys of Jupiter in 1973 and 1974. (Curiously enough, there is no mention whatsoever in this work that the papers resulted from an IAU colloquium, or *any* meeting for that matter.) The text, intended for scientists and graduate students, is divided into five parts: I, Introduction and Overview; II, Origin and Interior; III, Atmosphere and Ionosphere; IV, Magnetosphere and Radiation Belts; V, Satellites. The first section contains three articles summarizing both earlier studies and information presented later in this work. The origin and structure of Jupiter and its satellites, its atmosphere and ionosphere, and its particles and fields are outlined. This excellent presentation adequately capsulizes the current information available.

Destined to become a standard source for several years to come, *Jupiter* contains countless references to related literature, making it an even more valuable tool. Added features of note are abstracts, a list of contributors, glossary, and a good index. Thirty-six contributed papers from the IAU Colloquium held at the University of Arizona in May 1975 are not included here but can be found in special 1976 issues of *Icarus* and the *Journal of Geophysical Research.* Because 90 percent of the information in this volume would be too advanced for the layman, a not-too-technical presentation summarizing the important information here would be a welcome addition to the literature. Hopefully we'll see such a book someday. Mr. Gehrels was assisted in his work on this book by Mildred Shapley Matthews.

399. Glasstone, Samuel. **The Book of Mars**. Washington, D.C., NASA, 1968. 315p. illus. index. refs. $5.25. NASA SP-179. LC 68-62244.

Possibly the best-written, most comprehensive non-technical book on the red planet, this work holds the reader's interest from start to finish. Aimed at the general reader and college student, *The Book of Mars* covers a variety of topics from history to observation. Chapter two, on historical perspective, talks first about early observations and theories of the planets in general before beginning with the history of Martian study. This chapter is longer and more detailed than similar treatments in other books. After a discussion of Mars's place in the solar system, the author devotes a great deal of space to its physical characteristics, atmosphere, surface, clouds, etc. The treatment of these topics is generally from a historical viewpoint, tracing ideas on these subjects from the beginning to the late 1960s. The fact that the book lacks later findings from Mariner 9 and other probes does not lessen the value of these chapters. The emphasis on what we "now know" is minimal. Nearly 100 pages are devoted to an excellent dissertation on life on Mars, its possibilities and detection. Glasstone first examines how life formed on Earth and applies much of this to the Martian situation. It makes fascinating reading and is, without a doubt, the finest consideration of the topic to be found. The final chapter, on spacecraft exploration, is good but not lengthy. More comprehensive treatments can be found elsewhere. For any type of library, it includes dozens of references to related literature. It is hoped that a new edition will appear to include the Viking Mars lander mission.

400. Hartmann, William K., and Odell Raper. **The New Mars; The Discoveries of Mariner 9**. Washington, D.C., NASA, 1974. 179p. illus. index. bibliog. $8.75. LC 74-600084. NASA SP-337.

As time passes, the strange red planet, named after the Roman god of war, is becoming less mysterious and yet even more interesting. Since the earliest telescopic look at Mars, generally attributed to Galileo, in the early 1600s, our knowledge of the wind-swept desert world has grown in leaps and bounds. Before Viking, the greatest advances occurred in 1971 and 1972, when the spacecraft Mariner 9 sent back the best photographs ever of the Martian surface. This major achievement brought forth thousands of pieces of new and revised information, most of which are neatly summarized in this excellent book aimed at the general reader. Amply illustrated with black-and-white and color photographs, this work begins with a well-written summary of Martian research before Mariner 9 (i.e., from Galileo to present). Here the reader learns of the controversial canals, the naming of Martian surface features, and theories about the possibility of life. The majority of the book, however, is concerned with summarizing and describing, in layman's language, the information gathered and our "current" knowledge about the planet that sent the creatures in H. G. Wells *The War of the Worlds*. Dust storms, volcanic activity, the ever-changing polar caps, and much more are discussed clearly and intelligently. Martian geologic formations are often compared to similar situations on Earth. The final brief chapter is, in many ways, the most interesting. Here the authors list and discuss 27 major scientific discoveries and 23 new hypotheses based on the data gathered, many of which were verified or disproved with the Mars landing of the Viking spacecraft in July 1976. Highly recommended for home, public, and college libraries.

401.   Jackson, Joseph H. **Pictorial Guide to the Planets.** 2nd ed. New York, Thomas Y. Crowell Co., 1973. 248p. index. illus. $12.50. LC 72-7573. ISBN 0-690-62443-3.

One of the more up-to-date books on the solar system, this work is frequently interspersed with photographs from the NASA files. The arrangement is unlike that of similar books, devoting its chapters not to discussion of individual planets alone, but rather to general features like planetary atmospheres and surfaces, interplanetary space, and life on other planets. There are individual sections on the Earth, Moon, and asteroids, meteors, and comets, however, which help set up comparisons of the other planets later in the text. As might be expected, there is a decided emphasis on what we have learned about the solar neighborhood through spacecraft. In fact, the last three chapters are devoted exclusively to rockets and space vehicles, artificial satellites and space probes, and man in space. Further, a special supplement highlights recent space explorations and presents some of the better photographs of the Moon, Mars, and selected sky panoramas taken by spacecraft. The book's bibliography is better and longer than most and is arranged by subject for easier retrieval. The author, who revised the latest edition of Dinsmore Alter's *A Pictorial Guide to the Moon*, exhibits again here his excellent ability to explain and describe our Universe. Aimed at the general reader, this volume would be ideal for the public and secondary school library.

402.   Koenig, L. R., *et al.* **Handbook of the Physical Properties of the Planet Venus.** Washington, D.C., Scientific and Technical Information Division, NASA, 1967. 132p. illus. refs. glossary. $0.60pa. LC 66-61837. NASA SP-3029.

Like the Mars and Jupiter publications with the same titles, this volume summarizes our knowledge through the mid-1960s. Produced mainly for the use of space scientists and astronomers, the guide follows the same format as its companions covering similar subjects: orbit, mass, diameter, rotation, gravity, temperature, atmosphere, surface, and more. Containing 162 references, it should be updated for increased usefulness.

403. Michaux, C. M. **Handbook of the Physical Properties of the Planet Jupiter.** Washington, D. C., Scientific and Technical Information Division, NASA, 1967. 142p. illus. refs. glossary. $0.60pa. LC 66-61838. NASA SP-3031.

Although some of the information in this book has been or will be revised, it is still a good volume to have in the library. Like the Venus and Mars books, it was written mainly for space program scientists as a reference work to be used in project planning and in research and development. The author and four "contributors" present a concise yet comprehensive (for 1967) look at the physical characteristics of our largest planet. Drawing on data gathered by both optical and radio observations, the book includes facts and figures in textual and tabular format on mass, rotation, gravity, temperature, atmosphere, satellites, and other pertinent areas. An important section is the description of the radiation Jupiter emits, radiation that could affect spacecraft sent to study the planet. A list of 265 references accompanies the clearly written text, which now can be updated using the data gathered by Pioneer spacecraft.

404. Michaux, C. M. **Handbook of the Physical Properties of the Planet Mars.** Washington, D.C., Scientific and Technical Information Division, NASA, 1967. 167p. illus. refs. glossary. $0.70pa. LC 66-61839. NASA SP-3030.

A new edition of this handy guide would be welcome, since it lacks so much of the newer data gained since its publication. Nevertheless, most of the information is still valid, and this is one of the better sources for complete physical data on the Red Planet. Selected topics include orbit, mass, diameter, rotation, gravity, internal structure, temperature, atmosphere, life, satellites; in short, it is a storehouse of knowledge. In most cases, equations and methods of determining the physical properties of Mars are presented first in each chapter. Included are 314 references to the literature, a substantial bibliography for the ambitious researcher. The only illustrations are diagrams (which is unfortunate, since some photographs would have been useful). The lay reader could handle this book with little trouble, but there are much better volumes along those lines (e.g., *The Book of Mars*, by Samuel Glasstone; NASA, 1968). Best suited for the astronomical or space science researcher; not all libraries would want it.

405. Moore, Patrick, and Charles A. Cross. **Mars.** London, Mitchell Beazley; New York, Crown Publishers, 1973. 48p. illus. index. $8.95. LC 73-78847. ISBN 0-517-50527-4 (Crown).

Intended mainly for the general reader, but useful to students and scientists as well, this excellent book is an atlas, including maps and photographs made possible by Mariner spacecraft. The superb charts, drawn by Charles Cross using photographic mosaics supplied by NASA, show the landforms in painstaking detail. Features like craters, deserts, canyons, and volcanoes are labeled, and there is an index to the names of these. The text includes information on Mars's place in the solar system, its make-up, exploration (past, present, and future), observation, topography, and satellites. This beautifully illustrated, well-organized atlas belongs in nearly every type library: public, college, and observatory. Highly recommended for home libraries as well.

406. Moore, Patrick. **The Planet Venus**. 3rd ed. London, Faber & Faber, Ltd.; New York, Macmillan, 1961. 151p. illus. refs. indexes (subject and reference). LC 64-32148 (Faber); 64-9223 CD (Macmillan).

One of the few books devoted to the planet Venus, this interesting work for the layman and amateur astronomer emphasizes history and speculation. Venus, shrouded with thick clouds, remains somewhat a mystery, even after investigation by space probes. Much more has been learned since this book was written, though, and it is unfortunate that this new information is missing. Since there are so few books on Venus, this being the only one for the general reader, a new edition would be desirable. Topics include the "evening star," movement, "polar caps," rotation, atmosphere, transits and occultations, surface, life, and more. It is a good introduction but is of little use beyond that level.

407. Moroz, V. I. **Physics of Planets**. Washington, D.C., NASA, 1968. 412p. illus. index. refs. $3.00pa. NASA TT F-515.

This technical volume attempts to integrate data on the physical characteristics of the members of the solar system as known in the mid-sixties. Chapter I, on planetary astrophysics, is useful for the beginner who needs some preliminary knowledge before tackling the remainder of the volume. Chapters II, III, IV, and V consider the physics of Mars, Venus, Mercury, and the Giant Planets, respectively. The author covers temperature of surfaces and atmospheres, composition, density, nature of the atmosphere and surface, internal structure, and so on. Each section has a brief descriptive introduction about the planet(s), but for the most part this is an advanced volume, intended for the graduate student and scientist. Though it has become partially outdated because of recent interplanetary probes, a majority of the information in the book is still valid. Like many Soviet works, this has an extensive bibliography (635 citations) to journals and books on related topics; it is a translation of *Fizika Planet* (Moscow, 1967). For astronomy libraries.

408. Nicolson, Iain. **Exploring the Planets**. New York, Grosset & Dunlap, 1971. 159p. illus. index. bibliog. (All Color Guides). $3.95. LC 74-147096. ISBN 0-448-00868-8.

Suitable for young and old alike, this colorful little volume examines past, present, and future studies of the solar system. Especially good for younger students' papers and for browsing, this guide discusses in detail each planet and its moon, noting physical characteristics like temperature, atmosphere, rotation, orbit, etc. Tied in with these descriptions are details of some of the planetary explorations already completed or planned for the future. Illustrations of the planets, spacecraft, and related information are the high points of the book, along with the clearly written text. Definitely intended for the beginner, the volume really only hits the high spots but, for the intended readership, this is all that is necessary. Further, there is an excellent list of elementary or beginning texts for the reader who wishes to advance his studies. A chapter on amateur planetary observation is included for those with stargazing leanings. For all ages and all libraries (home, public, and school), this book comes highly recommended. Also available from Bantam Books as volume 42 of the "Knowledge Through Color" series ($1.95).

409. Nourse, Alan E. **Nine Planets.** rev. ed. New York, Harper and Row Publishers, 1970. 322p. index. illus. $10.95. LC 70-105234. ISBN 0-06-013222-1.

Not purely scientific fact and equation, nor solely a "gee-whiz!" look at the planets, nor even just a book that talks about how to get into space and study the planets, this fine work is a bit of all three, with some imagination and speculation thrown in. It is purely a narrative, devoid of mathematics, that describes our solar system and poses some interesting questions about the conditions on the planets and their satellites. The reader's attention will be held securely by this experienced science writer's style, and his description of past and future exploration by spacecraft. Each planet has its own descriptive chapter, including discussions of its surface and atmospheric conditions, the possibility of life, and more. The book ends on a stimulating and controversial note, posing the question of other solar systems far beyond ours. There are very few illustrations, though these are quite good; but this lack does not detract from the book, which is intended for the general reader and is highly recommended for the public library.

410. Page, Thornton, and Lou Williams Page, eds. **Wanderers in the Sky.** New York, The Macmillan Co., 1965. 338p. illus. index. bibliog. glossary. (Sky and Telescope Library of Astronomy, v.1). $7.95. LC 65-12722.

Written for the amateur astronomer and layman, this book focuses on two types of "wanderers"—natural (the planets) and man-made (artificial satellites). The latter material is somewhat outdated (it covers through the Mercury flights), but it nonetheless provides some historical perspective on the early space shots. The material on early planetary orbital studies is especially good, explaining Kepler's laws of planetary motion, Galileo's observations of the "wanderers," the Copernican theory, and more. The celestial motions of the five outer planets are detailed in Chapter 2, called "Newton's Mechanical System" in reference to his theory of gravitation and laws of physics, both of which are essential to an understanding of planetary movement. The following chapter includes several reports of early satellite launchings and the data gathered as a result. The hazards of interplanetary space travel are discussed in Chapter 4 on meteors, the Van Allen Radiation Belt, cosmic dust, solar particles, etc. A now-outdated last section discusses in detail our natural satellite, its orbit, surface characteristics, atmospheres, seismology, and much more. It is interesting to read this chapter, comparing it with later, post-Apollo literature. Like the other volumes in the series, this work is a collection of articles from past issues of *Sky and Telescope* written, for the most part, by famous astronomers. Readers should keep in mind that the emphasis in this book is on planetary *motion*, not on the planets themselves.

411. Peek, B. M. **The Planet Jupiter.** London, Faber & Faber, Ltd.; New York, Macmillan, 1958. 283p. illus. index. LC 58-14816.

Although a certain portion of this semi-classic work is now outdated, it remains an excellent source of information about the giant planet. Drawing mainly on his own experience observing Jupiter for many years, the author, in effect, has written a history of Jovian observation from the late-nineteenth to the mid-twentieth century. After a very brief introduction to the nomenclature used and other general remarks, Peek dives into the major chapter of the book: "Observations of Jupiter's Surface." Here in minute detail, the author describes each region, belt, and zone of the planet's atmosphere. The Great Red Spot is dealt with at length. Telescopes used, observational techniques, and related concerns are described first. Along with the text in this chapter are numerous sketches of each area. Various observations of each region by the author and others are dated and include rotation period and number of spots observed. It is one of the most complete descriptions ever presented for *any* astronomical phenomena. The serious and enthusiastic planetary observer should be inspired by the work presented here. Peek also highlights photographic, spectroscopic, and very early radio observations. Part III discusses theoretical considerations on the planet's composition, physical condition, and temperature. It should be interesting to compare his observations of 20 years ago with "current" information gained from radio telescopes and spacecraft. The book ends with a chapter on the Jovian satellites and four appendices on special topics. Intended for any interested reader, the book is not at all technical, yet it is also not elementary.

412.　Richardson, Robert S., and Chesley Bonestell. **Mars.** New York, Harcourt, Brace & World, Inc., 1964. 151p. illus. index. glossary. LC 63-17767.

Although some of the information in this book has been updated by recent Viking and Mariner exploration of Mars, most of it is still valid, and this volume is ideal for the general reader. Aided by the excellent paintings of Chesley Bonestell, the author describes the "red planet" and theorizes on how a trip might be made there in the future. Not an overly elementary text, neither is it too technical. A wide variety of subjects are covered, giving the Martian enthusiast much to devour. Among some of the more interesting topics are life on other planets, seasons on Mars, visible surface changes (vegetative and non-vegetative hypotheses), the canals (origins and theories), the moons, and more. Included are many helpful hints for observing the planet, a glossary, and Mars numerical data. Since this is one of the better volumes available, and since we have recently gained much new information on Mars, a new edition would be greatly welcomed.

413.　Sandner, Werner. **The Planet Mercury.** New York, The Macmillan Co., 1963. 94p. illus. index. LC 63-13485.

Ten years after this book was published, a NASA spacecraft flew by Mercury gathering some incredible photos and a great quantity of new data. Why then, is this little book still of use? Because it is not technical, and because it is a good description of and introduction to the planet closest to the Sun. There are good historical material and excellent illustrations, too, and the volume is well suited to the amateur astronomer. Topics discussed include rotation, surface features, atmosphere, transits, and other interesting subjects.

414.　Sandner, Werner. **Satellites of the Solar System.** London, Faber & Faber; New York, American Elsevier Publishing Co., Inc., 1965. 151p. illus. index. LC 65-22523.

Concerned mainly with the moons of Jupiter and Saturn, this volume for the amateur and layman is a good general treatment of a subject that is usually relegated to a small section of an introductory text. Sandner begins by presenting the views of the solar system since the sixteenth century, and then spends two lengthy chapters on the Jovian satellites and Saturn's rings and moons. In the former, he describes the discovery of the moons by Galileo, the determination of the speed of light by Rømer, surface features, atmospheres, etc. The rings around Saturn are of particular interest here, and the author goes into detail on the history of their study, discovery, and observation; the nature of the Saturnian satellites is covered, as well. The final third of the book is concerned with Uranus, Neptune, and Pluto, and the Earth, Mars, Venus, and Mercury, respectively. The Earth's moon is given brief coverage, but the reader can easily find dozens of books devoted to our natural satellite. There are a few illustrations and a handful of tables of statistical data which add to the work's generally historical and descriptive outlook. A good introduction, suitable for the public and college library; it should be updated some day to include the newly discovered moons of our two largest planets.

415. Slipher, Earl C. **A Photographic Study of the Brighter Planets.** Flagstaff, Ariz., Lowell Observatory; Washington, D.C., National Geographic Society, 1964. 125p. 62 plates. illus. notes. LC 64-18807.

Because there are now much better photographs of the planets taken by NASA spacecraft, this book is no longer as useful from the practical standpoint as it once was. Now mainly of interest as a comparison of early planetary observations, the volume contains some of the best photographic sequences taken by the author over a period of 56 years (1907-1962). Included also are some of the early planetary drawings by Slipher, an astronomer at Lowell Observatory, and other artists. Each sequence showing Mercury, Venus, Mars, Jupiter and Saturn is discussed in detail, explaining photographic techniques and analyzing and comparing each plate. Surface features in the photographs are discussed at length, and it is interesting to note the changes which take place over the long period covered. Mars and Jupiter hold the spotlight, accounting for over half the book.

416. Tricker, R. A. R. **The Paths of the Planets.** New York, American Elsevier Co., 1967. 240p. illus. index. LC 67-27999.

Aimed at the astronomy student and interested amateur astronomer, this work is concerned with the "geometry of the solar system." As might be guessed from the title, the emphasis is on the orbits of the planets and their satellites, and the author uses a multitude of diagrams and a few simple equations to illustrate this subject. The book begins with a descriptive overview of astronomy, the constellations, and what can be observed about the motions of the heavenly bodies. The following chapter is a brief look at the history of the heliocentric (Sun at the center) theory of the solar system, including mention of Ptolemy, Copernicus, and others. The Earth's orbit is considered at length, the author explaining the period, the seasons, and distance from the Sun. The majority of the rest of the book is concerned with the individual planets and their orbital geometry; tables and diagrams of the orbital elements abound. The scale of the solar system and gravity theory round out this non-technical practical book. Serious amateurs concentrating on planetary observations should study and read it.

Comets, Meteors, and Meteorites

417. Barnes, Virgil, and Mildred A. Barnes, eds. **Tektites**. Stroudsburg, Pa., Dowden, Hutchinson & Ross, Inc.; distr. Halstead Press, John Wiley and Sons, Inc., 1973. 445p. illus. indexes (subject, author, and author citation). bibliog. (Benchmark Papers in Geology). $22.00. LC 72-95942. ISBN 0-87933-027-9 (D, H & R); 0-471-05340-6 (Wiley).

Until the publication of this compendium, and except for *Tektites* (University of Chicago Press, 1963), edited by John A. O'Keefe, the majority of the literature on the subject has been scattered in over 1,000 articles in astronomy, geology, and general science journals. The editors of this collection have selected, arranged, and commented on 46 of those pieces in this overview of the strange, meteorite-like stones. Both historical and current perspective may be gained from this volume, which shows the growth and change of ideas about tektites, as well as an illustration of the controversy surrounding their origin. Photographically reproduced, the papers appear exactly as they did in the original sources; many contain abstracts and most include a substantial number of references. Although this advanced text is more geological than astronomical, the tektites' probable extraterrestrial origin allows this book to fit into the astronomical literature along with work on meteorites. For the astronomy and geology collection, this book is divided into the following sections: I, Geology, Petrology, and Mineralogy; II, Physical and Chemical Data for Tektites; III, Ages of Tektites; IV, Sculpture of Tektites; V, Microtektites; VI, Origin of Tektites; VII, Bibliography Prior to 1959.

418. Brown, Peter Lancaster. **Comets, Meteorites and Men**. New York, Taplinger Publishing Co., 1974 (c1973). 255p. illus. indexes (subject and name). $12.50. LC 73-16633. ISBN 0-8008-1734-6.

A fascinating book for the general reader and student, this non-technical volume contains both historical and current information on the less-permanent wanderers of the solar system. Two-thirds of the work is devoted to comets, and, besides describing their physical characteristics, the author tells of individual comets of note and of famous astronomers connected in some way with the discovery of those comets. The chapter on comet hunting is very good, too; this is obviously a topic that is of great interest to Brown. In fact, his enthusiasm pervades this book, making it a pleasure to read. Meteorites, too, are discussed at length, including what they tell us about the origin of the universe, and what clues they provide as to the origin of the chemical elements. Seven useful appendicies accompany the profusely illustrated text and include information on orbits, magnitudes, meteorite impact sites, major meteor showers, etc. Originally published in London in 1973 by Robert Hale, this book is highly recommended for all types of libraries, but especially college and public.

419. Chebotarev, G. A., E. I. Kazimirchak-Polonskaya, and B. G. Marsden, eds. **The Motion, Evolution of Orbits, and Origin of Comets.** Dordrecht-Holland, D. Reidel Publishing Co.; distr. New York, Springer-Verlag, 1972. 521p. illus. refs. $47.90. LC 73-179895. ISBN 90-277-0207-1 (Reidel); 0-387-91103-0 (Springer).

This conference volume contains 85 technical papers discussing recent research on comets, touching on some general topics as well as highly specialized concerns. The editors have arranged the papers into six logical groupings for easier searching (there is no index): I, Observation and Ephemerides; II, General Methods of Orbit Theory (Analytical Methods; Numerical Methods; Determination of Orbits); III, Motions of the Short Period Comets (Planetary Perturbations and Non-gravitational Effects; Determination of Planetary Masses); IV, Physical Processes in Comets; V, Origin and Evolution of Comets (Orbital Stability and Evolution; Theories of Cometary Origin); VI, Relationship with Meteors and Minor Planets (Orbital Evolution of Meteors and Minor Planets; Possibility of Common Origin). There are several papers on the "current status" of particular areas, but the book would be even more useful if an overall review paper were included. Perhaps the interested reader should consult *Comets, Scientific Data and Missions* (Kuiper and Roemer, eds., 1972) for a slightly less technical overview. From IAU Symposium no. 45 (Leningrad, August 4-11, 1970), this volume would be appropriate for the observatory and university library.

420. Gehrels, Thomas, ed. **Physical Studies of Minor Planets.** Washington, D.C., NASA, 1971. 687p. illus. index. glossary. refs. $3.00pa. LC 73-169176. NASA SP-267.

Before this volume of conference proceedings there was no single compilation of a comprehensive nature devoted to the planetoids. Information was scattered throughout the literature—a chapter here, an article there. The gap in the literature has been more than adequately filled by this excellent group of 79 papers, arranged in three major divisions: Observations; Origin of Asteroids; Possible Space Missions and Future Work. The coverage of individual topics is quite broad, and it's a good bet that anything important is here. The editor sees it as a textbook, but that point is debatable: the arrangement of subjects into three groups with no subdivisions, and the not-always-logical flow from one paper to another tends to counteract this view. It is an excellent book, however, and should be considered a standard guide for those in the field, to be used along with *Tables of Minor Planets* (Pilcher and Meeus, 1973). The well-written introductory chapter, dozens of references, and an index are the high points of the volume, comprised of review-type papers.

421.   Heide, Fritz. **Meteorites**. Chicago, University of Chicago Press, 1964. 144p. illus. indexes (subject, meteorite, crater). (Phoenix Science Series, PSS522). $2.25pa. LC 63-20906. ISBN 0-226-32339-0, P522.

The general reader or amateur astronomer who wants a good, short introduction to meteorites would do well with this book. Well illustrated and not at all technical, the text includes discussions of meteorite showers, meteorite craters, historical facts, chemical constitution, classification, origin, and tektites. The book explains some basics like the two types of meteorites (stone and iron), how to recognize a meteorite (few people can), and the significance of their (geologic) age. The chapter on chemical constitution gives the reader just enough information, avoiding the typically overboard treatment of this subject. The composition of meteorites is not insignificant, however, and is treated substantially in other texts. A few highlights are pictures of large meteorite craters and fireballs, and the stories of some of the larger meteorite finds. The reader's attention will be held by this interesting "first" book on the subject, and those wishing more detail should consider *Meteorites and Their Origins,* by G. J. H. McCall, which emphasizes classification, structure, and composition of the stones. Translated from the German (*Kleine Meteoritenkunde,* 1957) by Edward Andrews with Eugene R. DuFresne, this would be a good selection for nearly any library.

422.   Krinov, E. L. **Giant Meteorites**. Oxford, Pergamon Press, Ltd., 1966. 397p. illus. author index. bibliog. $20.00. LC 65-13142. ISBN 0-08-011121-1.

This fascinating book considers only those meteorites large enough to penetrate the Earth's atmosphere to form craters. The author presents a descriptive look at these rare and interesting occurrences, explaining the unusual circumstances surrounding the giant blocks of stone and iron. Divided into four major sections, the volume begins with a general chapter including a description of the different types of craters formed by meteorite impacts, and a look at 12 known and 11 suspected meteorite craters. The circumstances and material found in and near each site are detailed. The second and third chapters are quite long, concentrating on the Arizona (Barringer) Meteorite Crater and the Tunguska Meteorite (USSR). The former is the first giant meteorite crater to be discovered as such, and the latter caused widespread damage. The author goes into so much detail on these two, especially the latter, that the reader learns more than he ever wanted to know about either. The final section zeroes in on a famous meteorite shower in the Soviet Union in 1947, which provided many "stones" for study. Translated from the Russian by J. S. Romankiewicz and edited by M. M. Beynon; this book should be approached only after an introductory text has been read.

423.   Kuiper, G. P., and E. Roemer, eds. **Comets; Scientific Data and Missions**. Tucson, Lunar and Planetary Laboratory, University of Arizona, 1972. 222p. illus. index. refs. LC 72-619613.

Sending a spacecraft to study a comet close-up has been on the minds of astronomers for years, and NASA had hoped to carry out the first such mission in 1976, when the Comet d'Arrest returned and was in a favorable position for rendezvous. To discuss cometary science in general and to consider the planning of a space venture, the Tucson Comet Conference was held in April 1970. This volume is the proceedings of that meeting, and it includes more than two dozen papers, primarily concerned with current knowledge. Among the papers presented were discourses on models of cometary nuclei, infrared measurements of comets, nature of the cometary head, photometry of comets, various types of tails, spectra, orbits, and more. A second section on a comet rendezvous with a spacecraft discusses the Comet d'Arrest mission, trajectory requirements, and scientific criteria. A supplementary section includes an overview of cometary knowledge in 1972 and some interesting photographs of Comet Bennett.

424. Ley, Willy. **Visitors From Afar: The Comets**. New York, McGraw-Hill Book Co., 1969. 144p. illus. index. bibliog. $4.72. LC 69-17451. ISBN 0-07-037636-0.

The student or general reader who wishes a brief introduction to comets should consider this good text as a start. Mainly a description of some of the most famous "visitors from afar," it devotes only a small section to their composition and origin. Readers will be treated to descriptions of cometary orbits, naming of comets, some famous comets, comet groups and families, and a proposed space mission to study a comet close-up. Lengthy appendices cover Halley's comet, the Jupiter Family, meteor streams associated with comets, and famous astronomers who discovered and studied comets. This book is good, but others are better, and this is far too brief. Readers who wish to proceed should next consult *Comets, Meteorites and Men*, by Peter Lancaster Brown, or one of the others listed in this section.

425. McCall, G. J. H. **Meteorites and Their Origins**. New York, Halstead Press, John Wiley and Sons, Inc., 1973. 352p. illus. index. bibliog. $12.95. LC 72-7640. ISBN 0-470-58115-8.

Drawing on a number of scientific papers and previously published volumes, the author touches on a large number of areas of interest in the study of "stones from the sky." Included in this comprehensive volume are historical and physical aspects, theory of origin and planetological considerations. Meteoritics embodies a variety of scientific disciplines from astronomy to chemistry to geology, and the emphasis here is decidedly on the latter two. In the brief chapter on the astronomy of meteorites, McCall touches upon the origin of meteorites, followed by the section on meteorite flight, falls, and impacts. Included here are details of large meteorites and fireballs. The morphology, classification, mineralogy, and chemistry of meteorites are just a selection of the topics that make up a majority of the text. The illustrations of whole meteorites and cross-sections of the same are numerous and well chosen. One of the best books available on the subject, it is certainly also one of the most comprehensive. This is not an elementary book, however, and, although it is not technical, it would best suit serious amateurs, university students, and scientists.

426.   Millman, Peter M., ed. **Meteorite Research.** Dordrecht-Holland, D. Reidel Publishing Co.; distr. New York, Springer-Verlag, 1969. 941p. illus. refs. (Astrophysics and Space Science Library, v.12). $49.20. LC 70-415881. ISBN 90-277-0132-6 (Reidel); 0-387-91025-5 (S-V).

The geologist and astronomer will both likely find several papers of interest in this extensive conference volume which presents a good survey of "current" work in meteoritics. The major portion of the 73 technical papers is devoted to composition and structure, with the next most important topic being isotope studies and chronology. There is also a section on the early history of meteorites, relating mainly to solar nebulae and cosmogony discussions. Orbits, a topic often slighted in collective works, are discussed at some length. Although most papers are in English, a few are in Russian and French; most contain abstracts and references in the standard technical paper format. From a symposium on meteorite research held in Vienna, August 7-13, 1968, this volume unfortunately has no index, a serious drawback to its usefulness.

427.   Moore, Patrick. **The Comets: Visitors from Space.** Shaldon, Engl., Keith Reid, 1973. 94p. illus. index. glossary. £2.25 £0.75pa. LC 74-155652. ISBN 0-904094-00-6; 0-904094-01-4pa.

This short but informative book was written to coincide with the arrival of Kohoutek, the expected-to-be-brightest comet of the century. Unfortunately, Kohoutek, later known as the flop of the century, fizzled and was visible only with telescopes. Fortunately, Moore's book does not fizzle after starting, and the layman interested in comets will find an enjoyable combination of historical and descriptive information. This volume is up to Moore's usual excellent narration, taking the reader on a tour through space to meet some of the more famous comets (in expectation, of course, that Kohoutek would join their ranks). Moore describes such notables as Halley's Comet, Encke's Comet, and several others. The source of comets and their paths are two other aspects discussed. There are more detailed descriptions elsewhere of cometary composition, but the basics are here for the new reader. Appropriate for public and college libraries, this book is a good introduction to the subject.

428.   Nininger, Harvey H. **Find a Falling Star.** New York, Paul S. Eriksson, Inc., 1972. 254p. illus. index. glossary. $8.95. LC 72-83710. ISBN 0-8397-2229-X.

If Harvey Nininger did not appear on *What's My Line?*, he should have. Certainly his profession, hunting meteorites, is one of the most unique jobs ever undertaken. For more than 50 years, Mr. Nininger, his wife, and children hunted meteorites; they collected them, traded them, sold them, and exhibited them. This volume, an autobiographical account of this continuing search for rocks from the sky will surely interest many astronomers, both amateurs and professional. Written in narrative style, it recounts the strange adventures surrounding a strange quest. The author, in his work with meteorites, has educated layman and scientist alike about their importance and what they tell us about the Universe. No other individual has contributed more in this respect. For any type of library.

429. Richardson, Robert S. **Getting Acquainted with Comets**. New York, McGraw-Hill Book Co., 1967. 306p. illus. index. LC 66-24480.

One of the longer and better books on the subject, this work is written for amateur astronomers and students who want an introduction but also want to go beyond the bare minimum. Because it is aimed at an audience of general readers, the text is descriptive rather than technical, and the author's style makes it enjoyable reading. Besides the usual subjects (physical composition, orbits, and famous comets), the book has a substantial portion devoted to comet hunting, a project that the ambitious amateur can embark upon rather easily. A discussion of telescopes is included in this section, the author telling what types are best for comet searching, and how to use them to this end. He even tells what to do with a comet after it's been found. Interested readers should also consult *Comets, Meteorites and Men* (1974) by Peter Lancaster Brown for more information on comet hunting. Richardson also devotes a portion of the book to the comets' third cousins, the meteors. This book would be appropriate for public and school libraries.

430. Vsekhsvyatskii, S. D. **Physical Characteristics of Comets**. Jerusalem, Israel Program for Scientific Translations; Washington, D.C., Office of Technical Services, U. S. Department of Commerce, 1964. 596p. illus. indexes (name and comet). $6.00pa. LC 64-2766. OTS 62-11031. NASA TT F-80.

If this book is not well known, it should be. The author begins with general chapters on the nature of comets, their photometry, and a statistical analysis of cometary orbits. Next comes a list of comets and their absolute magnitudes, along with other important data. The remainder of this fine volume (about 500 pages) lists and describes every comet ever recorded up through 1955. Description includes discoverers, location in the sky, further references, and orbital information. A treasury of statistical data, this volume should not be missed; it ought to be in the reference section of every astronomy library. Translated from the Russian, it should, however, be updated to include the last 20 years.

431. Wasson, John T. **Meteorites: Classification and Properties**. Berlin, New York, Springer-Verlag, 1974. 316p. illus. subject index. refs. glossary. (Minerals and Rocks, 10). $31.20. LC 74-4896. ISBN 3-540-06744-2 (Berlin); 0-387-06744-2 (N.Y.).

This handbook is both an introduction to meteorite science and a guide to classification. Aimed at advanced students and non-specialists, it comprehensively covers the field of meteorites, a topic of interest to geologists and astronomers alike. Selected chapter headings include classification (silicate-rich and metal-rich), mineralogy and phase composition, petrology, trace elements, stable isotopes, origin and history, orbits, fall and recovery, morphology, and magnetic properties. The author also explains various physical studies to determine density and porosity, and mechanical, thermal, electrical, and optical properties. A review paper on chrondites illustrates how meteorite properties are interpreted. References are very extensive and constitute a substantial bibliography. Four lists of classified meteorites comprise a lengthy and useful appendix. For the astronomy or geology library, the book is a good beginning text; it should be used along with Mason's *Handbook of Elemental Abundances in Meteorites* (1971).

432. Wood, John A. **Meteorites and The Origin of Planets.** New York, McGraw-Hill Book Co., 1968. 117p. illus. index. refs. (Earth and Planetary Science Series). $6.50; $3.95pa. LC 68-13886. ISBN 0-07-071581-5; 0-07-071580-7pa.

What meteorites tell us about creation in the Universe, in particular, the origin of the solar system, is the theme of this brief, non-elementary book. The emphasis of subject matter is on the composition and age of the "stones," the writer contending that we study meteorites because they are so old, not because they "fall" out of the sky. A fair amount of space is devoted to the dating of the material in the meteorites, along with an analysis of the chemical make-up. Wood's last three chapters are the major portion of his theme on the relation of meteorites to planetary formation. Various theories of the origin of the solar system are presented, culminating with the author's ideas on the origin of meteorites, the oldest material available for study. For the non-specialist and science student, this book is not *the* source of information on meteorites; rather, it is one aspect put forth by one researcher.

Atmospheres

433. Craig, Richard A. **The Edge of Space: Exploring the Upper Atmosphere.** Garden City, N.Y., Doubleday and Co., Inc., 1968. 150p. illus. index. (Science Study Series). $4.50; $1.25pa. LC 68-10568. ISBN 0-385-00263-7; 0-385-06363-6, S55pa.

The interested amateur or student who desires a brief introduction to the Earth's upper atmosphere would do well to read this short but informative text. Not an astronomical topic in the truest sense, the upper atmosphere is a subject that every astronomer comes into contact with in one way or another—for example, in the study of the auroras. And certainly the space scientist, who is concerned with satellites and rockets, needs to know about the environment near the Earth. On this premise, and the fact that many great strides have been made recently in the study of the upper atmosphere, the author has written this text. He describes in clear, understandable language how the upper atmosphere is observed using rockets, satellites, and ground-based instruments, and the result of those studies. The various layers of the atmosphere are described, including chemical composition and temperature. Other subjects of interest are the ionosphere, atmospheric tides, the Earth's magnetic field, aurora and airglow, and the outermost atmosphere. This well-written book would be appropriate for college and public libraries.

434.   Dobson, G. M. B. **Exploring the Atmosphere**. 2nd ed. London, Oxford University Press, 1968. 209p. illus. index. $10.25; $8.00pa. LC 70-373980. B 69-02551. ISBN 0-19-851917-6; 0-19-851918-4pa.

Written for general readers and non-specialists, this book selectively explains certain topics in the realm of meteorology and geophysics. Each chapter discusses one topic falling into the range of atmospheric studies, describing present knowledge and current research. The first six chapters are related to meteorology, giving an overview of the atmosphere from top to bottom, including discussions of temperature and density, and touching on topics related to the weather: clouds, rain, hail, thunderstorms. Chapter six, on the ozone layer, takes on a great importance in this day of environmental outcry. The last five sections are concerned with astronomical subjects, in particular The Sun, Sunspots, and Solar Activity; The Ionosphere; The Aurora and Airglow; The Earth's Magnetic Field and the Upper Atmosphere; and The Magnetosphere and the Van Allen Belts. The treatment of these areas is unfortunately brief, but that can be expected in a general volume on the Earth's atmosphere, which this is. If further reading on astronomical topics is desired, the reader is referred to Craig's *The Edge of Space: Exploring the Upper Atmosphere* (1968).

435.   Goody, Richard M., and James C. G. Walker. **Atmospheres**. Englewood Cliffs, N.J., Prentice-Hall, 1972. 150p. illus. index. bibliog. (Foundations of Earth Science Series). $7.95; $2.95pa. LC 78-172279. ISBN 0-13-050096-8; 0-13-050088-7pa.

Recent advances in radio astronomy and successful data-gathering missions by interplanetary spacecraft have provided astronomers with much-improved values as well as new information on the planets. The authors have drawn upon and condensed this newly acquired data, combined with previously known information in this excellent volume on planetary atmospheres. Aiming at providing an understanding of the interactions of atmospheres with solar radiation, the authors present comparisons of expected results based on known physical processes and theory, with actual observations. Mainly presenting the cases for the Earth, Venus, and Mars, the book shows the differences among the atmospheres, interpreting the meanings of the results. Aimed at the student and the scientifically oriented layman, the work is a good general account of some of the more interesting aspects of atmospheric studies. After a summary/data chapter on the Sun and the planets as background material, the book discusses the effects of solar emissions on the chemical composition of atmospheres. The remaining chapters include looks at temperatures in the atmospheres, winds, condensation and clouds, and evolution of atmospheres. A good acquisition for the public and university library.

436.   McCormac, B. M., ed. **Atmospheres of Earth and the Planets**. D. Reidel Publishing Co., Dordrecht-Holland, 1975. 454p. illus. index. refs. glossary. (Astrophysics and Space Science Library, v.51). $65.00. LC 75-4954. ISBN 90-277-0575-5.

An up-to-date collection of papers, this volume of proceedings devotes about three-quarters of its text to the study of the Earth's atmosphere. After summary papers on optical observations and neutral chemistry in the Earth's atmosphere, the book turns to a discussion of physical processes, the papers touching on atmosphere physics, vertical transport, energy sources, and more. The structure and composition of the neutral and ionized atmosphere are covered in two articles reporting satellite and radar results. Rate coefficients are covered in part IV, which includes papers on ion, oxygen, and hydrogen chemistry. The book then turns to some theoretical considerations in a section on models of the atmosphere and ionosphere. Optical observations in various portions of the spectrum are described in part VI, which summarizes some spacecraft, rocket, and satellite missions. Part VII is the only section of the book which deals with atmospheres other than the Earth, beginning with a comparison of Mars and Venus. Recent space missions have contributed greatly to our meager knowledge of the atmospheres of other planets, and this part of the book is a good summary of recent results. The technical material here is the proceedings of the Summer Advanced Study Institute, held at the University of Liège, Belgium, July 29-August 9, 1974.

437.   McCormac, Billy, M., ed. **Aurora and Airglow**. New York, Van Nostrand Reinhold Publishing Corp., 1967. 689p. illus. refs. index. $35.00. LC 67-23737. ISBN 0-442-15577-8.

An emphasis on the observation of the upper atmosphere phenomena known as aurora and airglow is the theme of this lecture series volume taken from the NATO Advanced Study Institute at the University of Keele, Staffordshire, England, 15-26 August 1966. Included are 50 papers, edited by the foremost authority on the subject, aimed at the advanced student and specialist. The book is more than descriptions of the observations aurora and airglow, however; it begins with six lectures entitled "orientation," in which background is set for the new reader. History, auroral morphology, and implications of aurora studies are the highlights of this part. There are also two sections of papers falling under the headings of "theory," which are review essays surveying the research and explanations of aurora, dayglow, solar wind and magnetosphere, radio aurora, and more. One section considers artificial aurora and airglow, which is used in comparison studies. The volume ends with a conference summary paper and some conclusions. The aforementioned arrangement into "theory" and "observation" is an often-used alignment in a book like this, and it is unfortunate that more concrete subject divisions were not employed. This is admittedly difficult to do when there are so many papers on different topics, but it would have made the book a little more useful. Fortunately, Mr. McCormac's books always include indexes, something most conference and lecture volumes do not.

438.   McCormac, B. M., ed. **The Radiating Atmosphere**. Dordrecht-Holland, D. Reidel Publishing Co., 1971. 455p. illus. index. refs. glossary. (Astrophysics and Space Science Library, v.24). $30.20. LC 70-154742. ISBN 0-90-277-0184-9.

One of the more up-to-date volumes on the aurora and airglow phenomena, this book resulted from a series of lectures presented at the Summer Advanced Study Institute held at Queen's University, Kingston, Ontario, August 3-14, 1970. A combination of review papers and descriptions of observational work, this volume is divided into eight major categories for easy location of subject material: I, Atmospheric Airglow Emissions; II, Atmospheric Processes; III, Aurora; IV, Auroral Interpretations; V, Particle Precipitation; VI, Radio Observations; VII, Auroral Morphology; VIII, Summary and Conclusions. In all, there are 39 lectures aimed at the advanced student and astronomer, substantially updating *Aurora and Airglow* (B. M. McMormac, ed; 1967) and similar works. The volume ends with the statement that no concrete conclusions can be drawn on these changing topics; indeed, there is not even a totally accepted definition of aurora or airglow. There is no doubt that progress will continue to be made and that the editor will continue to be associated with excellent volumes such as this one.

439.   Omholt, A. **The Optical Aurora.** Berlin, New York, Springer-Verlag, 1971. 198p. illus. refs. subject index. (Physics and Chemistry in Space, v.4). $19.90. LC 79-163747. ISBN 0-540-054861-3 (Berlin); 0-387-05486-3 (N.Y.).

Unlike the several conference volumes on auroras which mainly report results of research, this technical book is concerned with a unified, detailed explanation of the phenomena caused by the interaction of solar particles and the upper atmosphere. The first chapter is a good introduction to the occurrence and causes of auroras, including photographs of the various forms they take. Chapters two and three consider the electron and proton aurora, respectively, the author explaining the role of these elementary particles in the production of the phenomena. The optical spectrum of the aurora is described next, including local variations due to the height of the emission or its relative position in an auroral form. The physics behind the emissions in the optical range are also considered, as well as supplementary topics on temperature determination, pulsing aurora, radio observations, and auroral x-rays.

440.   Petrie, William. **Keoeeit—The Story of the Aurora Borealis.** Oxford, Pergamon Press; distr. New York, Macmillan, 1963. 134p. illus. index. refs. LC 63-14800.

The uniqueness of this short but informative book is that it is aimed at the general public; nearly all works on the auroras are technical and are intended for the scientist. The text begins with "The Aurora in History," a brief potpourri of some of the dozens of legends surrounding the phenomena. Here we learn about the Eskimo story of the great light of "Keoeeit," a belief that auroras are torches held by spirits to guide the souls of the dead to the land of happiness. The author next explains why we study the aurora and how it is done. Included here are pictures of the aurora borealis by the special "all-sky" camera which takes in nearly the entire sky in one exposure. The various forms of the aurora are described in the following chapter: homogeneous arc, homogeneous band, pulsating arc, diffuse luminous surface, pulsating surface, feeble glow, rays, rayed arc, rayed band, drapery, corona, and flaming aurora. The author's wife painted excellent water colors of several of these forms, and they accompany this chapter. The range of the auroras is detailed as well, showing the parts of the globe they inhabit. Other interesting chapters follow the aforementioned basic material: variations in time and space, light and sound (many claim to hear a whispering sound when auroras appear), relatives of Keoeeit, and the causes (particles from the Sun interacting with the upper atmosphere). Even though this book is fairly outdated, most of the material is unchanged, and it would serve as a good introduction to the general reader with no science background.

441.   Roach, F. E., and Janet L. Gordon. **The Light of the Night Sky**. Dordrecht-Holland, D. Reidel Publishing Co., 1973. 125p. illus. refs. subject index. (Geophysics and Astrophysics Monographs, v.4). $16.50; $9.90pa. LC 73-83568. ISBN 90-277-0293-4; 90-277-0294-2pa.

Aimed at students in astronomy and aeronomy, this fascinating book examines the various phenomena which cause the night sky to twinkle and glow: airglow, aurora, zodiacal light, starlight, twilight, etc. Adequately illustrated with photographs and diagrams, the volume contains brief but excellent introductions to the aforementioned topics. In particular, the authors discuss star counting and the distribution of starlight over the sky, the interaction of solar particles with the upper atmosphere, the nightglow phenomena ("a constant Earth 'envelope' "), and dust-scattered starlight, called diffuse galactic light. Scattered throughout the book are several biographical sketches of scientists who played an important role in the studies of the light of the night sky. This excellent volume could also be appreciated by the prepared layman, and it would be appropriate for college, observatory, and public library.

Gravity

442.   Airy, George Biddell. **Gravitation: An Elementary Explanation of the Principal Perturbations in the Solar System**. Ann Arbor, Neo Press, 1969. 173p. illus. $5.00; $2.50pa. NUC 70-96253. ISBN 0-911014-02-0.

A reprint of a book first published in 1834 and revised in 1884, this particular work examines the gravitational attractions among the bodies of the solar system. The book remains a useful work, as the foreword of this facsimile reprint explains: "The exceptionally long life of this book, a rare [phenomenon] among scientific books of non-historical interest, must be due to its success in explaining simply a wide variety of gravitational perturbations of the planets and their satellites in our solar system. Airy uses only simple diagrams and the qualitative applications of Newton's laws of motion." A fascinating look at nineteenth century science, this non-mathematical work discusses the basic laws of gravitation, the effects of gravity on bodies in motion (e.g., the planets), perturbations, the effects of planets and satellites on each other, lunar theory (gravitational effects on other bodies and vice versa), planetary theory, Jupiter's satellites, and more. Of interest to students, astronomers, and historians, this reprint would be a good acquisition for most libraries.

443.   Misner, Charles W., Kip S. Thorne, and John Archibald Wheeler. **Gravitation**. San Francisco, W. H. Freeman & Co., 1973. 1279p. illus. exercises. bibliog. indexes (subject and name). $39.50; $19.95pa. LC 78-156043. ISBN 0-7167-0334-3; 0-7167-0344-0pa.

An incredible number of physical, mathematical, and astronomical topics are considered in this advanced volume on gravitation physics. The overall theme, the study of Einstein's "general relativity," or "geometrodynamics," is divided into what the authors call Track 1 and Track 2; the former is more basic and less advanced, making the book useful as either a one-semester or a two-semester text. Track 1 constitutes the basic core of gravitation theory for advanced physics students. Assuming only previous knowledge of vector analysis and partial differential equations, it is suitable for a one-semester course at the junior, senior, or graduate level. Track 2 is additional material, sometimes merely supplemental but often an extension of Track 1, which can be used if the course in question is a full-year class. For the astronomy student, the most valuable parts of the book, beyond the basic physics, are the astronomical applications of Einstein's theory, the discussion of cosmology, gravitation and black holes, and relativistic stars. A listing of the contents will help illustrate the scope of this massive text: Spacetime Physics; Physics in Flat Spacetime; The Mathematics of Curved Spacetime; Einstein's Geometric Theory of Gravity; Relativistic Stars; The Universe; Gravitational Collapse and Black Holes; Gravitational Waves; Experimental Tests of General Relativity; and Frontiers.

444.   Valens, E. G., and Berenice Abbott. **The Attractive Universe: Gravity and the Shape of Space**. Cleveland, New York, World Publishing Co., 1969. 187p. illus. index. bibliog. $6.21. LC 68-14702. ISBN 0-529-03910-9.

A layman's look at one of the universal forces of nature, this volume describes the physics and effects of gravity. A physics rather than astronomy book, the text is included here because of the importance of gravitational attraction between celestial bodies. The reader should consult a basic astronomy text for specific discussions of gravity and astronomy. Well illustrated with diagrams and excellent photographs, the book examines topics such as falling, curving (orbital motion), conserving (Kepler's laws), the tides, determining the gravitational constant, and others. An extremely interesting educational volume for junior high age and up, it is appropriate for the public library.

Celestial Mechanics

445.   Chebotarev, G. A. **Analytical and Numerical Methods of Celestial Mechanics.** New York, American Elsevier Publishing Co., Inc., 1967. 331p. refs. index. (Modern Analytical and Computational Methods in Science and Mathematics, no. 9). $31.25. LC 66-23936. ISBN 0-444-00023-2.

Celestial mechanics is a very rigorous subject, and this book is definitely oriented toward mathematics. Fortunately, the author relates the numbers to astronomical examples, an approach that is often neglected by the many theoretical texts. This aspect becomes evident when one examines the subjects of this book: Astronomical Coordinates and Time; Theory of Major Planets; Theory of Minor Planets; Satellite Theory; Lunar Theory; Theory of Cometary Motion. The volume covers well the six methods of celestial mechanics, on which nearly all modern theories of planetary motion are based: the Laplace-Newcomb method, Hill's planetary method, the method of the variation of arbitrary constants, Hill's lunar method, method of periodic orbits, and Cowell's method. Intended as a textbook for university students, this book can also be used by non-astronomer scientists who desire an introduction. Originally published in Russian (*Analiticheskiye i chislennyye metody nebeshoy mekhaniki*, Moscow-Leningrad, 1965), this practical work would be appropriate for university and observatory libraries.

446.   Fitzpatrick, Philip M. **Principles of Celestial Mechanics.** New York, Academic Press, Inc., 1970. 405p. refs. indexes (author and subject). $13.50. LC 77-119612. ISBN 0-12-257950-X.

An advanced text for astronomy majors and students of mathematics, this volume presents the basics of a naturally rigorous subject. Readers should be prepared well in mathematics, mastering calculus, differential equations, vector analysis, and complex variables; a knowledge of classical mechanics will be useful, too. The major portion of the book consists of standard, basic discussions of celestial mechanics theory, including the one-body and two-body problems, the equation of motion, Lagrangian and Hamiltonian mechanics, gravitational potential, and more. There is also a chapter describing the equatorial, ecliptic, and orbital coordinate systems; transformations among the three systems are amply explained. The last two chapters put theory into practice, applying previous explanations to artificial satellite situations. A good text for someone with the proper background, it would be suitable for the university or observatory library.

447.   Hagihara, Yusuke. **Celestial Mechanics.** Cambridge, Mass., The MIT Press. Vol. 1: **Dynamical Principles and Transformation Theory.** 1970. 689p. index. refs. $35.00. LC 74-95280. ISBN 0-262-08037-0. Vol. 2: **Perturbation Theory.** (2 pts). 1971, 1972. 919p. index. refs. $35.00ea. LC 74-95280. ISBN 0-262-08048-6.

The first of two volumes of a projected five-volume set, rigorously recapitulating "the results of the whole field of celestial mechanics and the associated branches of science during the past one hundred years," this work is a combination of mathematical theory and historical references. This truly ambitious venture examines the various theories and equations of the last century in great detail, resulting in one of the most comprehensive studies of celestial mechanics ever published. Future volumes will consider more advanced theories and solutions to the classical problems outlined in these first two books. The references are so extensive and complete that they could, and should, stand alone as an excellent bibliography for astronomers and librarians.

448.   Stiefel, E. L., and G. Scheifele. **Linear and Regular Celestial Mechanics.** Berlin, New York, Springer-Verlag, 1971. 301p. refs. index. (Die Grundlehren der Mathimatischen Wissenschaften, 174). $23.40. LC 72-133369. ISBN 0-387-05119-8 (N.Y.); 3-540-05119-8 (Berlin).

Two-body motion is described here using linear differential equations that have constant coefficients, in a non-standard approach to celestial mechanics. Besides their goal of approaching the subject in a non-classical manner, the authors also introduce independent variables other than time, and attempt to "formulate the perturbation theory in the language of canonical analytical mechanics." A typical example of how rigorous and theoretical a volume on celestial mechanics can become, it is divided into three major parts: *Basic and Numerical Theory* (Preliminaries, Regularized Theory, Kepler Motion, The Initial Value Problem, The Fundamental Differential Equations, Typical Perturbations, Refined Numerical Methods); *Canonical Theory* (General Canonical Theory, Classical Canonical Theory of the Perturbation of Elements, The Canonical Theory of the Oscillator Generated by the Perturbed Problem of Two Bodies); and *Geometry and Outlook* (Geometry of the KS-Transformation).

449.   Szebehely, Victor. **Theory of Orbits: The Restricted Problem of Three Bodies.** New York, Academic Press, Inc., 1967. 668p. illus. refs. indexes (subject and author). bibliog. $37.00. LC 66-30106. ISBN 0-12-680650-0.

This advanced celestial mechanics text is concerned specifically with the three-body problem, its theory and applications. Aimed at the graduate student, the book also can be used by the astronomer, mathematician, and space engineer, as it holds something for all these groups. The reader, of course, will already be familiar with the two-body problem before attempting this book, and will be prepared to meet with pages and pages of advanced mathematics. There are historical notes and comments, extensive references, and applications of the theory explained. Chapters include Description of the Restricted Problem; Reduction; Regularization; Totality of Solutions, Motion Near the Equilibrium Points; Hamiltonian Dynamics in the Extended Phase Space; Canonical Transformations of the Restricted Problem; Periodic Orbits; Numerical Explorations; Modifications of the Restricted Problem.

STARS

Like the literature of the solar system, the material on stars is quite voluminous. While the majority of the works are technical, intended for the advanced student and astronomer, there are several very well written items for the amateur and general reader. Books in this particular section, it should be pointed out, are concerned with stellar composition and physical characteristics; works on the constellations, for instance, are listed under Amateur Astronomy.

The references listed below are divided into three sections: 1) general works, 2) structure and evolution, and 3) variable stars. The first contains publications of a general nature and those which do not fit into the other two categories; these are intended for both general and advanced readers. The books include introductory works, treatises on spectroscopy, general texts on astrophysics, binary-star books, etc. Part two, on structure and evolution, is one of the most important areas in all of astronomical research; the books listed here include some very standard works, some classics in the field. Finally, the third section, on stars that change in brightness, mass, or other quantity, is but a small part of stellar studies. Yet it is one of the richest in the literature because of its popularity among amateurs and professionals. A substantial number of books on variables is listed.

Card catalog subject headings: *Library of Congress:* STARS; names of individual stars; STARS–ATMOSPHERES; STARS–SPECTRA; STARS–CLASSIFICATION; STARS–EVOLUTION; STARS–STRUCTURE; STARS, DOUBLE; NOVAE; STARS, VARIABLE; ASTROPHYSICS. *Sears:* STARS; names of individual stars; ASTROPHYSICS.

General Works

450.   Aitken, Robert G. **The Binary Stars.** New York, Dover Publications; distr. Goulcester, Ma., Peter Smith, 1964. 309p. illus. indexes (subject and name). refs. $5.00. LC 64-13456. ISBN 0-8446-1517-X (Smith).

Reprinted from the original 1935 edition (McGraw-Hill), this classic volume should be on the shelf of every observatory library and nearly any other library with astronomy books. From an historical survey of binary star research to the origin of the double stars, the author comprehensively covers the subject in an interesting and clear writing style. The historical sketches are fascinating; names like Jean Baptiste Riccioli, William Herschel, John Herschel, F. G. W. Struve, S. W. Burnham, and more crop up in the author's history chapters. He next turns to observing methods, explaining techniques that are essentially unchanged today. The orbits of binaries are dealt with at some length, in particular those of visual and spectroscopic doubles. The most technical portion of the book is the discussion of the orbital elements, the author resorting to extensive use of trigonometry. Eclipsing binaries are described, too; the change in magnitudes due to the eclipsing member of the system is explained. Other topics of interest found here are "especially interesting binaries," data on stars in the Northern Sky, and thoughts on the origin of these two-star systems. Two extensive tables on the orbits of certain visual and spectroscopic binary stars comprise the appendices. Highly recommended for anyone interested in binary stars, from layman to scientist.

451.   Aller, Lawrence H. **Atoms, Stars, and Nebulae**. rev. ed. Cambridge, Mass., Harvard University Press, 1971. 351p. illus. index. (Harvard Books on Astronomy). $13.00. LC 76-134951. ISBN 0-674-05264-1.

"Astrophysics for the layman" might be an appropriate alternate title for this fine text. Aimed at college students and advanced amateurs as well, this book is mainly a description of classical astrophysics and stellar evolution. Relying heavily on graphs, charts, and photographs, instead of on advanced mathematics, Aller presents the stars and all their mysteries in a clearly written volume that is the best of its kind. One of the most important aspects of the study of astrophysics is the analysis of starlight using a spectrograph; the author covers this early on and makes extensive use of spectral photographs throughout the book. Some basic physics of the atom and radiation are presented early, too, in an effort to prepare the reader for the rest of the stellar story. The different types of stars are defined (e.g., dwarfs, giants, main sequence, novae) along with descriptions of their spectral type, physical characteristics, and place on the evolutionary scale. With regard to the latter, a good explanation of the Hertzsprung-Russell Diagram is given, showing the reader the importance of this temperature vs. luminosity plot. There is also a chapter on the gaseous nebulae and interstellar matter and their roles in the formation of stars. The energy-producing processes are explained next, with not-too-technical descriptions of the nuclear reactions. Stellar evolution is described in "The Biography of a Star," an illuminating section showing the life patterns of stars on the H-R diagram, showing how these patterns differ according to mass and other factors. The book's last three chapters describe specialized but important related topics: (pulsating) variable stars, novae (exploding stars), and high-energy astronomy (in particular, quasars and pulsars). An excellent text for public, college, and observatory libraries, it should be read by astronomy librarians, too, after they have read a basic volume like Abell's *Exploration of the Universe*.

452.   Batten, Alan H. **Binary and Multiple Systems of Stars**. Oxford, Engl., Elmsford, New York, Pergamon Press, Inc., 1973. 278p. illus. indexes (general and star). bibliog. (International Series of Monographs in Natural Philosophy, v.51). $14.50. LC 72-88026. ISBN 0-08-016986-4.

Concerned with both the characteristics of individual double stars and the features of these systems as a group, this work is comprised of ten related review articles by a binary star specialist. Chapter one, an overview of binary systems, including classification and orbital elements, briefly introduces visual binaries, astrometric binaries, spectroscopic binaries, and spectrum binaries (all classified according to method of determination), and eclipsing and ellipsoidal binaries. Next is a discussion of the frequency of various types of double stars, followed by the book's only chapter on multiple star systems, the latter comprising 19 pages, the longest single treatment of the subject in one book, according to the author. A point of interest among specialists is the period of a double star system, a quantity that can be measured quite accurately; this topic is taken up in chapter 4. The following section on masses and radii may be the most important since double stars provide fairly easy methods for determining these quantities accurately. Other chapters consider apsidal motion, close binary systems, and circumstellar matter, the material in space between the two component stars. An interesting chapter on the origin and evolution of double star systems finishes this fine work, which has an excellent bibliography as well. For the college and observatory library.

453. Hack, Margherita, ed. **Mass Loss From Stars**. Dordrecht-Holland, D. Reidel Publishing Co.; distr. New York, Springer-Verlag, 1969. 345p. illus. refs. (Astrophysics and Space Science Library, v.13). $27.00. LC 77-480957. ISBN 90-277-0118-0 (Reidel); 0-387-91014-X (Springer).

This conference volume is mainly concerned with non-violent stellar mass loss—that is, reduction of mass due to "normal" inconspicuous processes in a star. The first of 44 papers here is an excellent review paper on mass loss from stars, outlining the state of current research and reviewing past investigations. Volumes of proceedings in general would be much improved if all would include such a lead-off review article for the benefit of the non-specialist. Another feature of use here is the editor's arrangement of papers into those concerned with direct observations of mass loss for single and binary stars, and those concerned with theoretical considerations for both types. Such a grouping provides a convenient method for searching in a volume that unfortunately does not have an index. A final chapter deals with mass loss from unstable stars. Intended for the astronomer, this volume is the proceedings of the Second Trieste Colloquium on Astrophysics, 12-17 September 1968.

454. Hack, Margherita, ed. **Modern Astrophysics: A Memorial to Otto Struve**. New York, Gordon and Breach; Paris, Gautier-Villars, 1967. 360p. illus. refs. $19.50. LC 74-403060.

This compendium of 29 research and review papers has been published in honor of one of the world's greatest scientists and astrophysicists. No one central theme can be found among the articles in this text devoted to the spectrum of astrophysical subjects, and this reflects the wide body of knowledge contributed by the book's dedicatee. The papers, written by astronomers associated with Professor Struve, range from stellar theory and observation to Kepler's third law to galactic studies; Struve's personal contributions include research on emission lines, interstellar matter, stellar rotation, binary stars, and stellar atmospheres. While this volume is not a particularly unique or significant contribution to the literature of astronomy, it is a good example of the memorial volume, a frequently customary and fitting gesture in honor of a significant researcher. Its principal drawback is the lack of an index, rendering it almost unusable by the researcher.

455.	Hack, Margherita, and Otto Struve. **Stellar Spectroscopy: Normal Stars.** Trieste, Italy, Osservatorio Astronomico di Trieste, 1969. 203p. illus. refs. index. $6.00. LC 74-446903.

Written for the graduate student, this book presents a review of the information obtainable about a star from the examination of its spectra. Only concerned with normal stars (i.e., those Population I and II stars on the main sequence), the authors begin this work with a short introductory chapter on stellar atmospheres and how stellar spectra play a role in their study. Inductive and deductive methods of study of spectra are briefly touched upon. Spectral classification is looked at in depth in the second chapter, in much greater detail than many other similar works. Topics along this line are classification based on photographic methods, classification by means of continuous spectrum gradients and color indexes, and photoelectric two-dimensional classifications. The HR diagram (the plot of effective temperature versus luminosity) is discussed also, but the treatment is too brief, and the HR diagram itself is not well defined initially. Chapter three explains the effects of rotation on the spectrum of a star. Quantitative analysis of stellar spectra is the final section, and the authors investigate the curve of growth, model stellar atmospheres, and the chemical composition of stellar atmospheres. Although this book would not serve alone as a text, it might be used as supplementary material to a text on stellar atmospheres. The writing is clear and understandable, unusual for many books of this level, but not enough definitions are given, and too much prior knowledge is assumed. For the university and observatory library.

456.	Hack, Margherita, and Otto Struve. **Stellar Spectroscopy: Peculiar Stars.** Trieste, Italy, Osservatorio Astronomico di Trieste, 1971. 317p. illus. refs. index. $7.00. LC 70-590009.

The follow-up to the authors' book on the spectroscopy of normal (main sequence) stars, this work consider Population I stars that are not on the main sequence, novae, and Ap and Am stars. A graduate-level text, this book does a good job of explaining the peculiar stars and their spectra, but it is assumed that the reader will have some background in astronomy (for example, the authors do not define a nova, even though there is an entire chapter on the subject here). The illustrations, in the form of diagrams, sketches, and photographs of spectra, are abundant and well chosen, and the text is easy to follow. Like the previous volume (*Stellar Spectroscopy: Normal Stars*), this should not be considered the final word on the subject, but even though it is not totally comprehensive, it is an excellent introduction. The three chapters are I, Emission Lines in Spectra of Hot Stars and Related Problems; II, Novae and Explosive Variables; III, Magnetic, Metallic-Line and Related Stars.

457.	Herbig, G. H., ed. **Spectroscopic Astrophysics: An Assessment of the Contributions of Otto Struve.** Berkeley, University of California Press, 1970. 462p. illus. refs. indexes (subject and author). $15.00. LC 69-15939. ISBN 0-520-01410-3.

This volume, assembled six years after Otto Struve's death, highlights the many contributions to astronomy made by one of its most eminent scientists. Struve's main impact on astronomy, contends the book's editor, was his "skilled and imaginative use of the spectrograph." This book gathers 10 important papers published by him on spectroscopy, along with 11 invited review articles by astronomer colleagues on those topics dealt with in his papers. The 22 pieces (including a personal and scientific appreciation by the editor) cover spectral classification, curves of growth, hydrogen lines, shell spectra, peculiar stellar spectra, T Tauri Stars, interstellar matter, Beta Canis Majoris variables, stellar rotation and spectroscopic binaries. This second memorial is far more useful to the astronomer and graduate student than the first (*Modern Astrophysics: A Memorial to Otto Struve*, edited by Margherita Hack, 1967) because it deals exclusively with one subject area and because it has an index. The papers, reproduced photographically, are among Struve's most important. A complete bibliography of his writings, though a monumental task, would have been a nice addition to this fine collection.

458.   Jefferies, John T. **Spectral Line Formation**. Waltham, Mass., Blaisdell Publishing Co., 1968. 198p. illus. refs. indexes (subject and author). (A Blaisdell Book in the Pure and Applied Sciences). LC 67-20145.

An advanced, rigorous volume for the scientist and graduate student, this book discusses the theory and analysis of spectral line formation of a hot gas. Directly related to certain astrophysical problems, in particular the stellar atmosphere, the studies of line spectra are also of interest to physicists working a laboratory situation. However, the former application is the main emphasis of this book, and there is a great deal of information here on the solar and stellar atmosphere. Subject matter specifically includes: equation of transfer, source function, the line absorption coefficient, emission coefficient, equations of statistical equilibrium, theory of the line source function, and analyses of spectral line profiles. The usefulness of this work is that it imparts an understanding of how and why certain lines appear in the hot gas spectrum and the resultant astrophysical implications.

459.   King, Henry C. **Pictorial Guide to the Stars**. New York, Thomas Y. Crowell Co., 1967. 167p. illus. index. bibliog. glossary. $8.95. LC 67-12404. ISBN 0-690-62513-8.

More than just a book about the stars, this well-illustrated volume is also concerned with nebulae, galaxies, the Milky Way, and the Universe as a whole. Emphasizing our "current" knowledge, the author's descriptive, non-technical text is written with the layman and amateur astronomer in mind. Some scientific background would be very useful, but not absolutely necessary, for getting the most out of this comprehensive volume. Specific topics covered include the Sun (the nearest star), stellar evolution and composition, binary and multiple star systems, open and globular clusters, variable stars, exploding stars, and more. Each section involves a bit of history for background, combined with modern theory and observation. Although, like most pictorial guides, this one emphasizes the illustrations, the text is also quite good, providing a basic introduction to the topic.

460. Kruse, W., and W. Dieckvoss. **The Stars**. Ann Arbor, University of Michigan Press, 1957. 202p. illus. index. (Ann Arbor Science Library). $5.00. LC 57-7745. ISBN 0-472-00101-9.

Unfortunately, there are few books on stars for the layman, and even fewer still in print. This good volume is still available, however, despite its age, and it is just right for the new astronomy reader who wants a non-technical introduction. The first of the two main parts deals with the position, brightness, and color of stars, explaining those qualities in which astronomers are most interested. A brief introductory section preceding this describes telescopes and tells how they are used to gather information about the stars. In discussing stellar positions, the authors tell how star maps are made and used, why "fixed" stars are not fixed, how far the stars are, and more. In the section on the brightness of stars, they explain magnitudes, brightness measurement, stars that change in brightnesses, and how magnitudes can help measure stellar distances. This part of the book ends with a lengthy discussion of stellar spectra, how they are analyzed, and what they tell us about the stars' physical characteristics. Chemical composition and energy production are explained, too. Part two, "The World of Stars," consists of discussions of star clusters and galaxies; in the process, the authors build a model of the Universe. A good book for the public library, this work was translated from the German (*Die Wissenschaft von den Sternen*, 2nd ed., 1954) by Ralph Manheim.

461. Kurth, Rudolf. **Introduction to Stellar Statistics**. Oxford, Pergamon Press, 1967. 175p. illus. index. refs. (International Series of Monographs in Natural Philosophy, v.10). $12.50. LC 66-24821. ISBN 0-08-010119-4.

More mathematical than astronomical, as might be expected, this advanced text considers the statistical problems related to stellar motion, distances, and distribution. There is little here in the way of astronomical information (for example, how stars are distributed in the plane of the Milky Way); instead the concepts of determining such a distribution are presented. After a brief historical introduction, the author discusses "the observational data," or the direction, quantity, and quality of starlight that can be received by a telescope, and what can be learned from them. Chapter III, a review of the elements of statistics, is a worthwhile section for the new or old statistician. At this point the text turns to the astronomical applications of statistics—in particular, the distribution of stars and galaxies, solar motion, stellar and galactic motions, stellar distances (including primary, secondary, and tertiary methods of estimation). The final rigorous chapter is an in-depth look at the so-called integral equations of stellar statistics. One of the few such works available, it contains several pages of excellent references for further study. For the college and observatory library, it is aimed at the graduate student and astronomer.

462. Mavridis, L. N., ed. **Stars and the Milky Way System**. Berlin, New York, Springer-Verlag, 1974. 368p. illus. refs. indexes (author and subject). $62.00. LC 73-9108. ISBN 3-540-06383-8 (Berlin); 0-387-06383-8 (N.Y.).

The second of three volumes of the Proceedings of the First European Astronomical Meeting (held in Athens, Greece, September 4-9, 1972), this book is a typical example of a conference volume on stellar studies. Containing 39 papers outlining some of the latest astronomical research, it is also one of the most up-to-date volumes on stars. Not strictly reports of original research, the book contains several invited and general lectures in the form of review papers as well. A great variety of individual topics are presented under the following headings: Variable Stars; Binary Stars; Space Distribution and Motions of the Stars; Interstellar Matter; Galactic Center; Chemical Evolution of the Galaxy; Infrared Astronomy; Instruments; Three-Body Problem; Galactic Dynamics. These papers on the overall theme of stars in the Milky Way system include abstracts, references, and selected discussion comments. Overall, the volume is not particularly unique, but it does present a wide variety of subjects, whereas most proceedings volumes concentrate on one specific aspect of an area. This book for the scientist and graduate student would be a good selection for the astronomy library, but its exorbitant price may eliminate it from many acquisition lists.

463.   Page, Thornton, and Lou Williams Page, eds. **Starlight: What It Tells Us About the Stars.** New York, Macmillan Co., 1967. 337p. illus. index. glossary. (Sky and Telescope Library of Astronomy, v.5). $7.95. LC 67-12798.

Material on stellar spectra abounds in the literature, but unfortunately most of it is far too technical for the average reader. Luckily, the *Sky and Telescope* Library of Astronomy includes this fine volume on the subject, which is just right for the layman and beginning astronomy student. Comprised of past articles from the world's best general astronomy periodical, the book examines at length what stellar spectra are and what they tell us about the myriad of suns in the sky. The introduction includes eight pieces of stellar miscellany from star charts to parallaxes to nuclear reactions within stars. Building up to the analysis of starlight in Chapter 4, the second and third chapters offer discussions of the Sun and stellar brightness and luminosity, respectively. How starlight is turned into spectra and then studied is the subject of the reprints in the fourth chapter. One of the highlights of this short section is the story of Annie J. Cannon and the *Henry Draper Catalogue*, the listing of almost 250,000 stars, incorporating Miss Cannon's spectral classification, a scheme still used. The following chapter then tells what we actually learn from the spectra about individual stars. A few derived quantities include mass, rotation, and chemical composition. The latter half of the book looks at individual types of stars and how their spectra appear; variables, novae, magnetic and flare stars, and double star systems are considered. The final two chapters discuss related topics: star counts and distribution, and star clusters. The illustrations are quite good, as usual, and readers who have never seen a photograph of a stellar spectra will get an eyeful. An excellent work, well edited (like all the other volumes in the series), it belongs in both public and astronomy libraries.

464. Payne-Gaposchkin, Cecilia. **Stars in the Making**. Cambridge, Mass., Harvard University Press, 1952. 160p. illus. index. bibliog. (Harvard Books on Astronomy). LC 52-9378.

Despite this fine volume's age, it remains one of the few (and one of the best) books for the layman on stars and stellar evolution. Not only is it chockful of facts on a variety of subjects, it is also highly readable—not at all dry, like so many books of its type. The author begins her fascinating text with an overall picture of stars and interstellar dust and gas, describing the nearest star (the Sun), variable stars, dwarfs, etc. The way dust and gas block the light from stars is discussed, along with the role they play in the formation of stars. The author next considers binaries, clusters, and galaxies, showing that most stars travel in pairs or groups. The following chapter considers "age" on a cosmic scale; for example, how old are the atoms, stars, galaxies? The life expectancy of various celestial objects is explained, too. Stellar and galactic evolution close out the book, telling the reader how stars and star systems are born and die, over a stretch of billions of years. Sixty-seven well-chosen photographs of stars and galaxies are included at the end. This excellent book should be issued in a revised edition or, at the very least, should be reprinted.

465. Rose, William K. **Astrophysics**. New York, Holt, Rinehart, & Winston, 1973. 287p. illus. index. refs. $15.95. LC 72-89470. ISBN 0-03-079155-3.

Stellar theory and physical processes are the main theme of this graduate text, which introduces a variety of astrophysical topics. Beginning the book is a chapter called "Matter and Energy in Space," an introduction to cosmology that gives the reader an overall view of the Universe and the interstellar medium. Descriptions of the density of interstellar space and the dust and gas lying between the stars are a major portion of this section, which also considers the overall makeup of the galaxy. Stellar theory, including the standard topics of the H-R diagram, stellar composition, the equations of stellar interiors, and nuclear reactions in stars, is taken up next. In a book this length, it is not surprising that these topics are glossed over, as they are here, and the reader should seek out one of the other works in this guide for a more complete picture. The birth of stars from clouds of interstellar gas, an important astrophysical problem, is approached next, followed by discussions of astrophysics related to variable stars, red giants, planetary nebulae, novae, galactic x-ray sources, white dwarfs, supernovae, pulsars, neutron stars, and galaxies. A more in-depth coverage of cosmology is the final chapter. One of the more up-to-date general works on astrophysics, it would be appropriate for the university library.

466. Sobolev, V. V. **Course in Theoretical Astrophysics**. Washington, D.C., NASA, 1969. 493p. refs. subject index. $3.00pa. NASA TT F-531.

Not limited to stellar considerations, as many astrophysical texts are, this rigorous and highly mathematical volume is based on lecture notes used by the author over a 20-year period at the University of Leningrad. Originally published in Moscow in 1967 (*Kurs teoreticheskoy astrofiziki*), the book reflects many of the significant changes in theoretical astrophysics in the last two decades. The text begins with a basic discussion of stellar processes in the photosphere, considering energy transfer, thermodynamic equilibrium, the absorption coefficient, and more. The following chapter on stellar atmospheres deals with the absorption line phenomena, the chemical composition and physical processes of atmospheres, and brief comparisons of stars of varying spectral classifications. The solar atmosphere is presented next in a lengthy section of general information, and discussions of the chromosphere and corona, and solar radio emission. The Sun's close proximity, of course, affords detailed studies of the astrophysical processes, and the author places heavy emphasis on these aspects. The chapter on planetary atmospheres, often neglected by astrophysical texts, takes a look at light scattering and optical properties of the atmospheres, and their structure. A lengthy chapter on gaseous nebulae follows, the author covering the emission spectra characteristic of these celestial objects, and the causes of the emission spectra. The book ends with sections on variable stars, interstellar space, and the internal structure of stars. The advanced student will best appreciate this "heavy" work.

467. Vorontsov-Vel'yaminov, B. A., *et al.* **Physics of Stars and Stellar Systems.** Jerusalem, Israel Program for Scientific Translations, 1969. 718p. illus. refs. indexes (subject and name). $3.00pa. NASA TT F-506. TT 68-50307.

A technical book for the advanced student, this is the second volume of *A Course in Astrophysics and Stellar Astronomy*, originally published in Moscow in 1962. A comprehensive work, its material is divided into six major parts: Absolute Stellar Magnitudes and Stellar Masses; Binary Stars; Variables and Novae; Diffuse Matter; Theory of Stellar Atmospheres and Gaseous Nebulae; Stellar Systems. Many chapters have brief historical introductions to provide perspective, and there are many useful references at the end of each section. Nearly every major type of star and star system is covered, giving the reader a complete basic text. Published by the IPST for NASA, it was translated by Z. Lerman from the Russian (*Kurs astrofiziki i zvezdnoi astronomii*, v. 2, edited by A. A. Mikhailov).

Structure and Evolution

468. Aller, Lawrence H., and Dean B. McLaughlin, eds. **Stellar Structure.** Chicago, University of Chicago Press, 1965. 648p. illus. index. refs. (Stars and Stellar Systems, v.8). $24.50. LC 63-16723. ISBN 0-226-45960-8.

Early studies in the field of stellar composition were concerned mainly with models of main-sequence type stars, chemically homogeneous suns in hydrostatic equilibrium (i.e., structural stability due to equal and opposite forces of gravity and pressure). Later research, including much of the work today, was concerned with stellar models away from the main sequence, where abnormal conditions affect a star's mass, energy, and stability. Both types of stellar structure problems are presented here, in an excellent volume of bibliographic essays by various astronomers. These review articles include observational evidence as well as theoretical stellar model analysis. The evolution of stars is covered in a comprehensive, logical, professional manner. For the astronomy library, this book, like the others in the series, is must reading for the astronomer and graduate student.

Contents: 1, The Origin of the Chemical Elements; 2, Stellar Energy Sources; 3, Stellar Absorption Coefficients and Opacities; 4, Stellar Models for Main-Sequence Stars and Subdwarfs; 5, The Theory of White Dwarfs; 6, Theory of Novae and Supernovae; 7, Supernovae; 8, Magnetic Stars; 9, Meridian Circulation in Stars; 10, Stellar Stability; 11, Stellar Evolution and Age Determination.

469.   Baade, Walter. **Evolution of Stars and Galaxies.** Cambridge, Mass., Harvard University Press, 1975. 321p. illus. index. $5.95. LC 63-9547. ISBN 0-262-52033-8.

One of the better general treatises available on stellar and galactic evolution, this volume is designed for college undergraduates and the laymen with some background in astronomy. In a rather non-technical fashion, the book comprehensively covers the Milky Way and other galaxies, emphasizing both observational considerations and theoretical views. The author examines the various types of stars that make up galactic systems, commenting on the various evolutionary patterns that emerge. Red giants, main-sequence stars like the Sun, dwarf stars, supergiants, and more are described. A substantial portion of the book is concerned with the evolution of galaxies, including the Milky Way; the spiral, barred-spiral, elliptical, and irregular systems are each considered at length. Prepared by its editor, Cecilia Payne-Gaposchkin, from tape transcriptions of lectures given by Dr. Baade in 1958, the volume effectively uses tables and diagrams to illustrate the text, which is suitable for any type of library.

470.   Brancazio, Peter J., and A. G. W. Cameron, eds. **Supernovae and Their Remnants.** New York, Gordon and Breach Science Publishers, 1969. 240p. index. refs. illus. $29.50. LC 78-84192. ISBN 0-677-13290-5.

A variety of observational and theoretical aspects of supernovae are presented in this conference volume containing 13 well-edited papers. One of the handful of books on the subject, this volume is now a bit outdated; the newer material, and in particular the relationship between supernovae remnants and pulsars, is missing. Like most volumes of proceedings, this work is not a basic text, but rather a collection of topics, many highly specialized, with a varying level of mathematical content. Designed for the astronomer, it is from the Conference on Supernovae held at NASA Goddard Institute for Space Studies on November 2-3, 1967. The contents include: Some Results of the International Search for Supernovae; Supernovae and Supernovae Remnants; Computation of Relativistic Gravitational Collapse; Supernova Hydrodynamics; Exploding Star Models and Supernovae; Heavy Element Synthesis in Supernovae; Properties of Neutron Stars; Gravitational Radiation from Collapsed Supernova Remnants; Fluorescence Theory of Supernova Light; X-ray Emission from Supernova Remnants; Evidence for Continued Activity in the Crab Nebula; The Crab Nebula.

471. Chandrasekhar, S. **An Introduction to the Study of Stellar Structure.** New York, Dover Publications, Inc., 1957. 509p. illus. index. bibliog. $4.50. LC 58-162 ISBN 0-486-60413-6.

One of the finest texts to be found on the subject of stellar make-up, this book is intended for the advanced graduate student of physics or astronomy who desires a rigorous treatment of the topic. Originally published by the University of Chicago Press in 1939, this introductory work contains discussions of thermodynamics, stellar equilibrium, the gas equations, radiation and equilibrium equations, as well as in-depth treatments of the H-R diagram, stellar envelopes, stellar models, quantum statistics, degenerate stars, and stellar energy. Written by the most eminent stellar theoretician, this classic ranks with Eddington's *The Internal Constitution of the Stars* and Schwarzschild's *Structure and Evolution of the Stars* as a definitive treatment of the complex field it describes. Must reading for any graduate student in astronomy.

472. Chandrasekhar, S. **Radiative Transfer.** New York, Dover Publications, Inc., 1960. 393p. illus. indexes (subject and definitions). $5.00pa. LC 60-3117. ISBN 0-486-60590-6.

Primarily a problem for the astrophysicist, the radiative transfer of energy has been extensively investigated since 1905, when it was used to explain absorption and emission lines in stellar spectra, and 1906, when it was connected with radiative equilibrium in stellar atmospheres. Professor Chandrasekhar discusses the subject in terms of plane-parallel atmospheres as a branch of mathematical physics using the general principles of invariance and non-linear integral equations. Although the main thrust of the subject deals with the production and transfer of energy within a star, the problem can also involve radiative transfer in planetary atmospheres, and the author rigorously deals with both problems. Now a classic text, it considers in detail the equation of transfer, isotropic scattering, principles of invariance, the general laws of scattering, diffuse reflection and transmission, and more. Intended for the advanced graduate student and scientist, this highly theoretical treatise was originally published by Oxford University Press in 1950.

473. Chiu, Hong-Yee, and Amador Muriel, eds. **Stellar Evolution.** Cambridge, Mass., The MIT Press, 1972. 812p. illus. refs. $20.00. LC 79-38325. ISBN 0-262-03043-8.

Probably the most massive volume on the subject ever published, this work is arranged roughly in the order of a star's development, from main-sequence to most of the advanced forms. Aimed at the practicing astronomer and graduate student, the book rigorously attacks stellar evolution on all fronts. Chapter 1, easily the longest and most comprehensive chapter in the book, is an excellent look at so-called "normal stellar evolution," the development of stars on the main sequence. One must first master this subject before considering advanced stages of stellar development. From this point, the collection of 22 sections, taken from a series of lectures at SUNY Stony Brook (June 18-July 16, 1969), moves gradually into the evolutionary pattern. Some selected topics are evolution near and away from the main sequence, variable stars, novae, neutron stars, stellar opacity, nucleosynthesis. Overall, it is a good text, with numerous diagrams, dozens of relevant citations to related literature, and advanced mathematics. Not a particularly good volume for the advanced student beginning stellar evolution studies, it is appropriate for the specialist. The book's major drawback is the lack of an index.

474.   Clayton, Donald D. **Principles of Stellar Evolution and Nucleosynthesis**. New York, McGraw-Hill Book Co., 1968. 612p. illus. index. notes. problems. $24.50. LC 68-12263. ISBN 0-07-011295-9.

The formation of the elements inside stars is the main thrust of this graduate-level text, which emphasizes the concepts of stellar interiors in a different approach to the evolution of stars. The author's background in nuclear physics explains this viewpoint, and it is refreshing to see a divergence such as this. Readers seeking a comprehensive, descriptive look at the various evolutionary stages of a star, however, will do better to look elsewhere (e.g., Schwarzschild's *Structure and Evolution of the Stars*). The major concerns here are the nuclear reactions that take place at various stages in a star's life. The major events discussed include the proton-proton reaction, PPI chain, PPII and PPIII chains, the CNO bi-cycle, etc. Other chapters on familiar topics precede and follow the sections on nucleosynthesis: stellar interiors, stellar structure, general introduction to the stars, and so on. A final chapter considers heavy-element synthesis. Interspersed with dozens of problems for the student, it would be suitable for a class of astronomy or nuclear physics undergraduates who have had modern physics and quantum mechanics.

475.   Cosmovici, Cristiano Batalli, ed. **Supernovae and Supernova Remnants**. Dordrecht-Holland, D. Reidel Publishing Co., 1974. 387p. illus. refs. (Astrophysics and Space Science Library, v.45). $37.00. LC 73-91428. ISBN 0-90-277-0427-9.

The first paper alone (*Review of the Research on Supernovae*, F. Zwicky) makes this up-to-date volume a good buy for the astronomy library. In this article, the author describes how he and Walter Baade first coined the word supernova and went on to do pioneering work in the field; it is a fascinating chapter in the history of astronomy. The book's main usefulness, however, is in providing an excellent survey of current research and observation. First, several separate and systematic supernovae searches are described here by astronomers, and there is even a paper on historical observations of the exploding stars, with a special emphasis on Tycho's star (the Crab Nebula). The majority of the 48 papers from the international conference held in Lecce, Italy, May 7-11, 1973, are concerned with various scientific analyses and theorization. Papers are arranged in the following sections: I, Results and Techniques of Supernovae Surveys; II, Photometric Studies of Supernovae; III, Spectra of Supernovae and their Interpretation; IV, Statistics of Supernovae; V, Supernova Remnants; VI, Theories on Supernovae and Supernova Remnants. An excellent choice for the observatory and university library, the book contains hundreds of useful references to the literature.

476.	Cox, John P., and R. Thomas Giuli. **Principles of Stellar Structure**. New York, Gordon and Breach, 1968. 1327p. 2v. illus. refs. indexes (subject and author). LC 68-26755.

Encompassing the broad topics of stellar interiors and atmospheres, this two-volume set is an attempt to bring together information from many sources into a comprehensive text complementing previous work on the subject. Following what the author calls "a fairly leisurely style" (of mathematics interspersed with pages of text), the books are logically divided into theory and application. Volume 1, *Physical Principles*, covers the characteristics of stellar interiors, radiation theory, including the infamous "equation of transfer," thermodynamic equilibrium, general thermodynamic theory, radiation pressure, various methods of energy transfer, including radiation and convection, stellar energy sources, and more. Applications of these concepts—in particular, stellar models—is the theme of the larger second volume. Homologous stars, computation of stellar models, survey of stellar evolution, and pulsating (variable) stars are just a few of the topics covered. Each book contains very long and fairly complete lists of references, further enhancing the usefulness of the work. Based on lecture notes used by the author over several years, the work would be suitable for advanced graduate students, who should first, however, read a basic text like Schwarzschild's *Structure and Evolution of the Stars*.

477.	Eddington, A. S. **The Internal Constitution of the Stars**. New York, Dover Publications, Inc., 1959. 407p. illus. index. refs. LC 59-65135.

Another classic in the field of astronomy, this book is far less rigorous than many succeeding volumes on the study of stellar evolution. Written at a time when the subject was in a state of growth and speculation (much remains speculation, of course), the work discusses the then-prevailing theories on stellar processes. The author realized that his book was not the final word—even as it went to press, a new quantum theory was emerging—and Eddington points this out early in the text. He goes on to discuss at length quantum theory, radiative equilibrium, the mass-luminosity relationship, opacity, stellar energy, and other basic topics. Originally published by Cambridge University Press in London in 1926, the book is also of great historical value in the study of the developments of stellar theory. Astronomers and graduate students should include this reprint on their "must" reading list.

478.	Greenstein, Jesse L., ed. **Stellar Atmospheres**. Chicago, University of Chicago Press, 1960. 724p. illus. refs. indexes (star and subject). (Stars and Stellar Systems, v.6). $20.00. LC 61-9138. ISBN 0-226-45958-6.

This highly technical work includes a series of contributions on the study of the outermost layers of stars, emphasizing the study of stellar spectra, the key to our knowledge of the composition of the suns. A lengthy first chapter on stellar atmosphere model theory begins a section of five chapters outlining the basis of standard methods of theoretical analysis of stellar atmospheres. The book then discusses sources of stellar energy, followed by a review of stars with high luminosity or extended atmospheres. Individual types of stars and their spectra, stellar mass loss, and isotopes are other subjects considered. A standard text for all advanced astronomy collections.

Contents: 1, The Theory of Model Stellar Atmospheres; 2, Stellar Energy Distribution; 3, Basic Theory of Line Formation; 4, Quantitative Analysis of Normal Stellar Spectra; 5, Interpretation of Normal Stellar Spectra; 6, Non-thermal Phenomena in Stellar Atmospheres; 7, Stellar Magnetic Fields; 8, Stellar Rotation and Atmospheric Turbulence; 9, The Spectra of Supergiants and Cepheids of Population I; 10, Early-type Stars with Extended Atmospheres; 11, Eclipses by Extended Atmospheres; 12, Composite and Combination Spectra; 13, Spectra of Long-Period Variables; 14, Physical Properties of the Red Giants; 15, The Loss of Mass from Red Giant Stars; 16, Isotopes in Stellar Atmospheres; 17, The Spectra of Novae; 18, Spectra of Dwarf Variable Stars; 19, Spectra of Stars below the Main Sequence.

479.   Menzel, Donald H., Prabhu Lal Bhatnagar, and Hari K. Sen. **Stellar Interiors**. London, Chapman & Hall; New York, John Wiley & Sons, Inc., 1963. 317p. refs. indexes (author and subject). (The International Astrophysics Series, v.6). $10.50. LC 64-9790. ISBN 0-470-59420-9 (Wiley).

The authors of this advanced text point out that it is impossible to separate completely the study of stellar structure from stellar evolution; therefore, they incorporate one chapter at the end on the latter subject. On the whole, however, the rigorous material here is concerned with the inner layers of stars, touching on radiation transfer, thermodynamics of perfect gases, energy production, the equations of stellar structure, and more. Chemically homogeneous stellar models in convective or radiative equilibrium are presented in the latter pages of the work, followed by homogeneous composite models with both types of equilibrium simultaneously. The non-astronomy graduate student will appreciate the first chapter, which explains what can be observed about stars and how these quantities lead into the question of stellar structure, and the section on stellar evolution, which ties everything together in the final analysis. A fairly good text, it would be appropriate for any advanced astronomy collection.

480.   Mihalis, Dimitri. **Stellar Atmospheres**. San Francisco, W. H. Freeman and Co., 1970. 463p. index. refs. glossary of physical symbols. (A Series of Books in Astronomy and Astrophysics). $16.00. LC 77-116897. ISBN 0-7167-0333-5.

A comprehensive and rigorous introductory text on the graduate level, this book provides good background for later research. The stellar atmospheres problem is presented, as it should be, in the abstract, with no references to particular types of stars. In fact, the author has limited these theoretical discussions to the "classical stellar atmospheres problem—i.e., atmospheres in hydrostatic, radiative, and steady-state statistical equilibrium." The non-classical case is another problem altogether and is discussed in other works. As could be expected, the author delves much more deeply into topics that are introduced on a more elementary level in other volumes such as Novotny's *Introduction to Stellar Atmospheres and Interiors*. The basic equation of energy transfer within a star is set up first, laying the groundwork for the rest of the book. Topics discussed by the author include the gray atmosphere; the equation of state, opacity, LTE and non-LTE model atmospheres; spectral line considerations; and more. Definitely not for the beginner, this text is suitable for astronomy libraries and their advanced users.

481.   Novotny, Eva. **Introduction to Stellar Atmospheres and Interiors.**
New York, London, Oxford University Press, 1973. 543p. illus. bibliog. indexes
(author and subject). problems. $19.50. LC 72-86302. ISBN 0-19-501588-6.

Written for the advanced undergraduate and beginning graduate, this text would be an excellent choice for an intermediate course in stellar evolution. The volume is arranged well, beginning with a basic chapter called Observational Data, where the author defines terms and discusses what can be observed about a star (energy output in terms of magnitude and spectra) and what characteristics can be subsequently deduced from the aforementioned quantities; these include mass, radius, luminosity, chemical composition, and population type. The H-R diagram and its various forms and uses are discussed as well. A science major who has no background in astronomy will find this chapter essential. The following two chapters are the meat of this book, discussing atmospheres and interiors, respectively. Topics in the former include radiative transfer, properties of gases, model atmospheres, and spectra. The latter chapter considers the stellar interior equations and the basic stellar evolution theory. These two chapters could be read in either order, or independently. The final section of the book shows the student how to calculate a model solar atmosphere and a model stellar interior. Projects follow for actual experience in a complicated but instructive exercise on modeling. Well illustrated with diagrams and sketches, the book also has many useful tables. For university and observatory libraries.

482.   Page, Thornton, and Lou Williams Page, eds. **The Evolution of Stars:
How They Form, Age, and Die.** New York, The Macmillan Co., 1967, c1968.
334p. illus. index. bibliog. glossary. (Sky and Telescope Library of Astronomy,
v.6). $7.95. LC 67-28468.

From Bart Bok's 1938 general article on stellar evolution to a short paragraph on x-rays emanating from the Crab Nebula, this book is a fascinating look at the development of knowledge about stars since the mid-1930s. This collection of edited *Sky and Telescope* articles gives the general reader and amateur astronomer an inside look at the discoveries and challenges of stellar evolution. We learn "What's Inside the Stars?", how they are born out of great, hot gas clouds, and what happens as they grow old and eventually die quietly, or with a bang. The late Otto Struve, a renowned astronomer and a frequent *Sky and Telescope* contributor, tells us how the chemical elements are formed inside stars and just how much of each ingredient is there. The reader will be fascinated, too, by the story of supernovae, the situation which occurs when certain types of stars are at the end of their evolutionary rope. Exotic topics in stellar development are spotlighted, too: x-ray astronomy, neutron stars, variable stars, etc. Although this obviously does not read as smoothly as a text by one person, it is an excellent volume, well arranged and carefully edited. For the college and public library, its contents include Inside the Sun and Stars; H-R and Color-Magnitude Diagrams; The Ages of Star Clusters; Star Formation; The History of Spinning Material; Stellar Explosions; Changes in Chemical Composition; and Peculiarities in the Lives of the Stars.

483. Schwarzschild, Martin. **Structure and Evolution of the Stars.** New York, Dover Publications, Inc., 1965, c1958. 296p. illus. index. refs. $3.00pa. LC 58-6109. ISBN 0-486-61479-4.

One of the classic works on stars, this fine book ranks with Eddington's *The Internal Constitution of the Stars* and Chandrasekhar's *An Introduction to the Study of Stellar Structure*. An excellent text for the advanced undergraduate, it presents, as no other work has, the basics of stellar make-up and evolution in clear, understandable language, with just the right amount of mathematics. Like the many volumes that followed it, this text begins with a discussion of what we can observe about the stars and what it all means: luminosities, radii, masses, stellar populations, and abundances of the elements are all mentioned. A stellar interior is described in the next chapter, the longest and possibly most important section; here hydrostatic and thermal equilibrium, energy transport, equation of state, opacity, and more are explained in detail. The majority of Schwarzschild's volume, however, is concerned with the various stages of evolution from stellar birth to death. Making extensive use of the H-R diagram and applications of the stellar interior equations, he expertly details the fascinating story of a star's lifetime, setting up theoretical models for each stage of growth. A final chapter summarizes and reviews for the benefit of the student. The book was first published by Princeton University Press in 1958. It is also currently available in hardcover from William Gannon, Santa Fe, New Mexico (ISBN 0-88307-247-5).

484. Shklovsky, I. S. **Supernovae.** London, New York, John Wiley and Sons, Inc., 1968. 444p. illus. refs. (Interscience Monographs and Texts in Physics and Astronomy, v.21). $28.00. LC 68-56972. ISBN 0-470-78650-7.

Translated from the Russian, this book is currently the only text devoted entirely to the study of supernovae (though there are volumes of conference proceedings on the subject). Fairly technical in content, the book begins logically with a section on general background information, including historical notes on the investigations (photometric and statistical) of supernovae, interpretation of their spectra, an overview of supernovae detected in our galaxy, and more. Much of the information we have on these stars that explode violently is gathered from observations of the aftermath, or remnants; section two deals with the study of these remnants from type II supernovae. The third chapter deals with the Crab Nebula, the last known supernovae in the Milky Way, and the one object that has provided the most information to astronomers. The final chapter looks at the relationships between supernovae and other astrophysical problems. Overall, the book does not read well and there is not enough background information or definitions. The term supernova is nowhere explicitly defined; the reader must read between the lines too much. One point in the book's favor is that it contains 360 references to related literature, providing an excellent bibliography. A better, more up-to-date work is needed, but for the time being, this book will suffice.

485. Slettebak, Arne, ed. **Stellar Rotation.** Dordrecht-Holland, D. Reidel Publishing Co.; distr. New York, Gordon and Breach, 1970. 355p. illus. refs. $45.00. LC 76-118131. ISBN 90-277-0156-3 (Reidel); 0-677-60170-0 (G & B).

The effect of a star's rotation on its interior, atmosphere, evolution, etc., is the subject of this conference volume, which contains 39 contributed papers. Included are several review-type articles on stellar rotation, as well as reports of original observational and theoretical research. One-third of the papers are concerned with how stellar rotation figures in the evolutionary pattern of several types of stars and how it affects the atmospheres and spectral lines of still others. Half the text contains pieces on rotation in binaries, clusters, and other special situations, including a small portion devoted to statistics related to the subject. The remainder of the book looks at solar rotation, the situation most easily observable and from which assumptions are applied to other stars. The volume is broad in topical coverage and represents one of the few works available on the subject. Unfortunately, it has no index and no introductory section explaining in general terms the problem at hand. For the astronomer and advanced student, the publication results from the IAU Colloquium on Stellar Rotation held at Ohio State University, September 8 to 11, 1969.

486. Stein, R. F., and A. G. W. Cameron, eds. **Stellar Evolution.** New York, Plenum Press, 1966. 464p. illus. refs. index. LC 65-25285. ISBN 0-306-30221-7.

This volume brings up to date the work of Eddington (1926), Chandrasekhar (1939), and M. Schwarzschild (1958) by presenting a series of papers summarizing recent work in the field. The book begins appropriately with an excellent introductory paper that condenses the basics of stellar evolution in analytical terms, covering the stellar structure equations and the various stages of evolution for homogeneous and inhomogeneous stars. Next is a set of four papers on the physics of stellar interiors, touching on topics of energy generation and transport. The next section, on aspects of stellar evolution, is divided into three major parts and one minor part: pre-main-sequence contraction, hydrogen-burning, advanced stages of evolution, and stellar evolution with varying G, respectively. Stellar variability and mass loss are the subjects of chapters 4 and 5. Following these survey-theoretical papers is a section of 13 articles related to observational evidence for stellar evolution. A brief summary paper follows. In all, there are 41 pieces providing an up-to-date survey of the field, emphasizing stellar models created in part with the use of computers. From an international conference sponsored by the Institute for Space Studies of the Goddard Space Flight Center, held November 13-15, 1963, this work is a good, advanced text for scientists and graduate students; it should be in the university and observatory library.

487. Swihart, Thomas L. **Basic Physics of Stellar Atmospheres.** Tucson, Pachart Publishing House, 1971. 86p. refs. index. (Astronomy and Astrophysics Series). (Intermediate Short Texts in Astrophysics). $9.95. LC 72-180256. ISBN 0-912918-04-7.

The non-specialist who wants a very brief introduction to the problem of stellar atmospheres should consider this advanced volume, which does not emphasize the mathematical basis of the subject. The reader should, however, have a solid background in differential equations and basic physics before attempting the book. Radiation transfer and the equation of transfer are dealt with first in laying the ground-work for the remainder of the volume. Other topics include the gray atmosphere, the non-gray atmosphere, and line formation. The student should shy away from this volume, which is at once too brief and too advanced for his needs. On the other hand, it will be useful for the astronomer who is not working in the area. Problems are included at the end of the book to help the reader learn and retain what was written.

488.   Swihart, Thomas L. **Physics of Stellar Interiors.** Tucson, Pachart Publishing House, 1972. 119p. index. problems. (Astronomy and Astrophysics Series). (Intermediate Short Texts in Astrophysics). $9.95. LC 72-94354. ISBN 0-912918-05-5.

The companion to the author's *Basic Physics of Stellar Atmospheres*, this volume emphasizes the physical concepts related to the problem instead of expounding the most recent research in stellar studies. Like the other volume, this text hits the high points on the way to explaining the complexities involved. Among the general topics covered are Radiation Theory, Gas in Thermodynamic Equilibrium, Polytropes, Stellar Energies, and Structure and Evolution of the Stars. The latter chapter is the least mathematical and consequently the most descriptive in the book. In just 23 pages, Swihart deftly condenses the theories and equations of stellar structure and evolution into an understandable, coherent essay; considering the volumes and volumes written on this topic, it is quite admirable that Swihart has presented this short chapter so well. The entire volume, in this respect, is a piece of craftsmanship. The non-specialist will best appreciate this book, which requires a knowledge of differential equations and basic physics.

489.   Tayler, R. J. **The Stars: Their Structure and Evolution.** London, Wykeham; distr. New York, Springer-Verlag, 1970. 203p. illus. index. (Wykeham Science Series for Schools and Universities, 10). $5.50. LC 73-116975. ISBN 0-085109-110-5 (Wykeham); 0-387-91054-9 (Springer).

Readers who want a basic beginning text on stars at the college level should consider this work, which is one of the best-written and most comprehensive volumes of its kind. The author avoids the trap of becoming overly involved with mathematics (in a subject that is admittedly very mathematical), and instead has written a book with more description and explanation than usual, and with just the right proportion of equations and symbols. The first chapter, which is strictly descriptive, introduces the topic of stellar structure and evolution, and the second chapter then begins to get to the heart of the matter by considering at length the observational properties of the stars. It is those quantities—luminosity, magnitude, radii and mass (sometimes observable), temperature, etc.—that help determine the structure of stars, which ultimately leads to the theories of evolution. Tayler goes into substantial detail on these topics, laying the important ground-work for the remainder of the volume. Next come the all-important equations of stellar structure, the laws that govern the physical processes in the star, holding it together, producing the energy, transmitting it, etc. The author then considers the physics of stellar interiors, delving into the creation of the elements and explaining how pressure, opacity, and rate of energy generation depend on temperature, chemical composition, and density. The remainder of the book deals with the various stages of stellar evolution from main sequence stars to the final stages of development: white dwarfs, neutron stars, etc. Each chapter contains a two-or three-paragraph summary which adds to the effectiveness of an already excellent work. The addition of chapter questions would have made it even better.

490.   Underhill, Anne B. **The Early Type Stars**. Dordrecht-Holland, D. Reidel Publishing Co.; distr. New York, Gordon and Breach, 1966. 282p. illus. refs. indexes (star and subject). (Astrophysics and Space Science Library, v.6). $27.50. ISBN 90-277-0141-5 (Reidel).

This review-type book considers O and B type stars, as well as Wolf-Rayet stars and certain subluminous stars. "Early" type stars refers to stars that are early in the spectral classification scheme, and these objects typically display absorption lines in their spectra. The first chapter of the work is a brief introduction to these special stars and the methods used in observing them; the spectrograph and classification of early type spectra are discussed. The following chapter reviews various "improved" systems of classification of spectra, emphasizing their applications to O and B type stars. Ms. Underhill goes on to discuss luminosities, photometric studies, distribution, spectrophotometric techniques, and more. One of the book's high points is the list of over 600 references; in all, it is a well-written, comprehensive treatment of the topic. Recommended for the observatory and university library.

Variable Stars

491.   Detre, L., ed. **Non-Periodic Phenomena in Variable Stars**. Dordrecht-Holland, D. Reidel Publishing Co., 1969. illus. refs. $39.50. LC 74-467583. ISBN 90-277-0115-6.

Unusual occurrences or abnormal situations pose more interesting, complex questions in any scientific field. In astronomy, scientists are frequently concerned with atypical situations in stellar evolution, radio astronomy, and so on. This volume, which contains the proceedings of the 4th Colloquium on Variable Stars, held in Budapest, 5-9 September 1968, is a good example of the type of research just mentioned. Sponsored by the IAU, the conference was concerned with variable stars that do not exhibit the typical "regular" characteristics found in a large percentage of such stars. Irregular and eruptive variables were the center of attraction of this show, which produced 67 papers, edited and arranged here for the interested astronomer. The nature of the abrupt changes in light, mass, magnetism, etc., and the resultant spectra of these stars is the main theme of this technical volume.

Contents: Part I: Statistical and Physical Interpretation of Irregularities; Modern Techniques of Observation and Analysis. Part II: Intrinsic Irregular Variables: Very Young Stars; Flare Stars; Of, Be, Shell Stars; Irregular Magnetic Variables; R CrB Variables; Nova Outbursts. Part III: Irregular Activity in Periodic Variables: β CMa, δ Sct Variables; Variations in δ Scuti Stars; Late Type Giants. Part IV: Non-Periodic Phenomena in Binary Systems: Hot, Very Short-period Eruptive Binaries: Old Novae, U Gem Stars, etc.; Symbiotic Stars; Contact Binaries and W UMa Stars; Convectional Binaries; Part V: Miscellaneous Problems.

492.   Fernie, J. D., ed. **Variable Stars in Globular Clusters and in Related Systems.** Dordrecht-Holland, D. Reidel Publishing Co., 1973. 234p. illus. refs. (Astrophysics and Space Science Library, v.36). $26.50. LC 73-83560. ISBN 90-277-0341-8.

This book contains both reports of original research and several review-type papers comprising the proceedings of an IAU Colloquium (no. 21) held at the University of Toronto (Aug. 29 to Sept. 3, 1972). Concerned with intrinsic variables, such as RR Lyrae stars, the colloquium papers touch on "the relationship of variables to non-variable cluster members, the position of the variables in the HR diagram and their importance for problems of stellar evolution, empirical data on the variables, periods and period changes, and the relevant parts of pulsation theory." The volume is divided into four major sections, each containing about seven to ten papers: I, General Problems of Variables in Population II Systems; II, RR Lyrae Variables in Population II Systems; III, Slow Variables in Population II Systems; IV, Theoretical Considerations of Population II Variables. Both theoretical and observational aspects of the problem are presented in this book, which represents only a part of the variable star problem. For the observatory and university library.

493.   Glasby, John S. **The Dwarf Novae.** New York, American Elsevier Publishing Co., Inc., 1970. 293p. illus. index. refs. $9.50. LC 74-125628. ISBN 0-444-19633-1.

A follow-up to the author's more general work, *Variable Stars*, this book zeroes in on those eruptive variables whose brightness increases about four magnitudes in novae-like bursts every few weeks. Because these stars present researchers with a wide variety of unsolved problems concerning their nature and evolution, they are of great interest to many astronomers. The first part of the volume speaks of the dwarf novae in general and their relationship to other variables with similar characteristics. The second part talks in detail about the novae themselves, in particular the U Geminorum variables and Z Camelopardalis variables. The observation of these objects and a discussion of cataclysmic variables make up the final two chapters of this book.

494. Glasby, John S. **The Nebular Variables.** Oxford, Engl., Elmsford, N.Y., Pergamon Press, Inc., 1974. 210p. illus. refs. indexes (subject, name, variable star). (Natural Philosophy Series). $22.50. LC 74-3354. ISBN 0-08-017949-5.

 The author's fourth book on variable stars examines those objects associated with both bright and dark nebulae. These variables are classed as "irregular" because of a lack of periodicity in light variations, and they pose many interesting questions for astronomers in the area of stellar evolution. The book explores in detail the four main categories of nebular variables which were spoken of in more general terms in Glasby's *Variable Stars* (1968): I, RW Aurigae Variables; II, T Orionis Variables; III, T Tauri Variables; IV, Peculiar Nebular Variables. Since the stars considered here are young, pre-main-sequence suns still connected to their parent clouds, they provide the astronomer with a first-hand look at the early evolutionary stages. The author's expertise in this field shows again as he proves his ability to write for all levels of readership. An excellent book for the student, advanced amateur, or non-specialist astronomer, it would be a good choice for the astronomy or university library.

495. Glasby, John S. **Variable Stars.** London, Constable & Co.; Cambridge, Mass., Harvard University Press, 1969, c1968. 333p. illus. index. $10.00. LC 79-405238 (Constable). NUC 69-47935 (Harvard). B 68-22920. ISBN 0-09-456200-8 (Constable); 0-674-93200-5 (Harvard).

 Intended for non-specialists, astronomy students, and interested amateurs, this volume is devoted entirely to stars that undergo some change in light, mass, or magnetic field strength. The most general of the author's four books on variable stars, this volume gives a good overall picture of the subject with details that are substantial enough to solve the needs of the intended readership. In all, 24 types of variables are discussed, classed into three major divisions: extrinsic, intrinsic, and eruptive. The first group refers to the so-called eclipsing binaries, where a double or multiple star system is the cause of change in apparent magnitude (i.e., one star passes in front of the other). In this section Glasby considers the Algol, β Lyrae, W Ursae Majoris, and "peculiar" eclipsing variables. The second type of variable star undergoes changes from within, and these stars are often referred to as "pulsating stars" because of the contractions and expansions of the outer layers. Examples are RR Lyrae Variables, Cepheid Variables, Long-Period Variables, and RV Tauri Variables. Finally, the author spends nearly half the book on eruptive variables, stars that undergo rapid and massive, erratic changes in light or mass or both. Some of the most interesting stars in the galaxy are described in this chapter: U Geminorum Variables, Z Camelopardis Variables, novae, supernovae, and more. Readers who are interested will wish to follow up with the author's other works, and similar books by other writers.

496. Kukarkin, B. V., ed. **Pulsating Stars.** Jerusalem, Israel Program for Scientific Translations; issued by Keter Publishing House Jerusalem Ltd.; distr. New York, John Wiley and Sons, 1975. 320p. illus. refs. author and subject indexes. (A Halstead Press Book). (IPST Astrophysics Library). $37.50. LC 75-17851. ISBN 0-7065-1342-8 (Keter); 0-470-51035-8 (Wiley).

An important addition to the literature of variable stars, this work consists of ten chapters by various Russian specialists. Mainly concerned with the various types of stars whose light changes due to regular physical pulsations, the articles are generally review-type essays, delineating work already done. An introductory chapter on the theory of stellar pulsations gives way to the discussion of the classical Cepheids, long-period variables that provide the best means for determining the distances to other galaxies. Other variables are reviewed at length, too: RR Lyrae stars (by V. P. Tsesevich, author of a book on the subject), Delta Scuti stars, RV Tauri stars, and so on. A final chapter is concerned with semi-regular and irregular variables. Aimed at the scientist and graduate student, this work should be considered must reading for those in the field, and especially for newcomers desiring an overview. Where previous volumes by J. S. Glasby (*Variable Stars*, 1967) and W. Strohmeier (*Variable Stars*, 1972) have devoted one or two chapters to pulsating variables, this volume goes into far greater detail and is slightly more technical. Translated (*Pulsiruyushchie zvezdy*) by R. Hardin.

497.   Strohmeier, W. **Variable Stars**. Oxford, Engl., Elmsford, N.Y., Pergamon Press, Inc., 1972. 279p. illus. index. refs. bibliog. (International Series of Monographs in Natural Philosophy, v.50). £7.90. LC 76-179913. ISBN 0-08-016675-X.

Written for the student and professional astronomer, this volume deals almost exclusively with intrinsic variables, excluding eclipsing systems and glossing over some of the eruptive variables like novae and supernovae. The author presents his definition of variable stars (those stars whose brightness fluctuations are due to geometrical or physical factors) in the preface and then very briefly discusses history, nomenclature, techniques for discovery and measurement in the introductory chapter. More detailed and mathematical than Glasby's *Variable Stars* (1968), the book explores at length the physical processes at work inside the stars. Five appendices discuss special related topics, and another lists books and other references in the field. Edited by A. J. Meadows, the book would be best suited for university and observatory libraries.

Contents: Variability and High-Energy Astrophysics. Variability in the Form of Lower-Energy Outbursts. Variability in Young Stars. Variability Due to Pulsation. Semi-regular and Irregular Variability. Variability with Extensive Convection. Variability Due to Geometrical and Physical Factors. Variability and Magnetism. Variability of an Entire Galaxy.

498.   Tsesevich, V. P., ed. **Eclipsing Variable Stars**. New York, Halstead Press, John Wiley and Sons, Inc., 1973. 310p. refs. indexes (name and subject). (Israel Program for Scientific Translations. Astrophysics Library). $28.50. LC 73-1343. ISBN 0-7065-1311-8(IPST); 0-470-89222-6(Wiley).

Several Soviety astronomers have contributed to this book, which deals with stars whose magnitudes vary extrinsically. The chapters examine in detail the typical close binary system whose brightness varies when one component eclipses the other. The study of the light curves (change in brightness with respect to time) tells astronomers much about the component stars and their orbits, and these aspects of the subject are explained in detail here. Selected topics include photometric phases of eclipses, limb darkening of spherical stars, eclipsing systems with deformed components, determination of elements using computers, and more. Translated from the Russian (*Zatmennye peremennye zvezdy*) by R. Hardin, the text is rather technical, and is aimed at the scientist and advanced student.

499. Tsesevich, V. P. **RR Lyrae Stars**. Jerusalem, Israel Program for Scientific Translations, 1969. 357p. illus. bibliog. index of stars. $3.00pa. NASA TT-F-562. TT 69-55025.

Summarizing the results of half a million observations of RR Lyrae variables, this volume is a very comprehensive look at those objects which pulsate regularly at frequencies ranging from 56 minutes to 40 hours. The main significance of these stars is that they provide good measures of distance both within and without the Milky Way, but the author, a well-known variable star expert (having conducted over 50,000 observations of RR Lyrae stars), does not spend an inordinate amount of space rehashing this fact. Rather, he discusses and describes at length dozens of interesting RR Lyrae objects that he and other astronomers have watched for several decades. Their characteristics, periods, radial velocities, light curves, etc., are detailed in the last four of five chapters: Stable and Non-Stable Oscillations; Regular Variation of Periods; Irregular Variation of Periods; "Extreme" and "Strange" Stars. An invaluable handbook for the variable star specialist, this work is the best and most comprehensive volume available. In light of the importance of RR Lyrae stars, its usefulness cannot be emphasized enough. Translated by Z. Lerman from the Russian (*Zvezdy tipa RR Liry*), this book would be a good purchase for the observatory library.

MILKY WAY AND INTERSTELLAR SPACE

Books on our own galaxy are few and far between. In fact, those included below represent a substantial segment of the volumes currently available. Representing general and technical works alike, but with more of the latter, these items are concerned with the properties of the Milky Way as a whole, rather than with the properties of, say, individual stars. The works often cover the distribution and populations of particular types of stars, the shape and rotation of the galaxy, and the evolution of our star system.

Also in this part are publications devoted to the matter between the stars in the Milky Way, known as interstellar gas and dust. These works on the nebulae and interstellar matter are almost exclusively aimed at the student and astronomer; all the books below are fairly technical, save one. The layman can obtain good descriptions of the clouds of dust and gas from a standard elementary text or general work.

Card catalog subject headings: *Library of Congress:* MILKY WAY; INTERSTELLAR MATTER. *Sears:* GALAXIES; SPACE ENVIRONMENT; OUTER SPACE.

General Works

500. Becker, W., and G. Contopoulos, eds. **The Spiral Structure of Our Galaxy**. Dordrecht-Holland, D. Reidel Publishing Co.; distr. New York, Springer-Verlag, 1970. 478p. illus. refs. $28.10. LC 75-115886. ISBN 90-277-0109-1 (Reidel); 0-387-91038-7 (S-V).

Eighty-six separate papers comprise this conference volume on the Milky Way for the astronomer and advanced student. The structure of the galaxy has been a prime target of astronomical research in this century, and the number and variety of topics in this book attest to that fact. Part I is a compendium of papers on spiral structure in galaxies in general, zeroing in on observed patterns in other stars systems for comparison with our own. The great galaxy in Andromeda (M31) is a good example of spiral structure, often compared to the Milky Way, and several papers in this section discuss its aspects. The following part, on radio and optical observations, is by far the longest section, and it provides ample support for the spiral structure hypothesis. General theory of spiral structure is considered in part III, with discussions of normal and barred spirals, and an outline of numerical experiments, including computer modeling. Finally, theory and observation are compared in two areas, gravitational and magnetic effects. The volume ends with a summary by Bart Bok and suggestions for future research. This advanced volume is the result of the IAU Symposium No. 38 held at Basel, Switzerland, August 29 to September 4, 1969.

501. Bok, Bart J., and Priscilla F. Bok. **The Milky Way**. 4th ed. Cambridge, Mass., Harvard University Press; distr. Cambridge, Mass., Sky Publishing Corp., 1974. 273p. illus. index. (The Harvard Books on Astronomy). $15.00. LC 73-83418. ISBN 0-674-57501-6.

An excellent overview of our galaxy, understandable to the layman, amateur, and beginning student, this standard work is written in a semi-popular style, with little technical language and notation. After presenting a general description of the Milky Way and the instruments used to study it, the authors discuss the stars that make up the system. Here the reader learns of the usual and unusual members of the galactic family: main sequence stars, red giants, dwarf stars, binary systems, and more. The motion of the galaxy and its stellar members are described, too, along with their spatial distribution. The chapter on radio astronomy and its use in the study of the Milky Way is quite good; the authors describe radio observations of dust and gas clouds, and the determination of the galaxy's spiral structure. This latest edition finds a new format (double columns), added illustrations, and new material on the recent research. Highly recommended for college, public, and special libraries.

502. Blaauw, Adriaan, and Maarten Schmidt, eds. **Galactic Structure**. Chicago, University of Chicago Press, 1965. 606p. illus. index. (Stars and Stellar Systems, v.5). $24.00. LC 64-23428. ISBN 0-226-05725-9.

The book on galactic astronomy, this compendium avoids examining the Milky Way as a whole, instead looking at the individual constituents and processes within the galaxy's structure. The 23 review papers for students and astronomers do not consider stellar evolution, for example, or characteristics of supernovae, etc. Instead, the authors are concerned with types of stars and other celestial objects as a group, how they are distributed throughout the galaxy, how they move with respect to other members of the star system, and so on. The first four chapters discuss space distribution and motions of the stars in the solar neighborhood—i.e., in our part of the galaxy. The next fifteen sections deal with the motions and space distributions of various members of the galactic population, and the remainder of the book discusses galactic dynamics, mass distribution, the dynamics of gas and magnetic fields, and spiral structure. The work was intended to fill the gap created by outdated and limited textbooks, and it succeeds quite well. Unfortunately, this volume is now somewhat outdated itself, but it contains a great deal of "permanent" information.

Contents: 1, Distribution of Common Stars in the Galactic Plane; 2, Classical Methods for the Determination of Luminosity Function and Density Distribution; 3, Distribution of Common Stars in Intermediate and High Galactic Latitudes; 4, Solar Motion and Velocity Distribution of Common Stars; 5, Motions of the Nearby Stars; 6, Moving Groups of Stars; 7, Distribution of Associations, Emission Regions, Galactic Clusters, and Supergiants; 8, Distribution of Classical Cepheids; 9, Distribution of Interstellar Hydrogen; 10, Galactic Structure and Interstellar Absorption Lines; 11, Continuous Radio Emission in the Galaxy; 12, Distribution and Motions of Late-Type Giants; 13, Distribution and Motions of Variable Stars; 14, Distribution of Novae in the Galaxy; 15, Planetary Nebulae; 16, High-Velocity Stars; 17, Subluminous Stars; 18, Blue Stars at High Galactic Latitudes; 19, Globular Clusters in the Galaxy; 20, The Concept of Stellar Populations; 21, Stellar Dynamics; 22, Rotation Parameters and Distribution of Mass in the Galaxy; 23, Dynamics of Gas and Magnetic Fields; Spiral Structure.

503.   Mavridis, L. N., ed. **Structure and Evolution of the Galaxy.** Dordrecht-Holland, D. Reidel Publishing Co.; distr. New York, Springer-Verlag, 1971. 312p. illus. refs. indexes (subject and name). (Astrophysics and Space Science Library, v.22). $28.50. LC 77-135107. ISBN 90-277-0177-6 (Reidel); 0-387-91072-7 (S-V).

While not really a text, this lecture series volume could almost be used as one. Containing an excellent selection of 17 articles of current theories about the Milky Way, the book updates, in part, Mihalis and Routly's *Galactic Astronomy* (1968) and other similar works. Taken from the NATO Advanced Study Institute held in Athens, September 8-19, 1969, the lectures begin with a look at the history of galactic study, highlighting early observations and later work in stellar statistics. (Readers especially interested in the history of this subject should see Whitney's *The Discovery of Our Galaxy*, 1971.) From this point the book considers topics that are standard in any volume on galactic studies: photometry, interstellar dust, space distributions of different galactic objects, galactic X-ray sources, stellar evolution, galactic evolution, and many more. This advanced volume contains hundreds of references to related work and is recommended for the university and observatory library.

504. Mihalis, Dimitri, and Paul McRae Routly. **Galactic Astronomy**. San Francisco, W. H. Freeman and Co., 1968. 257p. illus. index. refs. (A Series of Books in Astronomy and Astrophysics). LC 68-16760.

A textbook for undergraduates in astronomy or physics covering both theories and observations of the Milky Way. In the preface, the author promises to follow a philosophy of physical explanation of phenomena with a minimum of mathematics; upon inspection, however, it is seen that the volume contains a fair number of equations, and the student should be prepared in calculus and trigonometry. The book's first two chapters, on astronomical terminology and stellar physical properties, lay the groundwork for the remainder of the text, which is concerned mainly with spatial, statistical, and dynamic considerations. The author does a good job in these introductory sections of condensing and explaining things like coordinate systems, proper motion, magnitudes, distances, spectra, the H-R diagram and more. The last ten chapters are the meat of the book; they include discussions of stellar distribution, statistical parallaxes, galactic rotation, the galaxy's spiral structure, stellar and galactic dynamics, stellar orbits, star clusters, and more. The author's frequent references to other works make this a particularly worthwhile book. An excellent introduction, the natural follow-up to this would be *Galactic Structure*, edited by Blaauw and Schmidt (Stars and Stellar Systems, v.5, 1965).

505. Page, Thornton, and Lou Williams Page. **Stars and Clouds of the Milky Way**. New York, The Macmillan Co., 1968. 361p. illus. index. glossary. bibliog. (Sky and Telescope Library of Astronomy, v.7). $7.95. LC 68-27037.

Subtitled "the structure and motion of our galaxy," this book traces developments in the study of the Milky Way since 1933. Like the other volumes of the *Sky and Telescope* series, it is comprised of articles from past issues of the magazine, and its predecessors, *The Sky* and *The Telescope*. The introductory chapter provides some historical perspective, describing early studies by William Herschel (counting stars in various portions of the sky) and later work by J. C. Kapteyn (the astronomer who constructed one of the first correct models of our galaxy). The first chapter, on stellar distances, approaches one of the fundamental problems in determining the Milky Way's size, and the use of Cepheid variable stars as a distance indicator is explained. Globular clusters as another distance indicator are taken up in Chapter 2, followed by a discussion of the rotation of the galaxy. Other topics that help make this fine work a prime source for the amateur are interstellar material (dust and gas) and nebulae (planetary, bright, and dark). The chapter on radio studies of the Milky Way is excellent, and appropriately rather long, since research in this area has taught us much previously unknown about the composition of our star system. (The interested reader should also see *The Milky Way*, Bok and Bok, 1974, for another excellent treatment of this subject.) The final section, on the size and structure of the galaxy, shows, by means of diagrams and photographs of other systems, what the Milky Way would look like if viewed from far out in space. Introductory comments for clarification and explanation precede and conclude each section. Highly recommended for any library with a science collection.

Interstellar Space

506.   Dufay, Jean. **Galactic Nebulae and Interstellar Matter.** New York, Dover Publications, Inc.; distr. Gloucester, Ma., Peter Smith, 1968. 352p. illus. bibliog. notes. indexes (subject and name). (Dover Books on Astronomy and Astrophysics). $6.50. LC 68-23803. ISBN 0-8446-2004-1 (Smith).

   At irregular intervals between the stars lie great clouds of dust and gases, interstellar matter whose existence was first determined by its damping effects on starlight. Although the study of this material has taken a back seat to research on the more conspicuous members of the galactic system, it is nonetheless of great importance to many astronomers. One of the best books ever written on this subject is this fine volume, "an unabridged, republication of A. J. Pomerans' translation of *Nébuleuses Galactiques et Matière Interstellaire*, originally published by Hutchinson & Co., in 1957 . . . . " With the new reader in mind, Dufay begins with an introduction called "Some Fundamental Ideas of Astrophysics," in which he explains stellar luminosities and magnitudes, distances, absolute magnitudes, stellar spectral classification, stellar color-temperatures, the Hertzsprung-Russell Diagram, and much more. From this point the reader can delve into the remainder of the book with little difficulty. Part one explains the clouds of gas around stars called nebulae, as well as interstellar gas, while part two discusses the clouds of dust that dim and scatter starlight in and near the galactic plane. The next section considers common clouds of dust and gas, and the birth of stars from this diffuse matter. The final section briefly discusses diffuse matter outside the Milky Way. Clearly written and not bogged down in mathematics, this classic should be in most libraries that have an astronomy collection.

507.   Field, George B., and A. G. W. Cameron, eds. **The Dusty Universe.** New York, Neale Watson Academic Publications, Inc., 1975. 323p. illus. index. refs. $15.00. LC 75-15576. ISBN 0-88202-033-1.

   Thirteen review papers on interstellar and interplanetary dust comprise this volume, which resulted from a conference held in honor of Fred L. Whipple, former directory of the Smithsonian Astrophysical Observatory. The possible cosmogonic relationship between the two types of dust is explored in this work, which updates portions of several previous books on the subject, like *Nebulae and Interstellar Matter* (Middlehurst and Aller, eds., 1968). Briefly, the book explores the chemical composition of the dusts, their origin (and possible connections), the interaction of dust and gas between the stars, cometary debris, chrondritic meteorites, the size and shape of the dust particles, etc. The dust cloud, starlight extinction aspects of the topic are not emphasized. The volume ends with a paper by the honored guest, who summarizes and comments upon the state-of-the-art. A worthwhile purchase for the astronomy library, it is aimed at the advanced student and astronomer.

508. Gurzadyan, G. A. **Planetary Nebulae.** New York, Gordon & Breach, Science Publishers, 1969. illus. bibliog. $39.00. LC 69-11664. ISBN 0-677-20220-2.

Updated from the 1962 Russian edition, this English version contains reports of recent research results and includes various editorial revisions. The exact place in the stellar evolutionary process has not yet been firmly located for the planetary nebulae, there being many theories concerning these objects; the author discusses these hypotheses at some length, after first describing the nature and physics of the nebulae. Suitable for a graduate astronomy course, the book is fairly technical, yet it includes a good deal of descriptive information and is highly documented. Combining a review-type approach with the results of three decades of the author's personal research, this is the only present-day monograph devoted to the ring-like nebulae. Translated from the Russian (*Planetarnye tummannosti*) and edited by D. G. Hummer with the assistance of C. M. Varavsky and Z. Lerman, a revised translation has since been published by Reidel in 1971 ($28.50).

509. Hindmarsh, W. R., *et al.*, eds. **Magnetism and the Cosmos.** New York, American Elsevier Publishing Co., Inc., 1967. 436p. illus. refs. indexes (author and subject). $32.50. LC 67-16768. ISBN 0-444-19911-X.

This advanced text considers at length another of the basic forces that pervade the Universe. Divided into five sections (Geomagnetism, Stellar Magnetism, Solar Magnetism, Planetary Magnetism, Solar System Magnetic Fields), this advanced work explains to the reader how magnetic fields arise on celestial bodies, the effects of magnetism on the solar and stellar neighborhood, what the existence of a planetary magnetic field tells us about the interior of that planet, and much more. Both theoretical discussions and descriptions of observations comprise this volume of 39 papers from the NATO Advanced Study Institute on Planetary and Stellar Magnetism, held at the University of Newcastle upon Tyne, England, April 1965. While not a topic of broad interest among astronomers, the book is one of the few collective works on magnetism, so it should be in the astronomy library.

510. Kaplan, S. A., and S. B. Pikelner. **The Interstellar Medium.** Cambridge, Mass., Harvard University Press, 1970. 465p. illus. index. refs. $20.00. LC 70-85076. ISBN 0-674-46075-8.

Going beyond the standard topics of dust and gas clouds, this advanced text considers radiation fields, cosmic rays, magnetic fields, and non-thermal radio emission as well. The physics of these interstellar particles and fields is the primary concern of the book, which talks about (but plays down) the descriptive aspects of the subject. Consequently, topics like types of nebulae, interstellar grains, etc., are de-emphasized. The reader is referred to *Nebulae and Interstellar Matter* (Middlehurst and Aller, eds., 1968) for a more general advanced treatment in review form. Discussions here include interstellar hydrogen (ionization, emission, distribution), the physical state of the interstellar gas (spectral line formation, temperature, molecules in space), interstellar dust (distribution, polarization and absorption of starlight, properties of particles), interstellar magnetic fields and non-thermal radio emission, and interstellar gas dynamics and the evolution of the interstellar medium. Translated from the Russian (*Mezhzvezdnaya sreda*, Moscow, 1963), this volume is well suited for graduate students and astronomers.

511. Middlehurst, Barbara M., and Lawrence H. Aller. **Nebulae and Interstellar Matter.** Chicago, University of Chicago Press, 1968. 835p. illus. index. refs. (Stars and Stellar Systems, v.7). $32.50. LC 66-13879. ISBN 0-226-45959-4.

Sixteen lengthy review papers comprise this volume for the astronomer; the papers are concerned with the clouds of dust and gas that comprise about 20 percent of the galaxy's total mass. Without a doubt the best and most comprehensive volume on the subject, this book takes into account both optical and radio observations of the interstellar medium, as well as theoretical considerations related to its existence and physical processes. The first paper provides an overview of the subject, describing the observed properties of the dust and gas (including spatial distribution and composition), the dynamics of interstellar matter (energy dissipation, kinetic energy, etc.), and the formation of stars from the material. The remaining chapters discuss individual spectra, scattering of light, radio observations, and other aspects of interstellar space (x-rays, cosmic rays, and galactic magnetic fields). Like the other volumes of this series, the papers are well edited, clearly written, and appropriately illustrated. References abound, and the reader will readily find many citations to related literature. Like most of the other volumes of this series, however, the text badly needs to be updated.

Contents: 1, Dynamics of Interstellar Matter and Formation of Stars; 2, Diffuse Nebulae; 3, Dark Nebulae; 4, Flare Stars; 5, Interstellar Extinction; 6, Interstellar Grains; 7, Interstellar Absorption Lines; 8, Atomic Processes with Special Application to Gaseous Nebulae; 9, Planetary Nebulae; 10, Radio-Line Emission and Absorption by the Interstellar Gas; 11, Nonthermal Galactic Radio Sources; 12, The Theory of Synchronotron Radiation; 13, Discrete X-Ray Sources; 14, Dynamical Properties of Cosmic Rays; 15, Evidence for Galactic Magnetic Fields; 16, Primordial Stellar Evolution.

512. Osterbrock, Donald E. **Astrophysics of Gaseous Nebulae.** San Francisco, W. H. Freeman, 1974. 251p. illus. index. refs. (A Series of Books in Astronomy and Astrophysics.) $17.00. LC 74-11264. ISBN 0-7167-0348-3.

Summarizing the current theory on diffuse nebulae, planetary nebulae, and supernovae remnants, this up-to-date volume fills a gap in the literature in one of the more important areas of astronomy. Aimed at graduate students and astronomers, the book emphasizes the physical processes in the nebulae rather than spending a great deal of time on description. The author, however, does briefly cover the physical characteristics of the objects in the first chapter to prepare the new reader. Chapter topics include photoionization and thermal equilibrium, calculation of emitted spectrum, comparison of theory with observations, internal dynamics of gaseous nebulae, interstellar dust, H_{II} regions in the galactic context, and planetary nebulae. The work appears at a time when H_{II}

regions and other nebulae have become an integral part of both optical and radio astronomy research. The author underscores this in his preface by pointing out the growth of knowledge in these areas and its importance for understanding the Universe. A high point of the book is the bibliography, in the form of annotated references at the end of each chapter. It is rare, unfortunately, that citations to related literature are presented in this fashion; such an approach is extremely helpful to new readers and to librarians. Not a lengthy volume, this tome presents a succinct overview, and all astronomy libraries will want to have it.

513. Terzian, Yervant, ed. **Interstellar Ionized Hydrogen.** New York, W. A. Benjamin, 1968. 774p. illus. refs. $24.00. LC 68-56107. ISBN 0-8053-9268-8.

Somewhat dated, this volume of conference proceedings considers the interstellar H_{II} regions from the optical, radio, theoretical, and infrared points of view. Although the interstellar ionized hydrogen question has more recently been a problem for the radio astronomers, there have been related research projects of importance using other observational techniques, and these are explored here. Of the 32 papers from the meeting held in Charlottesville, Virginia, December 8-11, 1967, the following major subjects were discussed: observations related to regions of star formation, evolution of stars and H_{II} regions, optical and radio observations of H_{II} regions, radio recombination lines, the electron temperature of H_{II} regions, large scale distribution of ionized hydrogen in the galaxy, and OH emission. More recent papers have since appeared in the literature, but this book is one of the few treatises that offers a comparison of the various methods of H_{II} region study.

514. Wickramasinghe, N. C. **Interstellar Grains.** London, Chapman and Hall Ltd.; distr. New York, Barnes and Noble, Inc., 1967. 154p. illus. refs. indexes (author and subject). (The International Astrophysics Series, v.9). LC 67-113260. B 67-17357.

A comprehensive coverage of interstellar dust on a non-elementary level is presented in this text for the advanced student and scientist. Summarizing the various theories and observations of past researchers in this area, the author discusses all major grain theories but emphasizes the graphite theory, which he himself has been most closely associated with. The theoretical models covered include iron particles, dirty ice grains, Platt's complex molecules, graphite grains, and graphite core-ice mantle grains. Observation criteria (interstellar extinction, interstellar polarization, albedo, and backscatter criterion) are also summarized in this text, which is one of the better books available. Selected topics of interest include light scattering by spherical grains, interstellar reddening, interstellar polarisation, interstellar condensation theories, physical properties of grains, and optics of grains.

GALAXIES AND COSMOLOGY

Probably one of the most interesting and thought-provoking areas of astronomy is the study of the Universe as a whole. How large is it? Where, how, and when did it begin? What will happen in the future? These questions and many others have occupied the thought of cosmologists for several decades now, posing difficult problems for the astronomer. Consequently, the literature of cosmology contains some of the most fascinating and most speculative works in the entire realm of astronomical publication.

This section of the guide covers three subject areas, which are all related in that they concern the Universe at large: 1) galaxies (the vast star systems, like the Milky Way, which are islands in the almost limitless void); 2) cosmology (the structure and origin of the Universe); and 3) quasars (star-like emitters of vast amounts of radiation, among the most distant objects in the Universe). A few works are cited under each category, for both the general reader and the astronomer.

Card catalog subject headings: *Library of Congress:* GALAXIES; NEBULAE; COSMOLOGY; QUASARS. *Sears:* GALAXIES; UNIVERSE; QUASARS.

Galaxies

515.   Abetti, Giorgio, and Margherita Hack. **Nebulae and Galaxies.** London, Faber and Faber, 1964; New York, Thomas Y. Crowell Co., 1965. 264p. illus. index. bibliog. ₤4.25; $6.95. LC 65-17631 (Crowell); 65-8675 (Faber). ISBN 0-571-05699-7 (Faber).

This standard text presents a comprehensive, descriptive look at three types of objects in the sky referred to as nebulae: planetary nebulae, irregular bright and irregular dark nebulae, and the external galaxies. Logically, these topics are not related; their only common denominator is their nebulous appearance, a factor that led astronomers to believe they were like objects, even up to a hundred years ago. Logic aside, the treatment of the three subjects is quite good. While the book is not highly technical, it supposes that the reader has a bit of previous experience in the sciences, so it would best suit the serious amateur or student. Besides the three types of "nebulae" mentioned, the book also contains a discussion of our galaxy and the origin and evolution of the Universe. Several special features are worth mentioning, for they make this text special. First, the introductory chapter on nebulae gives some historical background on the astronomers who have studied them. Second, the arrangement of each chapter is well suited for both layman and student. Each object is described generally, noting physical characteristics, population frequency, distribution, chemical composition, spectrum, etc. Then, the most conspicuous and famous objects are listed, giving coordinates, historical background, peculiarities, etc. Serious amateurs and astronomy students will find this arrangement ideal for observation. Finally, most of the objects listed are accompanied by photographs. Translated from the Italian (*Le Nebulose e gli universi-isole*) by V. Barocas, this volume is highly recommended for anyone with an interest in the nebulae.

516. Page, Thornton, and Lou Williams Page, eds. **Beyond the Milky Way.** New York, The Macmillan Co., 1969. 336p. illus. index. glossary. bibliog. (Sky and Telescope Library of Astronomy, v.8). $7.95. LC 69-10504.

Galaxies and cosmology are the subject of this, the eighth volume in the excellent *Sky and Telescope* series. After some introductory articles on various topics (the expanding Universe, radio energy from galaxies, the local group of galaxies, etc.), the book begins its presentation of "large" numbers by talking about distances to other galaxies and the size of the Universe. The structure and content of galaxies is presented next, in a series of articles on the classification of galaxies, the Magellanic Clouds, the Andromeda Galaxy (M31), spiral structure, and several other topics. How galaxies develop is the subject of the next chapter, which examines the evolution of stars and the interstellar gas from which they form. Strange, peculiar galaxies are discussed next, with emphasis on topics like supernovae, exploding galaxies, dwarf galaxies, etc. Quasars, star-like emitters of large amounts of radio energy, are covered in the following two chapters (containing 17 articles on these fascinating, mysterious objects). Cosmology is the subject of the last section, which covers some of the highlights of the various theories over the past 30-odd years. It is not a very comprehensive section, but it includes some good reprints on some interesting topics. This book for the layman and amateur astronomer is comprised of articles from *Sky and Telescope* and its predecessors.

517. Sandage, Allan, Mary Sandage, and Jerome Kristian, eds. **Galaxies and the Universe.** Chicago, University of Chicago Press, 1975. 818p. illus. refs. indexes (author, subject, galaxy). (Stars and Stellar Systems, v.9). $45.00. LC 74-7559. ISBN 0-226-45961-6.

This long-awaited, much-heralded volume is both a major contribution to the literature and somewhat of a disappointment. Presenting a broad overview of the field, the book contains 19 review articles on topics ranging from classification of galaxies to redshifts. Unfortunately, a number of the papers are severely dated, some as old as 1965, leaving gaps in an otherwise up-to-date coverage. Quite a few reviews are fairly current, though, so the text is not completely outdated. Not surprisingly, the articles are of a high caliber, in keeping with the tradition of the previous volumes of the series.

Over half the book is devoted to characteristics and studies of individual galaxies and types of star systems. The subject of galaxies as they appear in the overall scheme of the Universe is covered in another third of the volume, followed by two chapters on cosmology. The latter treatment is unfortunately brief, but it does provide an acceptable introduction on the advanced level. The growth of radio astronomy and its importance in galactic studies is quite evident in this work, which explains many developments resulting from radio observations. Of special note are the hundreds of bibliographic references and the three fine indexes—in particular, the guide to galaxies cited in the text, which includes radio sources and quasars, groups, clouds, and clusters. Several useful tables are included as well: clusters (with brightest members listed), integral properties, redshifts and luminosities for 172 radio galaxies, etc. The editors note in the preface that the work done in this field has just begun, and that many new discoveries and theories are yet to come. A good record of the research done so far, this book is a must for the astronomy library; no comparable work is available.

      Contents: 1, Classification and Stellar Content of Galaxies Obtained from Direct Photography; 2, The Stellar and Gaseous Content of Normal Galaxies as Derived from Their Integrated Spectra; 3, The Masses of Galaxies; 4, Magnitudes, Colors, Surface Brightness, Intensity Distributions, Absolute Luminosities, and Diameters of Galaxies; 5, Integrated Energy Distribution of Galaxies; 6, The Identification of Radio Sources; 7, Strong Nonthermal Radio Emission from Galaxies; 8, Quasars; 9, Radio Observations of Neutral Hydrogen in Galaxies; 10, The Formation and Early Dynamical History of Galaxies; 11, Stellar Dynamics and the Structure of Galaxies; 12, The Extragalactic Distance Scale; 13, Binary Galaxies; 14, Nearby Groups of Galaxies; 15, Clusters of Galaxies; 16, Distribution of Galaxies; 17, Galaxy Clustering: Its Description and Its Interpretation; 18, Radio Astronomy and Cosmology; 19, The Redshift.

518.   Shapley, Harlow. **Galaxies**. 3rd rev. ed. Cambridge, Mass., Harvard University Press; distr. Cambridge, Mass., Sky Publishing Corp. 1972. 232p. illus. index. (Harvard Books on Astronomy). $10.00. LC 77-169859. ISBN 0-674-34051-5.

      A descriptive and lively text, excellent illustrations, and comprehensive subject coverage make this the best general text available on the external star systems. In this latest edition, revised by Paul W. Hodge, approximately one-third of the material is new and there are several dozen new illustrations. The layman and student will be held attentive by the author's style as he describes a variety of topics, drawing largely on his own experience as a galactic astronomer. The introductory chapter gives definitions of frequently used terms (useful for the neophyte), general comments, and a brief description of the types of galaxies (elliptical, spiral, irregular, etc.). The author next describes the two closest galaxies, neighbors of the Milky Way, the Magellanic Clouds. Used by navigators in the fifteenth century to find the south celestial pole, these galaxies have provided astronomers with a large body of information about galactic structure because of their proximity, so they hold a very important place in astronomical research. Appropriately, Shapley devotes a great amount of space to these two galaxies. The Milky Way as a galaxy is also explored, including discussions of distance determinations, size, and structure. (The reader is referred to *The Milky Way*, Bok and Bok, 1974, for a more detailed survey.) The so-called "local group" is described in "The Neighboring Galaxies," a chapter on the Andromeda Galaxy (M31) and several other nearby star systems. The physical characteristics of galaxies are also described (rotation, types of stars, interstellar matter, radio emissions, etc.). The "Surveys of Deep Space" chapter is one of the most interesting; here Shapley discusses the galactic census and clusters of galaxies. The final chapter, on the expanding Universe, takes a look at the motions of star systems and presents some cosmological theory. Highly recommended for all types of libraries.

519. Van den Bergh, Sidney. **The Galaxies of the Local Group.** Richmond Hill, Ontario, David Dunlap Observatory, 1968. 73p. illus. refs. (Communications from the David Dunlap Observatory, no. 195).

This short but interesting treatise is a reprint from the *Journal of the Royal Astronomical Society of Canada* (vol. 62, August and October, 1968), discussing in detail the cluster of nearby galaxies known as the "local group." Included in this short book is a variety of information on the Milky Way, the Magellanic Clouds, the Andromeda Galaxy (and its companion galaxies), Fornax, Sculptor, and the other members of our cluster. The book is written in the style of a long survey article, synthesizing the work of many astronomers rather than looking at one person's results in detail. An excellent coverage of a topic that has had little attention elsewhere, this book has a substantial number of useful references to related work. A new version of this paper should be written someday to include the newly discovered "Snickers" galaxy.

Cosmology

520. Dickson, F. P. **The Bowl of Night: The Physical Universe and Scientific Thought.** Cambridge, Mass., the MIT Press, 1968. 228p. illus. indexes (subject and name). notes. $13.95; $4.95pa. LC 71-78628. ISBN 0-262-04024-7; 0-262-54003-7, 126pa.

Man's view of the Universe is the subject of this non-technical volume written for the college student and serious layman. The author's treatment of cosmology is basically historical and philosophical, not physical. The book traces cosmological theory from the ancients through the Dark Ages on to the seventeenth century and up through the present. Among the many subjects discussed are Olber's paradox, relativity, theory of matter, early concepts of the physical universe, light, and various cosmological theories. The text is arranged well— a description of early views of the Universe sets the stage for later topics like relativity and expanding universes, etc. The book contains many references to related works for the serious student.

521. Evans, David S., Derek Wills, and Beverly J. Wills, eds. **External Galaxies and Quasi-Stellar Objects.** Dordrecht-Holland, D. Reidel Publishing Co.; distr. New York, Springer-Verlag, 1972. 549p. illus. refs. indexes (subject and name). $38.60. LC 77-154736. ISBN 90-277-0199-7 (Reidel); 0-387-91092-1 (S-V).

A very comprehensive look at current research and developments in the fields of galaxies and cosmology is presented in this conference volume of 82 technical papers. Dozens of topics are covered and one is at first taken aback by the lengthy table of contents, which contains only a straight list of papers, with no subject groupings. Fortunately, there is a good subject index, a not-too-frequent inclusion in volumes of conference proceedings, complete with NGC, M, and quasar numbers. The papers cover both optical and radio observations of galaxies, cosmological topics (theoretical and practical), and a large percentage of articles on quasars. Though the papers are mainly reports of original research, there are also 12 comprehensive survey papers, which nicely summarize the past and current state of galactic and cosmological studies. A very selected list of topics presented includes stellar populations in galaxies, gas and dust, galactic evolution, radio emissions, classification of galaxies, observation and theory of quasars, current views of the Universe, etc. An excellent compilation, the book is the result of IAU Symposium no. 44 held in Uppsala, Sweden, August 10-14, 1970.

522. Field, George B., Halton Arp, and John N. Bahcall. **The Redshift Controversy.** Reading, Mass., W. A. Benjamin, Inc., 1974, c1973. 324p. illus. index. refs. (Frontiers in Physics). $19.50; $11.00pa. LC 74-614. ISBN 0-8053-2512-3; 0-8053-2513-1pa.

Since the 1920s it has been generally accepted by astronomers that the redshifts observed in galaxies are due to the expansion of the Universe. Some reject this theory, however. This volume considers both sides of the question, presenting the two opposing views by astronomers Halton Arp and John N. Bahcall. This book resulted from a debate between the two scientists at an AAAS symposium on December 30, 1972, in Washington, D.C. Besides containing material from the formal arguments, the volume includes selected reprints illustrating both speakers' views. The stated purposes of the book are to stimulate further research in this area, and to illustrate for students of astronomy the ambiguity which sometimes besets astronomical research; the result is an interesting and thought-provoking book, worth reading by all types of scientists.

523. Gill, T. P. **The Doppler Effect: An Introduction to the Theory of the Effect.** London, Logos Press; distr. New York, Academic Press, Inc., 1965. 149p. index. refs. $10.00. LC 65-19285. ISBN 0-12-283350-3.

The Doppler effect, a topic usually introduced in freshman astronomy or in physics during discussions of relativity, is covered thoroughly in this book for the advanced student or scientist. Defined by the author as "the change in the apparent time interval between two events which arises from the motion of an observer together with the finite velocity of transmission of information about the events," the Doppler effect has several astronomical applications, as well as uses in navigation, rocket and satellite tracking, and studying blackbody radiation. All these topics are discussed at length, but the former will be of most interest to astronomers. Along these lines, Gill explains spectral line broadening, spectroscopic binaries, and the relativistic Doppler effect. These subjects are explained well, and with a great deal of rigor; the book, therefore, is intended for the advanced student. The only work to consider this subject at such length, it has an excellent introductory chapter.

524. John, Laurie, ed. **Cosmology Now.** London, BBC Publications, 1973. 168p. illus. index. ₤2.75. LC 74-179354. GB 74-14152. ISBN 0-563-12370-2.

In general, the current cosmological literature (except for works like Kilmister's *The Nature of the Universe*, 1971) is very technical and not at all suited for the general reader, an unfortunate situation since the subject is so interesting. A worthwhile addition to the astronomy collection of any library would be this good text, written for the layman in language that is not too technical and in a style that is captivating and descriptive. Emphasizing the Universe at it "currently" appears, the 12 separate lectures in this book exhibit the viewpoints of what the editor calls "the younger generation of cosmologists, who are writing in a language which would seem meaningless to an earlier generation—neutron stars, pulsars, and black holes are concepts of our own age." The articles are written by famous British cosmologers like Bondi, Sciama, and Taylor, to name three; they include discussions of cosmological models, the steady state theory, black holes, galaxies, etc. In short, an excellent, understandable picture of the Universe is presented, with observational evidence and theory in support, and previously mysterious topics are explained to nearly any reader's satisfaction.

525.   Kaufmann, William J., III. **Relativity and Cosmology**. New York, Harper and Row Publishers, 1973. 134p. illus. index. glossary. $4.50pa. LC 72-12002. ISBN 0-06-043568-2.

One of the best treatments of relativity for the layman is contained in this fascinating volume, which discusses many of the more exotic and puzzling aspects of current astronomical research. The author begins by presenting some background on gravitational theory, illustrating Kepler's laws of planetary motion and Newton's law of gravity. A discussion of space-time follows as Kaufmann moves toward the foundations of general relativity and the experiments made to prove its validity. These first chapters are very clearly written and are accompanied by excellent diagrams and drawings. The book then turns to cosmology and some interesting and related topics: the Doppler effect, the red shift, black holes, worm holes, galaxies and quasars, and the origin of the Universe. The chapter on the shape of space is very well presented and gives the reader a look at the Universe as a whole. Although the book is for general readers, those who have some background in science will find it easier to understand. Ideal for the public library, it would be appropriate for the university collection as well.

526.   Kilmister, Clive. **The Nature of the Universe**. New York, E. P. Dutton & Co., Inc., 1971. 216p. illus. index. bibliog. glossary. (The World Science Library). $3.95pa. LC 76-165333. ISBN 0-525-16430-8.

Many beautiful illustrations and a clearly written text comprise this fine book on the order and origins of the Universe, aimed at the general reader and student. Painting an excellent picture of the heavens, the author delves into some of the most interesting problems of space and time, relativity, and cosmology. From the first fascinating chapter describing the inconceivably enormous size of the Universe to some of the unusual cosmological theories, the reader will be captured by the descriptions of what the greatest scientific minds have theorized. The two chapters on mankind's concept of space and time trace these ideas from ancient Greece to the twentieth century, in an interesting story of how that viewpoint has changed so many times. Relativity, a mysterious and complicated-sounding topic, is clearly presented, as are some rival theories to Einstein's laws. New evidence supporting and counteracting certain cosmological theories is presented, along with some predictions for the future. Definitely one of the best books on cosmology, it is suited for all types of libraries and their readers.

527. Peebles, P. J. E. **Physical Cosmology**. Princeton, N.J., Princeton University Press, 1971. 282p. illus. refs. (Princeton Series in Physics). $12.50; $6.50pa. LC 74-181520. ISBN 0-691-08108-5.

A fairly technical survey of the theories of the Universe, this book would best serve the university student and non-specialist scientist. The author begins with a chapter title reminiscent of the phonograph record advertisements from television: "Golden Moments in Cosmology 1912-1950," a survey of ideas on the Universe and its beginnings. The expansion of the Universe, stead state cosmology, and more are reviewed in this section, which has many references to early twentieth century literature. The assumption that the Universe is homogeneous and isotropic on the large scale is the subject of Chapter II, which presents support for homogeneity and isotropy in the form of galaxy counts, Hubble's Law, and the radiation background. The cosmic time scale and the mean mass density of the Universe are also explored, the latter at some length. The Primeval Fireball Hypothesis and the Big Bang Theory are both discussed as a very likely origin of the Universe. Several cosmological models are presented in the latter stages of the book, followed by a "History of the Universe," a look at how it all got started. For the astronomy and university library; this book is probably too advanced for the layman.

528. Sciama, D. W. **Modern Cosmology**. Cambridge, Engl., Cambridge University Press, 1971. 212p. illus. indexes (subject and author). $12.95. LC 73-142961. ISBN 0-521-08069-X.

Lying between the popular book and the very technical treatise, this well-written volume presents a comprehensive look at our Universe and cosmological theory. Best suited for university students and educated laymen, the text requires a basic knowledge of mathematics and physics, but nothing too advanced. A description of the known Universe begins the book, as Sciama describes stars, galaxies, quasars, etc., discussing their physical characteristics, frequency, and distribution. The author next presents various models of the Universe, constructed according to several cosmological theories. Support for the various theories, mainly in the form of the study of the quantities of hydrogen and helium in the Universe, is advanced. Overall, this is an excellent treatment of the subject, and the book would be a good choice for the college, observatory, or public library.

Quasars

529. Bova, Ben. **In Quest of Quasars**. London, New York, the Macmillan Co., 1969. 198p. illus. index. bibliog. $6.95; $6.50pa. LC 77-83062.

One of the few popular treatises devoted mainly to quasi-stellar sources, this book is well illustrated with photographs and diagrams, and it covers all aspects of these interesting celestial objects. Putting quasars in perspective, the author also describes other astronomical phenomena like various types of stars and galaxies. The discovery of these star-like radio sources is covered first, followed by the discussions of stars and stellar systems. Included in the description of quasars are their size, distance, brightness, redshift, etc. A chronology of the history of quasars ends this fine book, which is subtitled, "An Introduction to Stars and Starlike Objects." The paperback edition (1975) is available from the New American Library.

530.   Burbidge, Geoffrey, and Margaret Burbidge. **Quasi-Stellar Objects.** San Francisco, W. H. Freeman and Co., 1967. 235p. (A Series of Books in Astronomy and Astrophysics). illus. indexes (subject and name). refs. $7.50. LC 67-17457. ISBN 0-7167-0321-1.

Written for astronomers and graduate students, this informative book is slightly more technical than (and twice as long as) Kahn and Palmer's *Quasars* (1967). Naturally, a great deal of new information has been obtained since this was published, but as a basic text it remains the best. Rather comprehensive, the book's 18 chapters cover the spectrum of subjects related to quasars. Selected headings include: identification, line spectra, radio emission, variations in flux, distribution, red shifts, models, local or distant phenomena?, energy output theories, etc. Especially useful is the list of approximately 150 quasars known in 1967; this table includes R.A., Dec., visual magnitude, redshift, and colors (B-V and U-B). Recommended for the astronomy and general science library, it concludes by admitting there are still many unanswered questions concerning the nature and place of quasars in cosmological theory.

531.   Kahn, F. D., and H. P. Palmer. **Quasars: Their Importance in Astronomy and Physics.** Manchester, Engl., Manchester University Press; Cambridge, Mass., Harvard University Press, 1967. 122p. $5.95. LC 67-93902. B 67-10954. ISBN 0-674-74100-5 (Harvard).

This short volume was published a few years after the quasistellar sources were first discovered, in an attempt to introduce the phenomena to scientists unfamiliar with their nature. Fortunately, it is not overly technical, and the layman could read this book, too, by skipping over the equations inserted at various points in the text. In a clear, fairly descriptive manner, the authors begin by discussing a previously known cousin to the quasar, the radio galaxy, stellar systems that emit a great amount of radio wave energy. The discovery of the strange star-like objects that emit 50 billion times more energy than the Sun is described next. Optical and radio properties of the objects are compared in the following two chapters. Other subjects which are sure to be of interest are quasars and relativity, quasars and stars, the production of fast particles, and a quasar model. A new edition of this excellent work would be welcome.

4

SPECIAL TOPICS

INSTRUMENTATION AND TECHNIQUES

Like many of the physical sciences, astronomy devotes a small but important section of its literature to equipment and its use. The works cited here deal with telescopes and their auxiliary devices, and with basic optics and optical engineering. Although most monographs on telescopes are designed for astronomers and technicians, there are also a few books for the general reader, and two of the best are listed below.

Volumes for the professional astronomer take the form of texts, review article books, proceedings, and more. A representation of each is included in this section. The books here deal only with optical instrumentation; books on radio telescopes are listed under Radio Astronomy. Further, those works on amateur telescope making and use are presented under Amateur Astronomy.

Card catalog subject headings: *Library of Congress:* TELESCOPE: ASTRONOMICAL INSTRUMENTS; TELESCOPE–POPULAR WORKS; TELESCOPE, REFLECTING. *Sears:* TELESCOPE; ASTRONOMICAL INSTRUMENTS.

532.   Barlow, Boris V. **The Astronomical Telescope.** London, Wykeham Publications (London) Ltd.; distr. New York, Springer-Verlag, 1975. 213p. illus. index. refs. glossary. (Wykeham Science Series, 31). $7.80pa. LC 74-78485. ISBN 0-387-91119-7; 0-387-91118-9pa. (Springer).

Best suited for the astronomy student and ambitious amateur, this most recent entry into the literature of instrumentation is an excellent text covering all aspects of the subject. The first two chapters are historical, discussing pre-telescope astronomy and the development of the telescope as an astronomical tool. Readers who find this well-written section too short should consult Asimov's *Eyes on the Universe* (1975) for a complete history. Basic optics is introduced next, with explanations of electromagnetic waves, reflection, refraction, diffraction, dispersion, aberration, and more; the important concepts of focal length, power, and aperture are also presented. The treatment of optics is substantial but not overwhelming. A related chapter on ray paths and light losses follows, showing the various types of telescope configurations and how light travels through each. Some of the steps and problems encountered in the engineering of large telescopes and their mirrors are outlined in Chapter 5, followed by a look at the instruments' drive mechanisms and controls. The latter is one of the most up-to-date and detailed treatments, covering the use of computers to direct the motions of the telescope. Other topics of interest are auxiliary equipment, observatory sites and buildings, extraterrestrial instruments, and some of the world's largest telescopes. Chapter 11 discusses the latest technological developments, including daytime viewing, non-visual observations, and some radical new observatory designs. Updating some old but good works like *Tools of the Astronomer* (Miczaika and Sinton, 1961), this fine book is sometimes rather technical, but not overly so. It would be ideal for the non-specialist who wants an introductory text with good background material. Ideal for the observatory, college, and public library, it is a welcome addition to the literature.

533.   Brown, Earle B. **Modern Optics**. Huntington, N.Y., R. E. Kreiger Publishing Co., 1974, c1965. 645p. illus. index. refs. $27.50. LC 73-92134. ISBN 0-88275-149-2.

Intended as a self-study text rather than a classroom textbook, this large volume gives a comprehensive overview of light, lenses, and optical systems, the knowledge of which is a prerequisite for the design of lenses for telescopes, a very complex task. Aimed at the optical engineer, whether working in physics, astronomy, or related fields, the book is one of the better available. The volume is divided into three major sections: Fundamentals; Optical Systems and Devices; and Lasers. The first includes such topics as the nature and properties of light, light as a wave motion, interaction of light and matter, geometrical optics, radiometry and photometry. The second section, concerned with applications, explains optical imaging systems, optical detection, optical measurement, communications theory, etc. The portion devoted to lasers is concerned with fundamentals, forms and characteristics, applications, and more. The author discusses the various types of telescopes, their lenses and mirror arrangements, uses, etc., but this is only a small portion of the book.

534.   Carleton, N., ed. **Astrophysics, Part A: Optical and Infrared**. New York, Academic Press, 1974. 587p. illus. refs. indexes (author and subject). (Methods of Experimental Physics, v.12). $43.50. LC 73-17150. ISBN 0-12-475912-2.

Intended for graduate students and others entering the field, this book of collected technical papers focuses on methods of observation and data collection in the optical and infrared portions of the spectrum. A "how-to" book for telescope auxiliary equipment, this effort is a significant contribution to the literature. It not only collectively explains the workings of the hardware of the astronomer, but it is very up to date as well. Papers give background and basic physical explanations as well as outlines of how the devices work. The types of data obtainable are discussed, but there is little here on results; however, that is not this book's purpose. A valuable text for the observatory library, it may be too expensive for some institutions. The book contains the following papers: Photomultipliers: Their Cause and Cure; Other Components in Photometric Systems; Observational Techniques and Data Reduction; Reshaping and Stabilization of Astronomical Images; Detective Performance of Photographic Plates; Two-Dimensional Electronic Recording; X-Ray and Gamma-Ray Detection by Means of Atmospheric Interactions: Fluorscence and Čerenkov Radiation; Polarization Techniques; The Instrumentation and Techniques of Infrared Photometry; Diffraction Grating Instruments; Fourier Spectrometers; Fabry-Perot Instruments for Astronomy.

535.   Evans, David S. **Observation in Modern Astronomy**. New York, American Elsevier Publishing Co., Inc., 1968. 273p. illus. index. refs. $16.75. LC 67-20394. ISBN 0-444-19941-1.

A combination astronomy text and professional observer's manual, this book intends to give the student who wants to be a working astronomer some idea of the types and methods of observation. Since most astronomical concepts are explained in the text, the student does not need to know any astronomy beforehand. Although the book explains methods of data gathering and analysis, students will probably not read this book and instantly become a practicing astronomer. They will learn by doing, not just reading. Therefore, this book should be considered as a secondary reference source, to be used for brushing up on some technique or as background before approaching new areas. The six chapter headings give an indication of the book's scope, which is not as wide as the title might indicate: Astronomy of Position; The Measurement and Analysis of Stellar Radiation; Interrelations between Observed Quantities; The Motions of the Stars; Variable Stars; Binary Stars and Multiple Stars; The Galaxy and the Galaxies. Sadly, there is no mention of planetary, lunar, or solar observation, often the first area encountered by the student. Telescopes and other equipment are introduced at various points, but this is not a definitive treatment of this topic. *Telescopes* (Kuiper, ed., 1960) and *Tools of the Astronomer* (Miczaika and Sinton, 1961) should also be read by the interested astronomer.

536.   Haug, Ulrich, ed. **The Role of Schmidt Telescopes in Astronomy**. Hamburg, W. Germany, European Southern Observatory, 1972. 160p. illus. refs. $7.00pa. LC 73-173917.

An example of a proceedings volume devoted to instrumentation, this book illustrates the wide range of uses of the so-called Schmidt telescope. Used frequently in sky surveys, Schmidt systems have provided astronomers with excellent photographic plates of the sky and are useful for studying faint, distant star systems. These aspects and more are covered in this collection of 25 papers presented at a meeting in Hamburg, March 21-23, 1972. The papers, which are generally shorter than most conference pieces, are mainly descriptions of particular applications of Schmidt instruments, but there are a few survey or general applications papers as well. Selected discussion comments are included in this volume for the astronomer.

537.   Hiltner, W. A., ed. **Astronomical Techniques.** Chicago, University of Chicago Press, 1962. 635p. illus. index. refs. (Stars and Stellar Systems, v.2). $22.50. LC 62-9113. ISBN 0-226-45954-3.

An extension of volume one (*Telescopes*, 1960) of the Stars and Stellar Systems series, this book presents review articles devoted to the design and use of various instruments. Primarily concerned with auxiliary equipment, rather than telescopes, the authors of the 23 papers mainly cover photometry, spectroscopy, and photography, three major areas of optical astronomical research. There is no mention of radio astronomy techniques and, because of the work's age, no mention of recent astronomical developments like infrared and x-ray astronomy. The thrust, then, is "optical Earth-bound astronomy," which, after all, is a prerequisite to the latest techniques. A new edition is badly needed; in the meantime, the reader is directed to *Astrophysics, Part A: Optical and Infrared* (N. Carleton, ed., 1974) for a good, up-to-date treatment of this field.

Contents: 1, The Detection and Measurement of Faint Astronomical Sources; 2, Spectrographs; 3, Radial-Velocity Determinations; 4, Spectrophotometry; 5, Measurement of Stellar Magnetic Fields; 6, Photomultipliers; 7, Photoelectric Photometers; 8, Photoelectric Reductions; 9, An Application of an Electronic Calculator to Photoelectric Reductions; 10, Polarization Measurements; 11, Instrumentation for Infrared Astrophysics; 12, Direct Recording of Stellar Spectra; 13, Image Detection by Television Signal Generation; 14, Application of the Image Orthicon to Spectroscopy; 15, Image Converters for Astronomical Photography; 16, Photographic Photometry; 17, Measuring Engines; 18, Techniques for Visual Measurements; 19, Astrometry with Astrographs; 20, Astrometry with Long-Focus Telescopes; 21, Orbit Determinations of Visual Binaries; 22, The Determination of Orbital Elements of Spectroscopic Binaries; 23, Orbit Determinations of Eclipsing Binaries.

538.   Horne, D. F. **Optical Production Technology.** New York, Crane, Russak and Co., Inc., 1972, c1962. 567p. illus. glossary. indexes (subject, name, and author). $42.50. LC 72-79282. ISBN 0-8448-0008-2.

Astronomers and technicians involved in designing and maintaining telescopes should be interested in this fine book, which explains in detail the applications and results of optical engineering. In particular, the author describes the production of lenses, prisms, mirrors, fibre optics, eye-glasses, and other optical materials, explaining the steps taken, materials used, and various tests to assure quality. Historical background frequently precedes the discussion of manufacture, which includes descriptions and photographs of the particular instruments being made. An example is the 98-inch Isaac Newton reflecting telescope made for the Royal Greenwich Observatory. There are quite a few references to astronomical optical production here, including large object glasses and mirrors, Schmidt cameras, mirror mountings, and testing of lenses and mirrors. Historical references as well as modern techniques are described. The subsection on the grinding and polishing of large mirror blanks (with many photographs) will be of interest to any astronomer, amateur or professional. Overall, the book covers dozens of optics applications, some of which are related directly to astronomy; almost all will be of interest to the telescope technician. These include, but are not limited to, grinding and polishing, optical tools, dioptric substances, prisms and flats, production control, testing optical components, mounting optical components, electro-optics and opto-electronics. This book does not cover basic optics, light, and lenses, and the reader will need to consult another source, like Brown's *Modern Optics* (1974, c1965), for a treatment of these basics.

539.   Kuiper, Gerard P., and Barbara M. Middlehurst, eds. **Telescopes.** Chicago, University of Chicago Press, 1960. 255p. illus. index. refs. (Stars and Stellar Systems, v.1). $12.00. LC 60-14356. ISBN 0-226-45953-5.

Written about the same time as Miczaika and Sinton's *Tools of the Astronomer* (1961), this review volume is slightly more technical and is aimed at the astronomer and graduate student. The text's main purpose is to describe various types of telescopes, along with their auxiliary equipment, including cameras, photoelectric devices, spectrographs, etc. An extensive discussion of astronomical "seeing" is also included, along with the connected problem of choosing an observatory site. The chapter devoted to radio telescopes describes the various types of instruments and lists many of the major telescopes. Two of the papers are concerned with two of the largest optical telescopes, written by astronomers affiliated with those observatories. There are good photographs of the 200-inch and 120-inch reflectors, and other telescopes as well, but the text is painfully dated, and a new, enlarged edition is needed.

Contents: 1, The 200-Inch Hale Telescope; 2, The Lick Observatory 120-Inch Telescope; 3, Design of Reflecting Telescopes; 4, Schmidt Cameras; 5, Telescope Driving Mechanisms; 6, The Transit Circle; 7, The Photographic Zenith Tube and the Dual-Rate Moon-Position Camera; 8, The Impersonal Astrolabe; 9, Astronomical Seeing; 10, Astronomical Seeing and Observatory Site Selection; 11, Radio Telescopes; 12, Radio-Astronomy Radiometers and Their Calibration.

540.   Miczaika, G. R., and William M. Sinton. **Tools of the Astronomer.** Cambridge, Mass., Harvard University Press, 1961. 294p. illus. index. (Harvard Books on Astronomy). LC 60-13299.

Telescopes and their uses are described in this book for general readers and astronomy students who want an in-depth look at these astronomical instruments, large and small. For those who wish an introductory treatment of light and basic optics, chapter one provides substantial detail. The electromagnetic spectrum is explained, as are reflection, refraction, and diffraction. A knowledge of these topics will better help the reader understand the workings of telescopes and their auxiliary equipment. Photography and its importance in astronomical research is the theme of chapter two, in which the authors discuss the photographic plate, exposure time, film sensitivity, and more. "Telescope Optics" is the book's most important section; here we learn about the types of telescopes (reflecting, refracting, and their variations), lenses, and eyepieces. The authors go far beyond just cursory descriptions, fortunately, discussing telescope performance, limitations, seeing, mirrors and their construction, resolution, distortion, etc. Additionally, the various astronomical cameras (astrographic, Schmidt, mirror) are described. All in all, it is a comprehensive, well-structured chapter. The construction of large telescopes, and their mirrors and lenses, is detailed next; included are discussions of the mountings, drives, and controls. The reader will quickly learn that planning and building a large instrument is not an easy task. Two special telescope applications, using special equipment, photometry and spectroscopy, are detailed, too. The former involves measuring the intensity and color of starlight, and the latter consists of spreading out starlight with a prism and analyzing the results. These descriptions are over-simplifications, but the authors go into great detail on these very important processes. Two special types of telescopes are discussed in the final two chapters: solar and radio instruments. This fine work should be revised and republished since it is one of the few not-too-technical books on instrumentation suitable for students, amateurs, and laymen.

541.   **Optical Telescope Technology.** Washington, D.C., NASA, 1970. 783p. illus. refs. $6.25pa. LC 79-605809. NASA SP-233.

Telescopes in space are the topic of this conference volume, which considers the current technology and the future of such ventures. Several successful orbiting telescope programs have been observing in visible, x-ray, infrared, and other regions of the spectrum, and many of these are described in this collection of 83 technical papers. These telescopes have been relatively small, since it is difficult to launch and successfully put into orbit a large instrument. An example of a technology and applications volume, this tome is a good description of our efforts in space astronomy through the end of the last decade. Included are papers on optical design, materials, instrumentation, logistics, testing, etc. There are also review papers on the past and future of space astronomy. Individual projects as well as general concepts are described. This state-of-the-art book is the result of a NASA workshop held at the Marshall Space Flight Center, April 29 to May 1, 1969.

542.   Smith, Warren J. **Modern Optical Engineering: The Design of Optical Systems.** New York, McGraw-Hill Book Co., 1966. 476p. illus. index. bibliog. refs. exercises. $24.00. LC 66-18214. ISBN 0-07-058690-X.

This advanced general text may be of use to the optical engineer working on telescopes, although there is not much here specifically on astronomical instruments. Not a text on basic optics, this book deals with the applications of light and lenses. Subjects include information on image formation (first-order and deviations), aberration, prisms, mirrors, stops and apertures, optical materials, basic optical devices, image evaluation, design of optical systems, and other important topics. Astronomical subjects include a cursory description of how telescopes work and a brief passage on the design of telescope systems. Not all astronomy libraries will want this, but those with optics shops probably will.

543.   Steel, W. H. **Interferometry**. London, Cambridge University Press, 1967. 271p. illus. bibliog. (Cambridge Monographs on Physics). $19.50. LC 67-12140. B 67-13353.

A theory of interferometry and a description of its techniques for all applications and in all regions of the spectrum where interferometry are used is the author's premise in this advanced book. In astronomical terms, interferometry refers to the measurement of stellar diameters or radio astronomy techniques. The author covers these applications, as well as other topics, in this rigorous text for scientists and advanced students. Chapter headings include Introduction; Mathematical Foundations; Optical Foundations; Coherence Theory; Theory of Two-beam Interferometers; Practical Two-beam Interferometers; Multiple-beam Interferometers; Measurements of Mean Phase; Measurements of Phase Validations; Interference Spectroscopy; Interference Imagery. The volume is a good overview of the subject, especially useful to radio astronomers.

SPACE SCIENCE AND ASTRODYNAMICS

Spawned by the race to the Moon, space science, an allied field including astronomy, physics, aeronautics, geology, chemistry, and more, is concerned with understanding and coping with the near- and outer-space environment. The boundary lines of space science as a discipline, and its literature, are not clearly definable, as it is a highly interdisciplinary field. Nevertheless, there are works that can definitely be "tagged" as space science, and a few of the more important of these are listed for the reader's information.

The literature of space studies is comprised almost exclusively of technical works, mainly in the form of student textbooks. There are also applications-type volumes describing past research and predicting the trend of future work; examples are astronomical observatories in space and probes to the planets. Several volumes on spacecraft exploration gathering astronomical data are presented under General Works.

Astrodynamics, a field closely related to space science, is included in this section. Also known as practical celestial mechanics, it is concerned with the orbits and trajectories of spacecraft, making it an important part of space research. A few sample volumes are presented here.

Finally, it should be re-emphasized that there are no books in this guide pertaining strictly to astronautics and aeronautics, since these subjects are not part of astronomy. There are a handful of items, though, that reflect both astronautics and astronomy, and several of these are scattered throughout the guide.

Card catalog subject headings: *Library of Congress:* SPACE SCIENCE; ASTRONAUTICS. *Sears:* Same headings.

544. Baker, Robert M. L. **Astrodynamics—Applications and Advanced Topics.** 2nd ed. New York, Academic Press, 1967. 540p. illus. refs. indexes (author and subject). $29.00. LC 67-14535. ISBN 0-12-075656-0.

The follow-up text and companion to the author's *Introduction to Astrodynamics*, this book is expectedly more mathematical than the first, with a good emphasis on space program applications. The majority of the work is concerned with the critical and highly important topic of orbit prediction and determination, including perturbations. The last two chapters consider the most advanced astrodynamic applications: lunar and interplanetary trajectories. Good problems and numerous references add to this book's usefulness. For the advanced student and aerospace engineer, this book ought to be updated to include advances made during the Apollo and Mariner years.

545. Baker, Robert M. L., Jr., and Maud W. Makemson. **An Introduction to Astrodynamics.** 2nd ed. New York, Academic Press, Inc. 1967. 439p. illus. indexes (author and subject). refs. $14.50. LC 67-14534. ISBN 0-12-075672-2.

This text is "meant to fill the gap between engineering handbooks, which tend rapidly to become outdated, and the more sophisticated texts, which tend to be over the head of most senior university students, or to lack practical value." One of the better treatments of practical celestial mechanics, this book is very readable and avoids becoming totally bogged down in mathematics. Often, equations are introduced to illustrate a point, rather than being an end in themselves. Excellent problem sets at the end of each chapter help the student apply what was learned. After some introductory material, the book is broken down into the following major portions, each directly related to astronomy: Minor Planets and the Moon; Comets, Meteorites, and Interplanetary Dust; Geometry, Coordinate Systems, and Ephemerides; Astrodynamic Constants; and Observation Theory. Several useful appendices and two glossaries are also included in this excellent book for the advanced student.

546. Haymes, Robert C. **Introduction to Space Science.** New York, John Wiley and Sons, Inc., 1971. 556p. index. illus. bibliog. problems (Space Science Text Series). $17.95. LC 78-140550. ISBN 0-471-36500-9.

The decided emphasis of this advanced text is astronomy, but there is a smattering of chemistry and geology and a large amount of physics as well. Haymes contends that space science is not concerned with technology (there are a few who might disagree), so no connection is made between the topics discussed and the space program, satellites, etc. Certainly, though, the subject matter here will be of great use to future space scientists who will have to understand the near and outer space environment. Aimed at students who know physics and ordinary differential equations, the book provides substantial problem sets with each chapter, giving the reader an opportunity to put learning into practice. Some selected subjects included are celestial coordinates and time, celestial mechanics, planets and their atmospheres, aurora, planetary interiors, comets, meteors, the interplanetary medium, stellar structure and evolution, radio astronomy, cosmology, and others. For the university or observatory library.

547. Herrick, Samuel. **Astrodynamics**. London, Van Nostrand Reinhold. v.1: 1971. 540p. index. bibliog. $18.50. LC 78-125199. B 71-14668. ISBN 0-442-03370-2. v.2: 1972. 348p. index. bibliog. $14.50. LC 78-125199. B 71-14668. ISBN 0-442-03371-0.

The celestial mechanician working in the space program will likely find a wealth of valuable information in this two-volume set, intended as a summary of the contributions made to the astrodynamic field. Scientists and engineers concerned with navigation, guidance, and control in the rapidly-changing field of space technology will likely want a copy of this work on practical celestial mechanics, which both reviews the classical concepts and applies them to the real world of space travel, satellites, and rockets. An advanced text in every sense of the word, its subject matter is divided as follows: Volume 1, Orbit Determination; Space Navigation; Celestial Mechanics; Volume 2, Orbit Correction; Perturbation Theory; Integration. The beginning student of astrodynamics may wish to consult *Introduction to Astrodynamics* (Baker and Makemson, 1967) before attempting this work.

548. Hess, Wilmot N., and Gilbert D. Mead, eds. **Introduction to Space Science**. 2nd ed., rev. and enlarged. New York, Gordon and Breach, 1968. 1056p. illus. refs. bibliog. indexes (subject and author). $79.00. LC 68-23418. ISBN 677-01450-3.

Space science is an incredibly varied field, encompassing astronomy, atmospheric science, aeronautics, computer science, and much more. Because of this variety, no two books on the subject ever include the same topics. This particular volume, a massive text of 23 chapters, is divided into three logical sections: The Earth and Its Environment; Space; and The Solar System and Beyond. In the first, the chapters deal with the near space environment and its problems, the Earth's magnetic field, the lower atmosphere, the ionosphere, the Earth's radiation belt, aurorae, meteorology from space, and the shape of the Earth. The (outer) space environment is the concern of the second section, a look at the interplanetary medium, the magnetosphere, cosmic rays, dust particles, cosmic chemistry, orbital mechanics, and man in space. The third portion of the book is devoted to astronomy, including the Sun, Moon, planets, stellar evolution, extragalactic radio sources, etc. Topics begin with historical background, move on to basic theory and description, and end up with current research. Unfortunately, "current" is now nearly 10 years old, and a new edition of this comprehensive text would be welcome. Intended for the student, this book would be appropriate for any astronomy or aeronautical library that can afford it.

549. Kopal, Zdeněk. **Telescopes in Space**. London, Faber and Faber Ltd.; distr. New York, Hart Publishing Co., 1970, c1968. 140p. illus. index. $12.50. LC 68-102661. B 68-07536. ISBN 0-571-08436-2 (Faber); 0-8055-4067-9 (Hart).

The importance of constructing and using telescopes beyond the Earth's atmosphere is explored in this informative book for the layman. Beginning with a history of the telescope, Kopal describes how this instrument, from its crude beginnings to present-day technology, has so greatly improved our knowledge of the Universe. And yet, and this is the point of the book, it has almost reached a limit on how much and how far it can "see." Its hindrance is the Earth's atmosphere, sometimes absorbing, sometimes distorting incoming starlight, in several areas of the electromagnetic spectrum. After describing the atmosphere and its negative effects, the author tells of efforts to get telescopes above it, using balloons, rockets, and satellites. The use of deep space probes to carry optical and other types of telescopes is considered, and efforts so far in this direction are described. The final brief chapter is a speculative one, looking into the future at possible space astronomy applications. For the public and college library, the book has good illustrations of various telescopes, earth- and space-bound.

550.   Labuhn, F., and R. Lüst, eds. **New Techniques in Space Astronomy.** Dordrecht-Holland, D. Reidel Publishing Co.; distr. New York, Springer-Verlag, 1971. 419p. illus. refs. $29.80. LC 75-159658. ISBN 90-277-0202-0 (Reidel); 0-387-91089-1 (S-V).

The study of electromagnetic radiation other than visible light by means of satellites, balloons, and rockets is the essence of this book for the astronomer. The sixty-six papers of this conference volume describe many of the methods of off-the-Earth observation of stars and other celestial objects in four separate sections: I, Gamma-Ray Astronomy; II, X-Ray Astronomy; III, UV Astronomy (New Results, Optical Systems, Detecting Systems, Calibration); IV, Radio Astronomy. But since astronomy, and especially space astronomy, changes so rapidly, much of this volume is no longer up to date. It is presumed that a similar conference will be held again to discuss "newer" techniques in space astronomy, including Skylab reports, etc. From IAU Symposium no. 41, held in Munich, August 10-14, 1970, this book includes abstracts and references but has no indexes.

551.   Morgenthaler, George W., and Howard D. Greyber, eds. **Astronomy from a Space Platform.** Tarzana, Calif., American Astronautical Society, 1972. 398p. illus. refs. (Science and Technology Series, v.28). $20.00. LC 73-159623. ISBN 0-87703-061-8.

The many positive aspects of the U.S. space program are brought out in this conference volume aimed at the layman and space scientist. Mainly held to counteract those who have criticized the space program as being worthless (i.e., no tangible results), the meeting presented the results of astronomical research carried out by NASA projects. Written in clear, understandable language, the 25 papers in this volume detail the objectives, methods, and results of astronomical data-gathering projects. Included are descriptions of orbiting observatories, Skylab experiments, specialized telescopes for observing in the non-optical portions of the spectrum, and much more. The conference, held in Philadelphia, 27-28 December 1971, was divided into the following sessions: Planetary and Solar Astronomy; Stellar and Galactic Astronomy; New Astronomy Areas and Very Large Space Telescopes; Advanced Applications and New Developments in Space Astronomical Instruments; and Strategies for Space-Based

Astronomy. This collective volume not only surveys current space astronomy research, but also proposes directions for the future and answers the prime question, "Was it all worth it?" The reader of this fine book will likely be convinced it was.

552.   Page, Thornton, and Lou Williams Page, eds. **Space Science and Astronomy; Escape from Earth.** New York, Macmillan Publishing Co., Inc. 1976. 467p. illus. index. bibliog. (Sky and Telescope Library of Astronomy, v.9). $13.95. LC 76-5879. ISBN 0-02-594310-3.

An informative group of articles on space exploration and related astronomical research, this book complements certain sections of *Wanderers in the Sky* (1965), the first volume of the *Sky and Telescope* series. The material spans nearly 20 years, outlining for the reader the birth and growth of space astronomy. Highlights include manned exploration of the lunar surface and the study of Moon rocks, flights to the three nearest planets (Venus, Mercury, and Mars), and optical and non-optical observations from spacecraft. Well-known astronomers and science writers, including Carl Sagan, Gerard Kuiper, and Raymond N. Watts, Jr., are the contributors to this collection of nearly 140 articles from *Sky and Telescope* magazine. Edited for easy reading, and frequently interspersed with expert commentary by the editors, it is illustrated with 150 outerspace photographs and various diagrams. From a 1957 excerpt on what astronomers can learn from space exploration to a 1975 article on colonizing space, this collection is sure to capture the attention of any armchair astronomer/astronaut. A space-age glossary and chronology round out the excellent text. Chapter headings include Concepts of the Space Age; Early Flights and the Hazards of Space; Exploring the Moon and Its History; Trips to Venus and Mercury; Preparations for Landing on Mars; Close-Up Views of Jupiter; Spacecraft Design and Workshops in Space; Optical Observations of the Sun, Earth, and Stars; X-Ray and Gamma-Ray Astronomy; The Frontier in Space.

553.   Papagiannis, Michael D. **Space Physics and Space Astronomy.** New York, Gordon and Breach Science Publishers, Inc., 1972. 293p. illus. index. bibliog. $22.50. LC 72-179021. ISBN 0-677-04000-8.

Readers interested in obtaining an overview of some of the "new astronomies" should consider this volume, which investigates the study of our upper atmosphere and beyond using spacecraft, rockets, and other devices. Introducing briefly many of the areas of interest in space science, the author whets the readers' appetite with his discussions of planetary atmospheres, the ionosphere, the magnetosphere, the "active Sun," interplanetary space, Earth-Sun interactions, and solar and galactic space astronomy. Aimed at the advanced undergraduate or beginning graduate student, the book begins with overviews of the topic to be discussed, with historical references and explanations of ground-based observations. The remainder of each chapter emphasizes information obtained from spacecraft, etc., highlighting important finds by specific space vehicles (for example, Mariner, Venera, etc.). Each topic is covered in enough detail to make the reader go on (if interested), and extensive bibliographies of of books and journal articles are included. It is not as comprehensive as some of the other space science volumes listed here, but it is one of the most up to date; although it is fairly technical, it is not as dry as the other volumes. For university and astronomy libraries.

554. Pecker, Jean-Claude. **Experimental Astronomy**. Dordrecht-Holland, D. Reidel Publishing Co.; distr. New York, Springer-Verlag, 1970. 105p. illus. indexes (name and subject). bibliog. (Astrophysics and Space Science Library, v.18). $11.60. LC 77-118378. ISBN 90-277-0157-1 (Reidel); 0-387-91042-5 (S-V).

In the same vein as his *Space Observatories*, the author has written another book intended to explore methods of astronomical research and at the same time to stimulate the reader to go beyond his treatment, which is nowhere complete. Showing "what space technology, by its very existence, can give to astronomers, and also what the requirements of the astronomers might be," Pecker deals with two types of investigation, one using spacecraft to gather data, the other involved with the direct exploration of the Universe. He discusses astronomy as an experimental science, experimental celestial mechanics (the use of artificial satellites), initiation in astronautics, experimental astrophysics, direct exploration of the extraterrestrial world, and the plurality of inhabited worlds. The book ends with some conclusions and predictions about space research. Many of the lunar and planetary space probes are briefly described, the author using these events to support his contention that space astronomy is a vital complement to Earth-bound research. His writing style is most enjoyable, a blend of narration and mathematics that is never dry and always interesting. Translated by R. S. Kabel from the French (*L'Astronomie expérimentale*), this volume will be of use to the non-specialist and student.

555. Pecker, Jean-Claude. **Space Observatories**. Dordrecht-Holland, D. Reidel Publishing Co.; distr. New York, Springer-Verlag, 1970. 120p. illus. bibliog. (Astrophysics and Space Science Library, v. 21). $12.00. LC 70-124847. ISBN 90-277-0168-7 (Reidel); 0-387-91066-2 (S-V).

Why we use rockets and satellites to observe the heavens and what information can be gathered using these techniques is the theme of this clearly written book for students, scientists, and laymen who have a scientific background. Translated by Janet R. Losh from the French (*Les Observatoires spatiaux*), the volume first looks at the Earth's atmosphere and tells how it limits what can be seen from ground-based observatories. Included is a detailed description of the electromagnetic spectrum and the individual classes of radiation, many of which are visible only above the Earth's atmosphere. The second half of the work lists, explains, and evaluates the type of information that can be gained by observing in space. It should be emphasized that this brief treatise is concerned with possible space science projects and not with reporting results of observations. Readers interested in facts and figures on specific research should consult the appropriate periodical and/or technical report literature. Among the possible areas of study, as delineated by the author, are solar colona, distant galaxies and other faint objects, the surface of the Sun, the planets, and double stars. All of these are, of course, observable from the Earth but are better studied above the atmosphere. Ultra-violet, x-ray, and other short wavelength radiation are additional areas available for research from rockets and satellites. Many of these phenomena are already being observed at this writing. This book's place in the space science literature is an important one, for it explains why we need to use satellites, etc., for astronomical research and makes sound suggestions for projects utilizing such techniques.

RADIO ASTRONOMY

Compared to the literature of optical astronomy, the number of books on radio astronomy is fairly small. This is not surprising, though, when one considers that this new field has only been around for about 30 years, compared to several millenia for its counterpart. Nevertheless, radio astronomical literature is growing in leaps and bounds, especially in the form of journal articles and published proceedings. The number of monographs is still relatively small, but even that is changing. The volumes listed here are representative and include some of the best works available. The great majority of radio astronomy books are technical, aimed at the astronomer and graduate student, but there are now a few popular books, the best of which is mentioned below. Books on amateur radio astronomy are found under Amateur Astronomy, despite the fact that such work is a very non-elementary venture.

Radio studies are varied and broad in scope; this section includes a sampling of the types of materials encountered: solar system studies, solar research, galactic and extra-galactic observation. Electrical engineering also plays an important role in the field of radio astronomy and astrophysics, and a good radio astronomy library will also contain publications from this vital area. The librarian should consult an engineering bibliography to identify some basic works.

Card catalog subject headings: *Library of Congress:* RADIO ASTRONOMY; RADIO ASTRONOMY–POPULAR WORKS. *Sears:* RADIO ASTRONOMY.

556.   Christiansen, W. N., and J. A. Hogbom. **Radiotelescopes**. London, New York, Cambridge University Press, 1969. 231p. illus. index. refs. (Cambridge Monographs on Physics). $18.50. LC 69-16279. B 69-23401. ISBN 0-521-07054-6.

Prior to the publication of this volume, most information on radio-telescope theory and design was scattered throughout the world's scientific journal literature, and no one monograph existed on the subject. This book, written by two designers of radio instruments, adequately fills that gap by consolidating much of the published and unpublished data on the subject.

Definitely not for the beginner, the book begins with an introductory chapter on the purpose of radio telescopes, a description of the objects observed, and an outline of the basic types of instruments. The theory behind the operation of radio telescopes is next, followed by chapters discussing in detail the types of instruments: the steerable parabolic reflector, other types of filled-aperture antennas, and unfilled-aperture antennas. Other sections consider aperture synthesis and sensitivity, with emphasis on "noise." Methods of observation and radio sources are generally not found in this text; the treatment here is strictly a theoretical discussion of the instruments, with occasional references to the uses of the various types of telescope. A rather dry treatment of an intrinsically dry subject, the book will be of interest to the scientist and graduate student concerned with radio telescope design and theory. Because of recent advances in technology, a newer, expanded version is desirable.

557.   Hey, J. S. **The Radio Universe**. 2nd ed. Oxford, Pergamon Press, 1975. 264p. illus. indexes (name and subject). (Pergamon Popular Science Series). $15.00; $9.50pa. £2.80. LC 75-23134. ISBN 0-08-018760-9; 0-08-018761-7pa.

Clear writing and comprehensive subject coverage make this excellent volume the best book available for the student and serious layman. The first third of the volume provides background and introductory information on history, radio waves, and radio telescopes. The reader is shown the difference between light and radio waves, sources of radio emission are discussed, and the nature of radio waves is explained. The chapter on instrumentation is quite good, with the author illustrating the various types and arrangements used; interferometry techniques are explained. The remainder of the book describes the various radio sources in some detail: the solar system, the Sun, the Milky Way, sources within the Milky Way, and radio galaxies and quasars. There are also chapters on radar astronomy and on cosmology from a radio viewpoint. This revised edition reflects several new techniques (e.g., long baseline interferometry) and discoveries (e.g., new types of radio stars) since the 1971 version. The book parallels, in part, Jennison's *Introduction to Radio Astronomy* (1966) but fortunatley does not become bogged down in mathematics as Jennison's work occasionally does. Although not specifically a text, it would be a good choice for an undergraduate course. A worthwhile acquisition for any type library, this work provides the literature of radio astronomy with another solid sourcebook, ranking with the author's *Evolution of Radio Astronomy* (1973) and Kraus's *Radio Astronomy* (1966).

558.   Jennison, Roger C. **Introduction to Radio Astronomy**. London, Newnes; distr. New York, Philosophical Library, 1966. 160p. illus. index. $4.75. LC 67-70600. B 66-24287.

Lying somewhere between the popular work and the textbook, this short volume concentrates on "how it works" and what is being investigated (or what *was* being investigated 10 years ago). Although much of the text is fairly descriptive, this book is rather technical and not for the neophyte. Occasional pages of long equations pop up, but some can be skipped without loss of meaning.

In "Tools of the Trade," Jennison explains the basic types of radio telescopes and schematically shows the components of receivers and interferometers. He next moves to the strongest source of extraterrestrial signals receivable, those from the Sun; these are categorized and summarized well. After a short discourse on solar system radio astronomy, the book next turns to galactic radio emissions and radio stars, describing Karl Jansky's pioneering work and the more recent studies by other scientists. Extragalactic radio sources like galaxies and quasars are covered, too. The last, longest, and most technical chapter covers the types of observing that can be performed by varying the number and configurations of aerials. This book definitely fills a gap in the literature, and it is certain to spur on those with a great interest in the subject.

559.   Kerr, F. J., and S. C. Simonson III, eds. **Galactic Radio Astronomy.** Dordrecht-Holland, D. Reidel Publishing Co., 1974. 654p. illus. refs. indexes. $78.00; $60.00pa. LC 74-81939. ISBN 90-277-0501-1; 90-277-0502-Xpa.

Covering much of the same subject matter as *Galactic and Extra-Galactic Radio Astronomy* (Verschuur, ed., 1974) this proceedings volume surveys current projects in the radio study of the Milky Way. Combining descriptions of original research with review papers, the book includes 73 pieces in the following six areas: The Interstellar Medium; Galactic H_{II} Regions; Supernova Remnants; Stellar and Circumstellar Sources; The Galactic Center; Large-Scale Galactic Structure. Happily, a third of the papers are survey articles, an unusually high number, giving the reader summaries of the most important work done. The index is very good, featuring NGC and similar numbers to assist galaxies or other objects. In general, the papers are fairly short (except for the review papers), reflecting the newness of the field and the rapid changes taking place. A good purchase for the astronomy library that can afford it and a must for the radio astronomy collection. These are the proceedings of IAU Symposium no. 60, held at Maroochydore, Queensland, September 3-7, 1973.

560.   Kraus, John D. **Radio Astronomy.** New York, McGraw-Hill Book Co., 1966. 481p. illus. indexes (subject and name). bibliog. problems. $19.50. LC 67-1537. ISBN 0-07-03592-1.

Intended as both a text and a reference work, this standard book takes a rigorous approach to the subject. A highly mathematical volume, it is intended for graduate students and astronomers. After a very brief introduction on what radio astronomy is and a few historical notes, Kraus presents two chapters on general astronomy fundamentals and radio astronomy fundamentals, respectively. In the former, the emphasis is on astronomical coordinate systems and basic quantities like time, distance, motion, magnitudes, etc. In the latter, the author explains electromagnetic radiation and its physics, the basis of radio astronomy; the basic quantities and related equations are spread out lavishly in this comprehensive section. Wave polarization and propagation are discussed in chapters 4 and 5. The principles of radio-telescope antennas and receivers are also detailed. After this thorough treatment of the basics, the book ends with a chapter on the types of astronomical radio sources and a description of some observational results. Extensive appendices on physical and numerical quantities and constants, lists of radio sources, and several other topics complete this excellent text. *The* book on the subject, it contains hundreds of citations to radio astronomy literature.

561. Kundu, Mukul R. **Solar Radio Astronomy**. New York, Interscience Publishers, John Wiley and Sons, Inc., 1965. 660p. illus. index. refs. $29.75. LC 65-19481. ISBN 0-470-51075-7.

Although radio astronomers actively study all portions of the sky and a multitude of celestial objects, the Sun, one of the earliest-known emitters of radio waves, is one of the most frequently observed radio sources. This volume takes a comprehensive look at the radio Sun, providing an excellent overview for the beginner and specialist alike. The author of this well-written volume begins with a short section of historical references, an outline of the classification of solar radio emmissions, and an outline of the problem of solar radio studies. A fairly unique second chapter reviews the physical characteristics of the Sun as viewed with optical telescopes; this part will enable the unfamiliar reader to better understand the radio and its relation to the optical counterpart. Few texts bother to include such a section. A rigorous and theoretical third chapter explains how radio waves are propogated and generated in the Sun's outer layers. An introduction to radio telescopes and techniques is included in the next chapter, which emphasizes solar observations. The majority of the text is concerned with the types of solar radiation and the various wave lengths of the emissions. Kundu explores both the quiet background radiation and slowly varying component radiation, and the more violent (and scientifically more interesting) bursts of radiation due to flares and other surface phenomena. The burst radiation is propagated on various wavelengths (centimeter, decimeter, meter), and these are all discussed. Various topics such as x-ray emission, solar cosmic rays, flares, and the irregular structure of the outer corona are also included. Newer data, gathered in the last decade, are missing, and a new edition of this now-standard work would be appreciated. In the meantime, the graduate student and astronomer who wish to do further reading should consult the current journal literature.

562. Meeks, M. L., ed. **Astrophysics, Part C: Radio Observations**. New York, Academic Press, 1976. 345p. illus. index. refs. (Methods of Experimental Physics, v.12C). $29.50. LC 73-17150. ISBN 0-12-475953-X.

Aimed at the astronomer, this advanced text consisting of 16 review papers deals with the current applications of radio instruments. The book's first section is concerned with single-antenna observations, touching on 1) observations of small-diameter sources; 2) spectral-line measurements; 3) using radio-frequency spectrometers; 4) measurements of galactic 21-cm hydrogen; 5) techniques of observing pulsars; 6) lunar occultation measurements; and 7) scintillation measurements. Occupying approximately one-half the volume, this portion, like the two that follow, is concerned not so much with results as with how they are obtained. The second part of the text takes the next logical step to a discussion of arrays of antennae and interferometers. Papers cover 1) two-element interferometer theory; 2) connected-element interferometry; 3) very long baseline interferometer systems; 4) frequency and time standards; 5) very long baseline interferometric observations and the reduction of data; and 6) estimation of astrometric and geodetic parameters. Finally, there are three articles on radio astronomy computer applications, an important topic about which little, so far, has been written. The first paper is an applications example, concerned with radial-velocity corrections for Earth motion. The second discusses the fast Fourier transform, its use in radio astronomy, and how it is programmed. The third paper considers data presentation techniques, in particular contour mapping, rule-surface mapping, and gray scale mapping. Like any review volume, this work covers only selected topics, but this does not detract at all from its usefulness. It fills a large gap in the literature; there is not much published in review format on radio astronomy.

563.   Meeks, M. L., ed. **Astrophysics, Part B: Radio Telescopes.** New York, Academic Press, 1976. 309p. illus. index. refs. (Methods of Experimental Physics, v.12B). $30.00. LC 75-34188. ISBN 0-12-475952-1.

Part of a series on current astronomical techniques and data gathering, this work covers instrumentation, how it functions, and related problems. The first six review papers zero in on radio telescopes, discussing how radiometric measurements are made, types of astronomical antennas, analysis of paraboloidal-reflector systems, feed systems for paraboloidal reflectors, antenna calibration, and problems of designing and constructing antenna arrays. The second section of the book contains five articles on how the Earth's atmosphere affects radio observations. The topics include the ionosphere, structure of the neutral atmosphere, absorption and emission by atmospheric gases, extinction by condensed water, and refraction effects in the neutral atmosphere. Finally, there are five papers on radiometers in general and special equipment: radiometer fundamentals, parametric amplifiers, maser amplifiers, multichannel-filter spectrometers, autocorrelation spectrometers. Like the other two volumes (*Part A: Optical and Infrared*, and *Part C: Radio Observations*), this work updates many of the "old" books on techniques of data collection. Further, there are so few works on radio methods, that Parts B and C are especially welcome additions to the literature of astronomy, despite their high prices. This advanced text is appropriate for observatory and university libraries.

564.   Smith, Alex G. **Radio Exploration of the Sun.** Princeton, N. J., D. Van Nostrand Reinhold Co., Inc., 1967. 159p. illus. index. bibliog. refs. (Van Nostrand Momentum Books, 15). $3.25. ISBN 0-442-08714-4.

The Sun is easily the most accessible and most-studied radio object, and its radio waves were first detected in 1942. The details of research since then are the subject of this book, whose only prerequisite is that the reader be familiar with beginning physics. Smith's book is, therefore, appropriate for college students and certain laymen; it is not comparable to Kundu's *Solar Radio Astronomy*, which is fairly technical and has a different audience. After some historical background, the author discusses general solar physics, an obvious prerequisite for radio studies; the physical structure and nuclear processes are explained. He describes solar radio telescopes and their principles before moving on to a description of the various types of radio signals being emitted. Other subjects include radar observations and Earth-Sun relations. An excellent introduction to solar radio study, it would be appropriate for public, university and astronomy libraries.

565. Verschuur, Gerrit L., and Kenneth I. Kellermann, eds. **Galactic and Extra-Galactic Radio Astronomy.** New York, Springer-Verlag, 1974. 402p. illus. index. refs. $37.80. LC 72-97680. ISBN 0-387-06504-0 (N.Y.); 3-540-06504-0 (Berlin).

An up-to-date compendium of some important research in non-solar system radio astronomy is the substance of this advanced text. Suitable for graduate students and astronomers, this collection of 13 papers is one of the better works available in the relatively small collection of radio astronomy monographs; whereas most books are mainly concerned with basic principles or one highly specialized topic, this volume happily considers a variety of subjects on a more advanced level. Each selection is well illustrated and contains quite a few references; the lists of citations are not necessarily complete but include those deemed most important by the authors and astronomers at the National Radio Astronomy Observatory. Contents: Galactic Non-thermal Continuum Emission; Interstellar Neutral Hydrogen and Its Small-Scale Structure; The Radio Characteristics of H_{II} Regions and the Diffuse Thermal Background; The Large-Scale Distribution of Neutral Hydrogen in the Galaxy; Supernova Remnants; Pulsars; Radio Stars; The Galactic Magnetic Field; Interstellar Molecules; Interferometry and Aperture Synthesis; Mapping Neutral Hydrogen in External Galaxies; Radio Galaxies and Quasars; and Cosmology.

566. Verschuur, Gerrit L. **The Invisible Universe: The Story of Radio Astronomy.** New York, Berlin, Springer-Verlag, 1974. 173p. illus. index. (Heidelberg Science Library, v.20). $5.90pa. LC 73-22202. ISBN 0-387-90078-0 (N.Y.); 0-540-90078-0 (Berlin).

Until this excellent work appeared, the relatively small but growing literature of radio astronomy was devoid of a good, popular treatment of the fascinating new branch of the world's oldest science. There have been a few other works for the layman, but they tend to be either dry or too technical. Not so here. The author, a well-known radio astronomer and author of many technical papers on the subject, has succeeded in writing an extremely interesting and non-technical overview of radio astronomy with an emphasis on the more important discoveries. Verschuur's writing ability shines as he explains in layman's language what radio astronomy is, what radio astronomers do, and what kinds of useful information have been gathered with radio telescopes. Especially good is his description of pulsars, whose accidental discovery opened an entirely new area of study and which provided the first evidence of neutron stars. The reader will be educated and entertained at the same time as the author discusses the history of radio astronomy, exploding stars, molecules between the stars, quasars, radio stars, radio telescopes, and life in the Universe. Highly recommended for any library, this is the best such volume available.

567. Zheleznyakov, V. V. **Radio Emission of the Sun and Planets**. Oxford, Pergamon Press, 1970. 697p. illus. index. refs. (International Series of Monographs in Natural Philosophy, v.25). $60.00. LC 75-76797. ISBN 0-08-013061-5.

Published at approximately the same time as Kundu's *Solar Radio Astronomy* (1965), this translation (of the 1964 work) is broader in scope than the former title. A long, review-type book, it considers not only solar radio studies but lunar and planetary research as well. The former topic, however, as expected, makes up about 75 percent of the book. A comprehensive survey of research through the mid-1960s, the book begins with some general background information on the physical conditions of the Sun, Moon, and planets, and an overview of celestial radio emissions and the techniques for their study. Chapters two and three parallel Kundu's work, discussing at length the "quiet" Sun and its radiation, and the sporadic radio emissions characterized by sudden bursts on various wavelengths. Next come descriptions of observations of planetary and lunar radio emissions, contrasting in particular the sporadic Jovian outbursts with the continuous radiation of the other planets. After these observational considerations, the author turns to theoretical aspects like propagation and generation of electromagnetic waves in the solar corona, theory of solar thermal and non-thermal radiation, and origin of planetary radio emissions. On the whole, much of this text duplicates topics in *Solar Radio Astronomy*, but both books include subjects not considered by the other. Therefore, both works should be read to get a more complete picture. Of note in this text is table IV, which succinctly summarizes the various types of solar emissions in three convenient pages, a 24-page bibliography, and a multitude of illustrations. Like the Kundu volume, a new edition would be desirable to add recent data.

THE NEW ASTRONOMIES

The obvious difficulty with the phrase "new astronomies" is that it will not be valid (i.e., "new") for long. Nevertheless, for the time being, this terminology is an apt description of the recent, exciting work going on in the field. In particular, "new astronomies" refers to non-visual studies of the Universe, using telescopes to see beyond the visible light portion of the electromagnetic spectrum to see cosmic rays, x-rays, gamma rays, infrared radiation, and more. Among the many discoveries of this special astronomy are strong x-ray sources, neutron stars and black holes, pulsars, clouds of interstellar molecules. One tangible result, besides a better understanding of the Universe, is the non-optical maps of the galaxy, showing stars and other objects that are strong emitters of non-visual radiation. Knowledge here is changing so rapidly that there are few textbooks available; those that exist go out of date far more quickly than the average general text. Discussions, mainly on the technical level, of these exciting new fields are found below in the sample items. The lay reader can find descriptions in fairly some of the general works elsewhere in this book.

Card catalog subject headings: *Library of Congress:* ELECTROMAGNETIC THEORY; COSMIC ELECTRODYNAMICS; GAMMA-RAYS; X-RAYS; RADIATION. *Sears:* ELECTROMAGNETIC WAVES; COSMIC RAYS; X-RAYS; SPECTRUM.

568.   Allen, David A. **Infrared, the New Astronomy.** New York, John Wiley & Sons, 1975. 228p. illus. index. refs. (A Halstead Press Book.) $12.00. LC 75-16584. ISBN 0-470-02334-1.

A solid overview of research and advances in infrared astronomy, this is the only such book available at present. A previous volume, *Infrared Astronomy* (Brancazio, Cameron, eds., 1968), a collection of conference papers, did not attempt to give a good overview. Infrared studies, a relatively new and unpublicized field, holds many possibilities for exciting new discoveries in a portion of the spectrum not yet fully investigated; the author surveys this unusual field with clear text and comprehensive coverage. Chapter one describes how the infrared portion of the electromagnetic spectrum was discovered, along with a general review of radiation. The detection of infrared waves, impossible with normal photographic techniques, is explained in the following section, with descriptions of materials used, photometric methods, atmospheric effects, telescopes, and data reduction. This extremely important summary chapter sets the stage for the more advanced topics that follow. Next is an outline of the early years of infrared astronomy, with a discussion of the development of infrared telescopes and the difficult task of choosing observing sites. The bulk of the text consists of detailed descriptions of the various celestial objects observed in the infrared spectrum: the Sun, Moon, planets, cool and hot stars, young stars, galaxies. Allen also presents some specialized topics: infrared spectra, infrared polarization, and infrared photography. A look to the future rounds out this fine effort for the astronomer. Of special note is the lengthy bibliography, which includes nearly all papers written on the subject.

569.   DeWitt, C., and B. S. DeWitt, eds. **Black Holes.** New York, Gordon and Breach, 1973. 552p. illus. refs. $39.00. ISBN 0-677-15610-3.

A rigorous and highly technical treatment of the black hole phenomenon is contained in this volume of seven edited lectures given at the Summer School of Les Houches in France, 1972. Mainly theoretical, but with some strong observational evidence as well, the authors' papers present a fairly solid case for the existence of these strange "objects" from which no radiation can escape. Currently the prime source of information in the field, it consists of the following papers: The Event Horizon; Black Hole Equilibrium States; Timelike and Null Geodesics in the Kerr Metric; Rapidly Rotating Stars, Disks, and Black Holes; Observations of Galactic X-ray Sources; Black Hole Astrophysics; and On the Energetics of Black Holes. The book unfortunately has no index, but it does have a fairly detailed table of contents that will be useful. Of special interest will be the description of the observations of x-ray sources from the Uhuru satellite, which points strongly toward Cygnus X-1 as the most likely candidate for black hole status. For the science and astronomy library, it is intended for the advanced student and scientist.

570.   Giaconni, Riccardo, and Herbert Gursky, eds. **X-Ray Astronomy.** Dordrecht-Holland, D. Reidel Publishing Co., 1974. 450p. illus. index. refs. (Astrophysics and Space Science Library, v.43). $50.00; $29.00pa. LC 74-79569. ISBN 90-277-0295-0; 90-277-0387-6pa.

The only volume to deal exclusively with this "new" field, this work presents ten chapters written by scientists directly involved in x-ray astronomy research. This highly technical book begins with an introductory chapter examining the role and history of the subject, which would not have come into being without this country's space program, for extra-terrestrial x-rays are obscured by the Earth's atmosphere and are invisible to ground-based instruments. Chapter 2 takes a look at techniques for observing x-rays, paying particular attention to methods and instrumentation used on Uhuru, the satellite used to gather a substantial portion of present data. Next a review of the physical processes that produce the rays (including synchrotron radiation, Thomson scattering, Coulomb *bremsstrahlung*, and line emission) is presented. The remainder of the book is slightly less mathematical (and more descriptive) than Chapters 2 and 3, presenting first an overview of present knowledge of how the x-ray sky appears. A schematic diagram in this chapter shows the currently known x-ray sources to lie mainly in or near the equatorial plane of the Milky Way. The remaining six chapters describe x-ray sources such as the Sun, supernovae remnants, extragalactic sources, etc. The highlight of the book, however, is the catalog of x-ray sources, listing 161 "objects" in all. Finally, the lists of references are excellent and will lead the reader to original literature on the subject. For the university and observatory library.

571.   Greisen, Kenneth. **The Physics of Cosmic X-Ray, Gamma-Ray, and Particle Sources.** 2nd ed. New York, Gordon and Breach, 1971. 115p. index. refs. (Topics in Astrophysics and Space Physics, 8). $15.00; $7.50pa. LC 78-135063. ISBN 0-677-03380-X.

Aimed at the science student and non-specialist, this short volume is an introduction to high-energy particles originating in outer space. A good overview is presented in the beginning chapter covering the sources of these particles, naming in particular the Sun, planetary and interplanetary sources, interstellar space, supernovae, radio galaxies, and quasars. Chapter two, on energetic solar particles (ESP), describes the solar wind phenomena, considering the composition of the ESP, time profiles, and the effects on the Earth. The author next considers discrete courses of charged particles remote from the Sun (like the supernovae, radio galaxies and quasars, all of which must be identified by indirect methods as originators of particles). A short chapter on the interactions in the interstellar medium follows. The origin of the so-called primary electrons is discussed next, before Greisen examines cosmic gamma-rays and x-rays. The study of highly charged particles is on the increase, with the radiation emitted by pulsars, quasars, and so on, of particular interest; therefore, this monograph fills an important place in the literature. Although this is a good introduction, it is no more than that, and the reader will want to consult the seven-page bibliography in order to find guidance for further study.

572.   Gursky, Herbert, and Remo Ruffini, eds. **Neutron Stars, Black Holes and Binary X-Ray Sources.** Dordrecht-Holland, D. Reidel Publishing Co., 1975. 441p. illus. subject index. (Astrophysics and Space Science Library, v.48). $54.00; $38.00pa. LC 75-15716. ISBN 90-277-0541-0; 90-277-0542-9pa.

The physics of collapsed objects is succinctly summarized and brought up to date in this fine compilation of nine review papers. The volume begins with a brief introductory article (by the editors), which reviews for the reader the basics of gravitational collapse and the discovery of neutron stars and black holes. The other eight advanced papers cover supernovae, pulse astronomy, cosmic gamma ray bursts, physics of gravitationally collapsed objects, observational properties of pulsars, galactic x-ray sources, optical observations of binary x-ray sources, black holes, and neutron stars. The importance of this volume to the literature of astronomy is its comprehensive and very up-to-date coverage of the above topics which, until now, were scattered in journals and monographs. Of special note is the "current" catalog of pulsars discovered as of 1974; a list of 92 galactic x-ray sources is also included. References to related literature abound; the article on pulsars alone has 334 citations. Alone, these nine papers would have made an outstanding book. However, the editors have also included reprints of seven classic papers in the field by such greats as Chandrasekhar, Landau, Baade, Oppenheimer, Weber, and Zwicky; further, they added a dozen more articles from the fairly recent literature which outline the search for gravitationally collapsed stars. A must purchase for the astronomy library; many other science collections may want it as well.

573.   Hewish, A., ed. **Seeing beyond the Visible.** London, English Universities Press; New York, American Elsevier Publishing Co., Inc., 1970. 150p. illus. index. $8.25. LC 74-547619 (U.K.); 74-122634 (U.S.). B 70-25249. ISBN 0-340-09894-5 (U.K.); 0-444-19649-8 (U.S.).

This interesting volume contains essays that were originally a series of talks broadcast by the BBC under the title "At The Speed of Light." More physics than astronomy, this book looks at the various portions of the electromagnetic spectrum, a topic of great importance to astronomers. The first two chapters are introductory discussions on the nature of light, and the particle and wave theories of light, a necessary prerequisite for the remainder of the book. Starlight (its source and what information it holds) is the major theme of chapter 3, a brief discussion of optical astronomy. The book then turns to a discussion of electromagnetic radiation from space in the form of radio waves, detailing their discovery, describing radio telescopes and how they are different from optical instruments, and going on to describe the "radio sky" and the objects radio astronomers study. Moving away from astronomy, the book explores x-rays, very short wave length radiation which allows us to see "inside" things; this section is one of the more technical (without resorting to mathematics) chapters. Still shorter radiation, microwaves, is taken up next, with descriptions of physical characteristics and communications and radar applications. The final chapter is a summing up and a discussion of some exciting developments in electromagnetic radiation research, like masers and lasers. Mainly descriptive, the book is not too technical, but neither is it elementary. For the layman and student.

574. Lenchek, Allen M., ed. **The Physics of Pulsars.** New York, Gordon and Breach, 1972. 173p. illus. refs. (Topics in Astrophysics and Space Physics). $22.00. $10.50. LC 76-150793. ISBN 0-677-14290-0; 0-677-14295-1pa.

One of the few books devoted to the pulsating radio stars, this volume is a compilation of lectures given at the University of Maryland. Since the topic is one of the newer astronomical discoveries, little has been settled finally on their nature. The 14 lectures transcribed here, then, represent reports of the first research that was done, little of it in final form (i.e., accepted as "known to be true" information). Indeed, more questions are posed in this compilation than are answered. Radio observations of pulsars, optical observation, x-ray observations, measurement of pulsar periods, searching for pulsars, and neutron stars are just a few of the topics explored in this interesting volume. The material is technical, suitable for advanced students and scientists, so the work is best suited for astronomy and college collections. Includes a list of 55 pulsars known as of 1970.

575. **Pulsating Stars.** London, Macmillan; New York, Plenum, 1969, 1970. v.1: 92p. v.2: 116p. illus. refs. (A Nature Reprint). v.1: $20.00. v.2: $20.00. ISBN 0-306-30401-5 (v.1); 0-306-37092-1 (v.2).

The title of this two-volume set is slightly misleading. The term "pulsating stars" usually refers to the intrinsic variable stars that show changes in brightness due to internal physical processes. These volumes, however, are about a different kind of pulsating star, a star which very regularly emits a rising and falling radio signal. Commonly known as pulsars, the astronomically "small" objects were first detected with a radio telescope in Britain in 1968. Since then, of course, pulsating radio stars have become one of the hottest astronomical topics in decades. Their discovery, and the subsequent early research on pulsars, was reported first in *Nature* magazine in a number of technical articles, which are reprinted here in two narrow volumes. An excellent record of recent astronomical history, they are appropriate for college and observatory collections. Titles in volume one include Discovery; Signal Characteristics; Optical Measurements; Theories; and Applications. Volume two contains PSR 0833-45 and NP 0532: The Fast Pulsars; Distribution and Distances; Theories; Observations; Radio, Optical, Gamma-ray, X-ray.

576. Shipman, Harry L. **Black Holes, Quasars, and the Universe.** Boston, Houghton Mifflin Co., 1976. 309p. illus. index. bibliog. glossary. $12.95; $5.95pa. LC 75-19535. ISBN 0-395-24342-4; 0-395-20615-4pa.

One of several volumes recently published on "the violent universe," this book is one of the most detailed and readable. The result is a non-elementary, but not overly technical, picture of the discoveries that have put astronomy in the headlines in the 1960s and 1970s. Divided equally among the topics mentioned in the title, the book presents a pleasing combination of fact, history, and theory, with an emphasis on the latter. The most theoretical object of all, the black hole, is described first, with Shipman outlining stellar evolution as background information. Emphasizing especially the final stages of stellar life, one of which, for certain stars, is probably the black hole, the author also tells about white dwarfs, novae, neutron stars (pulsars). The book takes the reader on an imaginary journey into a black hole, a fascinating tale of the physics and history. Two possible candidates for first confirmed black hole, Epsilon Aurigae and Cygnus X-1, are described, too.

Part two of the work concentrates on galaxies, quasars, radio galaxies, and similar phenomena. Among the controversial subjects here are the enormous amounts of energy emitted by the radio galaxies and the highly debatable red shifts of the quasars, along with the resultant cosmological implications. All are expounded in depth, yielding a thought-provoking essay. The last portion deals with an always-interesting subject, cosmology, and this treatment is no different. Clear explanations of the Big-Bang theory, the cosmic time scale, and the evolution of the Universe are given. Highly recommended for the educated layman and for college and observatory libraries.

577.   Taylor, John. **Black Holes: The End of the Universe?** New York, Random House, 1974. 175p. indexes (subject and name). $5.95. LC 73-20572. ISBN 0-394-49086-X.

This thought-provoking and totally non-technical book explains to the layman and non-astronomer scientist one of the most important scientific developments in recent years. Written in essay format, the work is a blend of astronomy, philosophy, religion, and cosmology; in short, it's an unusual volume, quite different from the conventional astronomy book. After some introductory or scene-setting chapters on Man's traditional concept of the Universe and himself, the author turns to a presentation of the concept of the black hole, explaining the physics of the strange phenomenon. The observational evidence for their existence is covered next, followed by a consideration of how we could (or should) harness the enormous amounts of energy associated with these objects. The changes in astrophysical theories and cosmological thought caused by the possible existence of black holes finishes this volume. It is hard to decide who this book is intended for, and even more difficult to recommend it for any particular type of library.

578.   Weekes, Trevor C. **High-Energy Astrophysics.** London, Chapman and Hall; distr. New York, Halstead Press, John Wiley & Sons, 1969. 209p. refs. subject indexes. $9.50. LC 70-6409. ISBN 0-470-93310-0. (Wiley); 0-412-09340-5 (Chapman).

One of the currently "exciting" subsets of the field of astrophysics is explored in this book aimed at the advanced student and non-specialist scientist. Written in a fairly descriptive but non-elementary manner, this volume deals with "high energies relative to the rest mass of the object (supernovae, radio galaxies, quasars), individual quanta possessing high energies (cosmic rays, x-rays, gamma rays), possible large cosmic energy densities (neutrinos, microwaves)." In particular, the author explains electromagnetic phenomena (the "synchrotron" and Compton radiations), variable stars (including novae and supernovae), the origin of the cosmic radiation, radio galaxies (and how they are observed), quasars, x-ray, astronomy, gamma ray astrophysics, and much more. Readers should consider this only as a survey or introductory volume, however; those who have a serious interest in high-energy astrophysics should peruse the literature for up-to-date articles and monographs.

APPENDIX
BIBLIOGRAPHY OF BASIC REFERENCE MATERIALS IN ASTRONOMY

KEY: A—Astronomy Libraries
P—Public Libraries
C—College/University Libraries

GUIDE TO THE LITERATURE

1. Kemp, D. A. **Astronomy and Astrophysics: A Bibliographic Guide.** London, MacDonald Technical and Scientific; distr. Hamden, Conn., Archon Books, Shoestring Press, Inc., 1970. A, P, C.

ABSTRACTS AND INDEXES: JOURNAL ARTICLES AND MONOGRAPHIC WORKS

1. **Astronomischer Jahresbericht.** Leipzig, W. de Gruyter. v.1-68, 1899-1968. Annual. A, C.

2. **Astronomy and Astrophysics Abstracts.** Berlin, New York, Springer-Verlag. v.1– , 1969– . Semi-annual. A, C.

3. **Astronomy and Astrophysics Monthly Index.** Sierra Madre, Calif., Olivetree Associates. Jan. 1976– . Monthly. A, C.

4. **Bulletin Signalétique 120: Astronomie, Physique Spatiale, Geophysique.** Paris, Centre de documentation du C.N.R.S. v.32– , 1971– . Continues *Bulletin Signalétique 120: Astronomie et astrophysique; physique du globe.* A, C.

5. **CSIRO Radio Astronomy Abstracts.** Sydney, Commonwealth Scientific and Industrial Research Organization, 1965-77. Weekly. A, C.

6. **Current Contents: Physical and Chemical Sciences.** Philadelphia, Institute for Scientific Information. v.1– , 1961– . Weekly. A, P, C.

7. **Current Physics Index.** New York, American Institute of Physics. v.1– , 1975– . Quarterly. A, C.

8. **International Aerospace Abstracts.** New York, American Institute of Aeronautics and Astronautics. v.1– , 1961– . Semi-monthly. A, C.

9. **Physics Abstracts.** London, Institution of Electrical Engineers. v.69– , 1966– . Bi-weekly. Continues in part *Science Abstracts, Series A*, and continues its numbering of volumes. A, C.

10. **Referativnyi zhurnal: Astronomiia.** Moscow, VINITI. 1963– . Monthly. A, C.

ABSTRACTS AND INDEXES: TECHNICAL REPORTS

1. **Scientific and Technical Aerospace Reports (STAR).** Washington, D.C., NASA. v.1– , 1965– . Semi-monthly. A, C.

DIRECTORIES

1. **Bibliography of Non-Commercial Publications of Observatories & Astronomical Societies.** rev. ed. Utrecht, The Netherlands, "Sonnenborgh" Observatory, 1973. A.

2. **Directory of Physics and Astronomy Staff Members.** New York, American Institute of Physics. 1959/60– . Annual. A, C.

3. **1976-77 Graduate Programs in Physics, Astronomy, and Related Fields.** New York, American Institute of Physics, 1977. A, C.

4. **International Physics & Astronomy Directory, 1969-70.** New York, W. A. Benjamin, 1969. A, P, C.

5. Kirby-Smith, Henry Tompkins. **U.S. Observatories; A Directory and Travel Guide.** New York, Van Nostrand Reinhold, 1976. A, P, C.

6. **List of Radio and Radar Astronomy Observatories.** Washington, D.C., National Academy of Sciences, National Academy of Engineering, Committee on Radio Frequencies. 1970– . Irregular. A, P, C.

7. Page, Thornton. **Observatories of the World.** Cambridge, Mass., Smithsonian Astrophysical Observatory, 1967. A, P, C.

Appendix: Bibliography of Basic Reference Materials / 285

DICTIONARIES AND ENCYCLOPEDIAS

1. Bizony, M. T., ed. **The New Space Encyclopedia; A Guide to Astronomy and Space Exploration.** New York, E. P. Dutton & Co., Inc., 1973. P, C.

2. Chiu, Hong-Yee, ed. **Chinese-English, English-Chinese Astronomical Dictionary.** New York, Consultants Bureau, 1966. A, C.

3. Hopkins, Jeanne. **Glossary of Astronomy and Astrophysics.** Chicago, University of Chicago Press, 1976. A, P, C.

4. Kleczek, Josip. **Astronomical Dictionary: In Six Languages.** Praha, Nakladatelství Československé Akademie Věd; distr. New York, Academic Press, 1961. A, C.

5. Muller, Paul. **Concise Encyclopedia of Astronomy.** Glascow, London. William Collins Sons & Co.; Chicago, Follett Publishing Co., 1968. P, C.

6. Satterthwaite, Gilbert E. **Encyclopedia of Astronomy.** New York, St. Martin's Press, Inc., 1971. A, P, C.

7. Wallenquist, Åke. **Dictionary of Astronomical Terms.** Garden City, N.Y., Natural History Press, 1966. A, P, C.

8. Weigert, A., and H. Zimmerman. **A Concise Encyclopedia of Astronomy.** New York, American Elsevier, 1968, c1967. A, P, C.

HANDBOOKS, MANUALS, AND ALMANACS

1. Akademiia Nauk SSSR. Institut Teoreticheskoi Astronomii. **Efemeridy malykh planet.** (Ephemerides of minor planets.) Leningrad, "Nauka." 1947– . Annual. A, C.

2. Allen, C. W. **Astrophysical Quantites.** 3rd ed. London, The Athlone Press, University of London; distr. Atlantic Highlands, N.J., Humanities Press, Inc., 1974, c1973. A, C.

3. Astronomisches Rechen-Institut. **Apparent Places of Fundamental Stars.** Heidelberg, W. Germany. 1940– . Annual. A, C.

4. Eichhorn, Heinrich. **Astronomy of Star Positions.** New York, Frederick Ungar Publishing Co., Inc., 1974. A, C.

5. **The Handbook of the British Astronomical Association.** London, Burlington House. 1921– . Annual. A, P, C.

286 / Appendix: Bibliography of Basic Reference Materials

6. Jones, Kenneth Glyn. **Messier's Nebulae and Star Clusters.** New York, American Elsevier Publishing Co., Inc., 1969, c1968. A, P, C.

7. Lang, Kenneth. **Astrophysical Formulae: A Compendium for the Physicist and Astrophysicist.** Berlin, New York, Springer-Verlag, 1974. A, C.

8. Mason, Brian, ed. **Handbook of Elemental Abundances in Meteorites.** New York, Gordon and Breach Science Publishers, Inc., 1971. A, C.

9. Moore, Patrick, ed. **Yearbook of Astronomy.** London, Sidgwick & Jackson Ltd.; New York, W. W. Norton & Co., Inc. 1962– . Annual. A, P, C.

10. Nautical Almanac Office. U.S. Naval Observatory. **The American Ephemeris and Nautical Almanac.** Washington, D.C., Government Printing Office. 1852– . Annual. Supplement: 1975. A, P, C.

11. Nautical Almanac Office. U.S. Naval Observatory. **Astronomical Phenomena for the Year 19– .** Washington, D.C., Government Printing Office. 1950– . Annual. A, P, C.

12. Percy, John R., ed. **The Observer's Handbook.** Toronto, Royal Astronomical Society of Canada. 1– , 1907– . Annual. A, P, C.

13. Pickering, James S. **1001 Questions Answered about Astronomy.** rev. ed. New York, Dodd, Mead & Co., 1975. P, C.

14. Pilcher, Frederick, and Jean Meeus. **Tables of Minor Planets.** Jacksonville, Ill., F. Pilcher, Illinois College, 1973. A, C.

15. Robinson, J. Hedley. **Astronomy Data Book.** New York, Halstead Press, John Wiley and Sons, 1972. A, P, C.

16. Roth, G. D., ed. **Astronomy: A Handbook.** 2nd ed. Berlin, New York, Springer-Verlag, 1975. A, P, C.

17. Voight, H. H., ed. **Landolt-Borstein Numerical Data and Functional Relationships in Science and Technology.** Group VI, Volume I: *Astronomy and Astrophysics.* Berlin, New York, Springer-Verlag, 1965. A, C.

ATLASES AND CATALOGUES

1. Alter, G., and B. Balazs, and J. Ruprecht. **Catalogue of Star Clusters and Associations.** 2nd ed. Budapest, Akademiai Kiado; distr. New York, Adler's Foreign Books, 1970. A, C.

2. Bečvář, Antonín. **Atlas Australis 1950.0.** Praha, Nakladatelství Československé Akademie Věd; distr. Cambridge, Mass., Sky Publishing Corp., 1964. A, C.

3. Bečvář, Antonín. **Atlas Borealis 1950.0.** Praha, Nakladatelství Československé Akademie Věd; distr. Cambridge, Mass., Sky Publishing Corp., 1962. A, C.

4. Bečvář, Antonín. **Atlas Coeli 1950.0.** Praha, Nakladatelství Československé Akademie Věd; distr. Cambridge, Mass., Sky Publishing Corp., 1956. A, C.

5. Bečvář, Antonín. **Atlas Coeli–II. Katalog. 1950.0.** Praha, Nakladatelství Československé Akademie Věd; distr. Cambridge, Mass., Sky Publishing Corp., 1964. A, C.

6. Bečvář, Antonín. **Atlas Eclipticalis 1950.0.** Praha, Nakladatelství Československé Akademie Věd; distr. Cambridge, Mass., Sky Publishing Corp., 1958. A, C.

7. "Cambridge Catalog of Radio Sources," in several parts in **Royal Astronomical Society. Memoirs.** (v.67, 1955; v. 68, 1959; v. 68, 1961; v. 69, 1965; v. 71, 1967) and **Royal Astronomical Society. Monthly Notices.** (v.134, 1966; v. 139, 1968; v. 144, 1969; v. 145, 1969; v. 151, 1970; v. 171, 1975). A.

8. Cannon, Annie J., and E. C. Pickering. **The Henry Draper Catalogue.** Cambridge, Mass., Harvard College Observatory, 1918-24. 10v. *Annals of the Harvard College Observatory*, v.91-99, 112. A, C.

9. de Vaucouleurs, Gerard, and Antoinette de Vaucouleurs. **Reference Catalogue of Bright Galaxies.** Austin, University of Texas Press, 1964. A, C.

10. de Vaucouleurs, Gerard, Antoinette de Vaucouleurs, and Harold G. Corwin, Jr. **Second Reference Catalogue of Bright Galaxies.** Austin, University of Texas Press, 1976. A, C.

11. **ESO/SRC/Atlas of the Southern Sky.** Geneva, ESO Sky Atlas Laboratory, CERN, 1976. A.

12. Gutschewski, Gary L., et al. **Atlas and Gazetteer of the Near Side of the Moon.** Washington, D.C., NASA, 1971. A, P, C.

13. Hoffleit, Dorrit. **Catalogue of Bright Stars.** 3rd rev. ed. New Haven, Conn., Yale University, 1964. A, C.

14. Jeffers, H. M., W. H. van den Bos, and F. M. Greeby. **Index Catalogue of Visual Double Stars.** (Lick Catalogue). Mount Hamilton, Calif., Lick Observatory, 1963. *Lick Observatory. Publications.* no.21 2v. A, C.

15. Kopal, Zdeněk. **A New Photographic Atlas of the Moon.** New York, Taplinger Publishing Co., 1971. A, P, C.

16. Kukarkin, B. V. **Obshchii katalog peremennykh zvezd.** (General Catalogue of Variable Stars.) 3rd ed. Moscow, "Nauka," 1968. Suppl.: 1971, 1972, 1974, 1976. A, C.

17. Massachusetts Institute of Technology. **Wavelength Tables.** rev. ed. Cambridge, Mass., The MIT Press, 1969. A, C.

18. "Master Source List." Columbus, Ohio State University Radio Observatory. (Originally published in **Astrophysical Journal. Supplement Series.** v.20 no.8, 1970, under the title "A Master List of Radio Sources.")

19. Moore, Patrick. **Atlas of the Universe.** London, Mitchell Beazley Ltd.; distr. Chicago, Rand McNally and Co., 1970. A, P, C.

20. Moore, Patrick. **Color Star Atlas.** London, Mitchell Beazley Ltd.; distr. New York, Crown Publishers, 1973. A, P, C.

21. Nilson, Peter. **Uppsala General Catalogue of Galaxies.** Uppsala, Sweden, Uppsala University, 1973. (Series: Uppsala. Universitet. Astronomiska observetoriet. Annaler. Bd.6.) A, C.

22. Nilson, Peter. **Catalogue of Selected Non-UGC Galaxies.** Uppsala, Sweden, Uppsala Astronomical Observatory, 1974. (Observatory Report no. 5). A, C.

23. Norton, Arthur P. **Norton's Star Atlas and Reference Handbook (epoch 1950.0).** 16th ed. Edinburgh, Gall & Inglis; distr. Cambridge, Mass., Sky Publishing Corp., 1973. P, C.

24. "Ohio Survey." Columbus, Ohio, Ohio State University Radio Observatory. 1967– . In seven parts, with continuing supplements. A, C.

25. **Palomar Sky Atlas.** National Geographic Society–Palomar Observatory Sky Survey. Pasadena, Calif., Mt. Wilson and Palomar Observatories, 1954. Fifth printing: 1977. A.

26. "Parkes Catalogue of Radio Sources; Declination Zone +20° to -90°'" in **Australian Journal of Physics. Astrophysical Supplement.** no.7. April 1969. A.

27. Seitter, Waltraut Carola. **Atlas for Objective Prism Spectra.** Bonn, West Germany, Ferd. Dummlers Verlag. v.1: 1970; v.2: 1973. A, C.

28. Smithsonian Astrophysical Observatory. **Star Atlas of Reference Stars and Nonstellar Objects.** Cambridge, Mass., The MIT Press, 1969. A, C.

29. Smithsonian Astrophysical Observatory. **Star Catalog.** Washington, D.C., Smithsonian Institution, 1966. 4v. A, C.

30. Sulentic, Jack W. and William G. Tifft. **The Revised New General Catalogue of Nonstellar Astronomical Objects.** Tucson, University of Arizona Press, 1973. A, C.

31. Zwicky, Fritz. **Catalogue of Galaxies and Clusters of Galaxies.** Pasadena, Calif., CalTech, 1961-68. 6v. A, C.

REVIEW LITERATURE

1. **Advances in Astronomy and Astrophysics.** New York, Academic Press. v.1-9, 1962-72. Irregular. A, C.

2. **Annual Review of Astronomy and Astrophysics.** Palo Alto, Calif., Annual Reviews. v.1– , 1963– . Annual. A, C.

3. Flügge, S., ed. **Handbuch der Physik.** (Encyclopedia of Physics.) Berlin, Springer-Verlag. A, C.

    v.50. **Astrophysics I: Stellar Surfaces and Binaries** (1958).
    v.51. **Astrophysics II: Stellar Structure** (1958).
    v.52. **Astrophysics III: The Solar System** (1959).
    v.53. **Astrophysics IV: Stellar Systems** (1959).
    v.54. **Astrophysics V: Miscellaneous** (1962).

4. Kuiper, G. P., and Barbara Middlehurst, eds. **Stars and Stellar Systems.** Chicago, University of Chicago Press. 8v. A, C.

    v.1: **Telescopes** (1960).
    v.2: **Astronomical Techniques** (1962).
    v.3: **Basic Astronomical Data** (1963).
    v.4: Clusters and Binaries (cancelled).
    v.5: **Galactic Structure** (1965).
    v.6: **Stellar Atmospheres** (1960).
    v.7: **Neublae and Interstellar Matter** (1968).
    v.8: **Stellar Structure** (1965).
    v.9: **Galaxies and the Universe** (1975).

5. **Vistas in Astronomy.** London, New York, Pergamon Press. v.1-19, 1955-75. Irregular. Became a review journal in 1976. A, C.

AUTHOR-TITLE INDEX

AAAS Science Books & Films, 7
ABC der Astronomie, 49
AGK, p. 56
Abbott, B., 444
Abell, G. O., 324, 325
Abetti, G., 156, 157, 515
Abriss der Astronomie 340
Abt, H. A., 72
Account of the Revd. John Flamsteed, 260
Acta Astronomica, 107
Adamczewski, J., 237
Advances in Astronomy and Astrophysics, 143
Air Almanac, 50
Airy, G. B., 442
Aitken, R. G., 450
Akademii Nauk SSSR, 51
Albert Einstein: Creator and Rebel, 241
Alexander A. F. O'D., 394, 395
All about Telescopes, 207
Allen, C. W., 52
Allen, D. A., 568
Aller, L. H., 451, 468, 511
Alter, D., 73, 148, 380
Alvan Clark & Sons: Artists in Optics, 254
Amateur Astronomer's Handbook, 229, 231
Amateur Astronomy, 197
Amateur Telescope Mirror Making, 208
Amazing Universe, 150
American Astronomical Society. Bulletin, 108
American Ephemeris and Nautical Almanac, 63
American Men and Women of Science, 24
Analitecheskiye i chislennyye metody nebeshoy mekhaniki, 445
Analytical and Numerical Methods of Celestial Mechanics, 445

Ancient Astronomical Observations and the Accelerations of the Earth and Moon, 305
Annals of the IQSY, 378
Annual Review of Astronomy and Astrophysics, 144
Annual Review of Earth and Planetary Sciences, 145
Apfel, N. H., 314
Apollo 11 Lunar Science Conference. Proceedings, 386
Apparent Places of Fundamental Stars, 53
Arp, H., 522
Arthur, D. W. G., 90
Asimov, I., 151, 271, 396
Astrodynamics, 547
Astrodynamics—Applications and Advanced Topics, 544
Astrographic Catalogue, p. 57
Astronomers, 248
Astronomers Royal, 249
Astronomical Atlases, Maps and Charts, 81
Astronomical Dictionary: In Six Languages, 37
Astronomical Ephemeris, 63
Astronomical Institutes of Czechoslovakia. Bulletin, 109
Astronomical Journal, 110
Astronomical Objects for Southern Telescopes, 210
Astronomical Phenomena for the Year 19– , 64
Astronomical Revolution, 242
Astronomical Society of Australia. Proceedings, 136
Astronomical Society of Japan. Publications, 111
Astronomical Society of the Pacific. Publications, 112
Astronomical Techniques, 537
Astronomical Telescope, 532
Astronomical Telescopes and Observatories for Amateurs, 228

Astronomicheskii Zhurnal, 135
Astronomie expérimentale, 554
Astronomie Populaire, 160
Astronomische Gesellschaft Katalog, p. 56
Astronomische Nachrichten, 113
Astronomischer Jahresbericht, 10
Astronomisches Rechen-Institut, 53
Astronomiskt Lexikon, 39
Astronomy, 152, 154, 287, 326
Astronomy: A Handbook, 230
Astronomy: Activities and Experiments, 328
Astronomy and Astrophysics: A Bibliographic Guide, 2
Astronomy and Astrophysics: A European Journal, 114
Astronomy and Astrophysics: A European Journal, Supplement Series, 115
Astronomy and Astrophysics Abstracts, 11
Astronomy and Astrophysics for the 1970s, 190
Astronomy and Astrophysics Monthly Index, 12
Astronomy Data Book, 68
Astronomy from a Space Platform, 551
Astronomy: Fundamentals and Frontiers, 316
Astronomy: Observational Activities and Experiments, 312
Astronomy of Birr Castle, 245
Astronomy of Star Positions, 54
Astronomy of the 20th Century, 292
Astronomy One, 314
Astronomy: The World's Most Beautiful Astronomy Magazine, 343
Astronomy with Binoculars, 199
Astrophysical Concepts, 327
Astrophysical Formulae, 59
Astrophysical Journal, 116
Astrophysical Journal. Supplement Series, 117
Astrophysical Letters, 118
Astrophysical Quantities, 52
Astrophysics, 119, 465
Astrophysics and Space Science, 120
Astrophysics and Space Science Library, p. 89
Astrophysics and Stellar Astronomy, 322
Astrophysics of Gaseous Nebulae, 512
Astrophysics, Part A: Optical and Infrared, 534
Astrophysics, Part B: Radio Telescopes, 563
Astrophysics, Part C: Radio Observations, 562

Atanasijević, I., 334
Atlas and Gazetteer of the Near Side of the Moon, 86
Atlas Australis 1950.0., 74
Atlas Borealis, 1950.0., 75
Atlas Coeli 1950.0., 76
Atlas Coeli–II. Katalog. 1950.0., 77
Atlas Eclipticalis 1950.0., 78
Atlas for Objective Prism Spectra, 101
Atlas of Cometary Forms, 98
Atlas of Deep-Sky Splendors, 105
Atlas of the Universe, 95
Atmospheres, 435
Atmospheres of Earth and the Planets, 436
Atoms, Stars, and Nebulae, 451
Attractive Universe: Gravity and the Shape of Space, 444
Aurora and Airglow, 437
Australian Journal of Physics, 121
Aveni, A. F., 94

Baade, W., 469
Backyard Astronomer, 201
Bahcall, J. N., 522
Baker, R. H., 205, 311, 326
Baker, R. M. L., 544, 545
Barlow, B. V., 532
Barnes, M. A., 417
Barnes, V. 417
Basic Astronomical Data, 69
Basic Astronomy: With Projects for Amateurs and Students, 211
Basic Physics of Stellar Atmospheres, 487
Batten, A. H., 452
Baxter, W. M., 206
Baum, R., 272
Beauty of the Universe, 155
Becker, W., 500
Bečvár, A., 74, 75, 76, 77, 78
Bedini, S., 257
Beer, A., 141
Beet, E. A. 224
Berendzen, R., 273
Bergamini, D., 149
Berlage, H. P., 351
Berman, L., 308
Beyond the Known Universe, 172
Beyond the Milky Way, 516
Bhatnagar, P. L., 479
Bibliography of Non-Commercial Publications of Observatories & Astronomical Societies, 140
Bibliography of Stellar Radial Velocities, 72
Biggs, E. S., 72
Binary and Multiple Systems of Stars, 452

Binary Stars, 450
Bizony, M.T., 40
Blaauw, A., 502
Black Holes, 569
Black Holes, Quasars, and the Universe, 576
Black Holes: The End of the Universe?, 577
Blank, D., 203
Bok, B. J., 501
Bok, P. F., 501
Bonestell, C., 412
Bonner Durchmusterung, p. 56
Book of Mars, 399
Boss, B., 79
Boundaries of the Universe, 161
Bova, B., 158, 529
Bowker, D. E., 80
Bowl of Night: The Physical Universe and Scientific Thought, 520
Boyd, L. G., 279
Brancazio, P. J., 470
Brandt, J. C., 309, 352, 364
Bray, R. J., 365, 366
British Astronomical Association. Journal., 122
Bronshten, V. A., 397
Brown, B. J. W., 81
Brown, E. B., 533
Brown, P. L. 191, 418
Brown, S., 207
Bulletin Signalétique 120: Astronomie, Physique Spatiale, Géophysique, 13
Burbidge, G., 530
Burbidge, M., 530
Buttmann, G., 238

Cadrans solaires, 235
Cambridge Catalog of Radio Sources, p. 57
Cameron, A. G. W., 470, 486, 507
Cannon, A. J., 82
Canon of Solar Eclipses, 61
Carleton, N., 534
Caspar, M., 258
Catalog of Emission Lines in Astrophysical Objects, 94
Catalog of the Naval Observatory Library, 6
Catalogue of Bright Stars, 87
Catalogue of Galaxies and of Clusters of Galaxies, 106
Catalogue of Selected Non-UGC Galaxies, 97
Cattermole, P. J., 392
Celescope Catalog of Ultraviolet Stellar Observations, 83
Celestial Mechanics, 447

Celestial Mechanics; An International Journal of Space Dynamics, 123
Celestial Navigation, 342
Celestial Objects for Common Telescopes, 223
Challenge of the Stars, 153
Chandrasekhar, S., 471, 472
Chebotarev, G. A., 419, 445
Cherrington, E. H., Jr., 192
Chinese-English, English-Chinese Astronomical Dictionary, 35
Chiu, H.-Y., 35, 473
Christiansen, W. N., 556
City of the Stargazers, 240
Clayton, D. D., 474
Clemence, G. M., 341
Cleminshaw, C. H., 148
Coblans, H., 1
Cole, F. W., 310
Coleman, P. J. Jr., 377
Color Star Atlas, 96
Comet of 1577: Its Place in the History of Astronomy, 298
Comets, Meteorites and Men, 418
Comets: Scientific Data and Missions, 423
Comets: Visitors from Space, 427
Commentariolus, 266
Comments on Astrophysics: A Journal of Critical Discussion of the Current Literature, 124
Concepts of Contemporary Astronomy, 313
Concepts of the Universe, 164
Concise Encyclopedia of Astronautics, 43
Concise Encyclopedia of Astronomy, 45, 49
Consolidated Lunar Atlas, 90
Contopoulos, G., 500
Cordoba Durchmusterung, p. 56
Corliss, W. R., 159
Coronal Expansion and Solar Wind, 370
Corwin, H. G., Jr., 85
Cosmology Now, 524
Cosmovici, C. B., 475
Cotter, C. H., 295
Course in Theoretical Astrophysics, 466
Cousins, F. W., 233, 353
Cox, J. P., 476
Craig, R. A., 433
Craters of the Moon; An Observational Approach, 392
Crawford, D., 239
Crime of Galileo, 267
Criswell, D. R., 388
Cross, C. A., 405
Current Contents: Physical and Chemical Sciences, 14
Current Physics Index, 15

Daumas, M., 296
Davis, R. J., 83
De Jager, C., 367
de Vaucouleurs, A., 84, 85
de Vaucouleurs, G., 84, 85, 318
Detre, L., 491
Deutschman, W. A., 83
DeWitt, B. S., 569
DeWitt, C., 569
Dialogue Concerning the Two Chief World Systems, Ptolemaic and Copernican, 263
Dicks, D. R., 297
Dickson, F. P. 520
Dictionary of Astronomical Terms, 39
Dictionary of Scientific Biography, 27
Dieckvoss, W., 460
Directory of Physics and Astronomy Staff Members, 25
Directory of Published Proceedings, 137
Directory of Special Libraries and Information Centers, 34
Discovering the Universe; A History of Astronomy, 288
Discovery of Our Galaxy, 293
Dobson, G. M. B., 434
Donath, F. A., 145
Donn, B., 98
Donnelly, M. C., 274
Doppler Effect; An Introduction to the Theory of the Effect, 523
Doyle, R. O., 186
Drake, S., 259
Dryer, M., 372
Dufay, J., 506
Dukas, H., 241
Dusty Universe, 507
Duveen, A., 330
Dwarf Novae, 493

ESO/SRC/Atlas of the Southern Sky, p. 57
Early Greek Astronomy to Aristotle, 297
Early Solar Physics, 302
Early Type Stars, 490
Earth and Extraterrestrial Sciences, 125
Earth and Planetary Science Letters, 126
Earth, Moon, and Planets, 363
Eclipse Phenomena in Astronomy, 357
Eclipsing Variable Stars, 498
Eddington, A. S., 477
Edge of Space: Exploring the Upper Atmosphere, 433
Edmond Halley: Genius in Eclipse, 250
Efemeridy malykh planet, 51

Eichhorn, H., 54
Einführung in die Astronomie, 169
Ellison, M. A., 368
Emerging Universe: Essays on Contemporary Astronomy, 178
Emerson, M. N., 208
Encyclopaedic Dictionary of Physics, 47
Encyclopedia of Associations, 26
Encyclopedia of Astronomy, 46
Encyclopedia of Atmospheric Sciences and Astrogeology, 41
Encyclopedia of Physics, 42
Ephemerides of Minor Planets, 51
Esplorazione dell'universo, 156
Essentials of Astronomy, 330
Euler-Mayer Correspondence (1751-1755), 261
Evans, D. S., 521, 535
Everyman's Astronomy, 184
Evolution of Radio Astronomy, 278
Evolution of Stars and Galaxies, 469
Evolution of Stars: How They Form, Age, and Die, 482
Experimental Astronomy, 554
Exploration of the Universe, 156, 324
Exploration of the Universe: Updated Brief Version, 325
Explorer of the Universe: A Biography of George Ellery Hale, 255
Exploring the Atmosphere, 434
Exploring the Cosmos, 308
Exploring the Moon through Binoculars, 192
Exploring the Planets, 408
External Galaxies and Quasi-Stellar Objects, 521
Eyes on the Universe: A History of the Telescope, 271

Fairbridge, R. W., 41
Fernie, J. D., 492
Field, G. B., 507, 522
Field Guide to the Stars and Planets, 196
Find a Falling Star, 428
Fisk, M., 26
Fitzpatrick, P. M., 446
Fizika Planet, 407
Fizika Solnecho: Korony, 376
Flammarion, G. C., 160
Flammarion Book of Astronomy 160
Flamsteed, J., 260
Flügge, S., 42
Forbes, E. G., 261, 262
Franzgrote, E., 253
Frederick, L. W., 311, 326
Friedman, H., 150
From Stonehenge to Modern Cosmology, 166

294 / Author-Title Index

From the Black Hole to the Infinite
  Universe, 162
Frontiers in Astronomy, 187
Fundamental Astrometry; Determination
  of Stellar Coordinates, 337
Fundamental Astronomy: Solar System
  and Beyond, 310

Gainer, M. K., 312
Galactic and Extra-Galactic Radio
  Astronomy, 565
Galactic Astronomy, 504
Galactic Nebulae and Interstellar
  Matter, 506
Galactic Radio Astronomy, 559
Galactic Structure, 502
Galaxies, 518
Galaxies and the Universe, 517
Galaxies of the Local Group, 519
Galiana, T., 43
Galilei, G., 263
Galileo, 251
Galileo: A Philosophical Study, 268
Galileo Studies: Personality, Tradition,
  and Revolution, 259
Galileo's Intellectual Revolution:
  Middle Period, 1610-1632, 269
Gahrels, T., 398, 420
General Catalogue, 79
General Catalogue of Variable Stars, 91
Getting Acquainted with Comets, 429
Giaconni, R., 570
Giant Meteorites, 422
Gibson, E. G., 369
Gill, T. P., 523
Gillispie, C. C., 27
Gingerich, O., 187, 188
Giuli, R. T., 476
Glasby, J. S., 161, 209, 493, 494, 495
Glass Giant of Palomar, 294
Glasstone, S., 399
Glossary of Astronomy and Astrophysics,
  36
Goldberg, L., 144
Goldsmith, D., 162
Goldstine, H. H., 55
Goody, R. M., 435
Gordon, J. L., 441
Government Reports Announcements
  & Index, 20
Grant, R., 275
Graubard, M., 276
Gravitation, 443
Gravitation: An Elementary Explanation
  of the Principal Perturbations in the
  Solar System, 442
Gray, D. A., 3
Gray, R. A., 3
Greenstein, J. L., 478

Greisen, K., 571
Greyber, H. D., 551
Griffith Observer, 344
Groei Van Ons Wereldbeeld, 286
Grosjean, C. C., 61
Guide to Reference Books, 4
Guide to Reference Material, 5
Guideposts to the Stars, 202
Gursky, H., 570, 572
Gurzadyan, G. A., 508
Gutschewski, G. L., 86

Hack, M., 453, 454, 455, 456, 515
Hagihara, Y., 447
Handbook for Planet Observers, 221
Handbook of Elemental Abundances
  in Meteorites, 60
Handbook of the British Astronomical
  Association, 56
Handbook of the Constellations, 203
Handbook of the Physical Properties
  of the Planet Jupiter, 403
Handbook of the Physical Properties
  of the Planet Mars, 404
Handbook of the Physical Properties
  of the Planet Venus, 405
Handbuch der Physik, 42
Handbuch für Sternfreunde, 230
Haramundanis, K. L., 83, 321
Hardy, D. A., 153
Hartmann, W. K., 354, 400
Hartung, E. J., 210
Harvard Books on Astronomy, p. 89
Harvard College Observatory: The First
  Four Directorships, 1839-1919, 279
Harwitt, M., 327
Haug, U., 536
Hawkins, G. S., 163, 277
Haymes, R. C., 546
Haysham, H., 211
Hazard, D. R., 323
Heide, F., 421
Heiserman, D., 225
Hellman, C. D., 298
Henry Draper Catalogue, 82
Herbig, G. H., 457
Herrick, S., 547
Hess, W. N., 548
Heuer, K., 240
Hewish, A., 573
Hey, J. S., 278, 557
Heymann, D., 388
Heywood, J., 226
High-Energy Astrophysics, 578
High Firmament: A Survey of Astronomy
  in English Literature, 303
Highlights in Astronomy, 167
Highlights of Astronomy, 189
Hiltner, W. A., 537

Author-Title Index / 295

Hindmarsh, W. R., 509
History of Astronomy, 286
History of Japanese Astronomy: Chinese Background and Western Impact, 304
History of Nautical Astronomy, 295
History of Physical Astronomy, 275
Hodge, P. W., 164, 165, 313, 352
Hoff, D., 328
Hoffleit, D., 87
Hoffman, B., 241
Hogbom, J. A., 556
Hopkins, J., 36
Horne, D. F., 538
Howard, N. E., 212, 213
Howarth, H. E., 289
Hoyle, F., 166, 167
Hubble Atlas of Galaxies, 100
Hughes, J. K., 80
Hundhausen, A. J., 370
Hyde, F. W., 227
Hynek, J. A., 314

Icarus: An International Journal of Solar System Studies, 127
In Quest of Quasars, 529
Index to Scientific Reviews, 142
Infrared, the New Astronomy, 568
Inglis, R. M. G., 193
Inglis, S. J., 315
Instruments scientifiques aux XVII et XVIII siécles, 296
Interferometry, 543
Internal Constitution of the Stars, 477
International Aerospace Abstracts, 21
International Astronomical Union, 189
International Astronomical Union. Transactions., 138
International Auroral Atlas, 88
International Conference on Education in and History of Modern Astronomy, 273
International Physics & Astronomy Directory, 1969-70, 28
Interstellar Grains, 514
Interstellar Ionized Hydrogen, 513
Interstellar Medium, 510
Introduction to Astrodynamics, 545
Introduction to Astronomy, 311, 321
Introduction to Planetary Physics: The Terrestrial Planets, 355
Introduction to Radio Astronomy, 558
Introduction to Space Science, 546, 548
Introduction to Stellar Atmospheres and Interiors, 481
Introduction to Stellar Statistics, 461

Introduction to the Solar Wind, 364
Introduction to the Study of Stellar Structure, 471
Introductory Astronomy, 320
Introductory Astronomy and Astrophysics, 331
Invisible Universe: The Story of Radio Astronomy, 566
Irish Astronomical Journal, 345

Jackson, J. H., 401
Jacobs, K. C., 178, 331
Jaki, S. L., 299
Jastrow, R., 168, 316
Jeffries, J. T., 458
Jennison, R. C., 558
Johannes Hevelius and His Catalog of Stars, 253
John, L., 524
John Herschel: Lebensbild eines Naturforschers, 238
Jones, B. Z., 279, 280
Jones, K. G., 57
Joseph, J. M., 194
Journal for the History of Astronomy, 346
Jupiter: Studies of the Interior, Atmosphere, Magnetosphere and Satellites, 398
Jupiter: The Largest Planet, 396

Kahn, F. D., 531
Kaler, J. B., 333
Kaplan, S. A., 510
Kaufmann, W. J., 525
Kaula, W. M., 355
Kazimirchak-Polonskaya, E. I., 419
Keene, G. T., 214
Kellermann, K. I., 565
Kelsey, L., 328
Kemp, D. A., 2
Keoeeit—The Story of the Aurora Borealis, 440
Kepler, 258
Kerr, F. J., 559
Kienle, H., 169
Kiepenheuer, K., 371
Kilmister, C., 526
King, E. A., Jr., 388
King, H. C., 459
King, I. R., 317
King's Astronomer: William Herschel, 239
Kinsler, D. C., 86
Kirby-Smith, H. T., 29
Kitamura, M., 58
Kleczek, J., 37
Kleine Meteoritenkunde, 421

Koenig, L. R., 402
Koestler, A., 281
Kopal, Z., 89, 143, 170, 282, 356, 381, 382, 383, 549
Koyré, A., 242
Kraus, J. D., 560
Krinov, E. L., 422
Kristian, J., 517
Kruse, W., 460
Kuiper, G. P., 90, 423, 539
Kukarkin, B. V., 91, 496
Kundu, M. R., 561
Kurs astrofiziki i zvezdnoi astronomii, 467
Kurs teoreticheskoy astrofiziki, 466
Kurth, R., 461
Kyed, J. M., 139

Labuhn, F., 550
Land, B., 243
Landolt-Borstein Numerical Data and Functional Relationships in Science and Technology, 71
Lang, K. R., 59
Lapedes, D. N., 38
Larsen, A. D., 253
Lattin, H. P., 171
Layzer, D., 144
Legacy of George Ellery Hale, 256
Lenchek, A. M., 574
Letter Against Werner, 266
Levinson, A. A., 384, 386, 387
Levitt, I. M., 172
Levy, D., 162
Ley, W., 283, 424
Life of Benjamin Banneker, 257
Light of the Night Sky, 441
Lighthouse of the Skies: The Smithsonian Astrophysical Observatory, 280
Lincoln, J. V., 378
Linear and Regular Celestial Mechanics, 448
Link, F., 357
Lippincott, S. L., 194
List of Radio and Radar Astronomy Observatories, 30
Livingston, D. M., 244
Long-Range Program in Space Astronomy, 186
Loughhead, R. E., 365, 366
Lovell, B., 173, 284, 300
Lucas, J. W., 385
Lunar Atlas, 73
Lunar Orbiter Photographic Atlas of the Moon, 80
Lunar Rocks, 391
Lunar Science: A Post-Apollo View, 393

Lunar Science Conference. Proceedings. Second: 387; Third: 388; Fourth: 389; Fifth: 390
Lüst, R., 550
Luyten, W. J., 92

McCall, G. J. H., 425
McCall, R., 151
McCormac, B. M., 436, 437, 438
McGraw-Hill Dictionary of Scientific Terms, 38
McGraw-Hill Encyclopedia of Science and Technology, 44
McGraw-Hill Modern Men of Science, 31
McGucken, W., 301
McIntosh, P. S., 372
McLaughlin, D. B., 468
McNally, D., 335
Macris, C. J., 373
Magnetism and the Cosmos, 509
Makemson, M. W., 545
Making Your Own Telescope, 222
Malinowsky, H. R., 3
Man and His Universe, 170
Manuel, F. E., 264
Maps of Lunar Hemispheres, 99
Maran, S. P., 309
Mars, 405, 412
Marsden, B. G., 419
Mason, B., 60, 391
Mass Loss from Stars, 453
Master of Light: A Biography of Albert A. Michelson, 244
Master Source List, p. 57
Matarazzo, J. M., 139
Mathematical Astronomy for Amateurs, 224
Mavridis, L. N., 462, 503
Mayall, M., 195, 215, 234
Mayall, R. N., 195, 215, 234
Mead, G. D., 548
Meadows, A. J., 265, 302, 303
Meeks, M. L., 562, 563
Meeus, J., 61, 67
Megalithic Lunar Observatories, 307
Mein Messier Buch, 105
Meinel, A. B., 94
Melson, W. G., 391
Menzel, D., 152, 196, 318, 374, 479
Mercury, 347
Messier's Nebulae and Star Clusters, 57
Meteorite Research, 426
Meteorites, 421
Meteorites and the Origin of Planets, 432
Meteorites and Their Origins, 425
Meteorites: Classification and Properties, 431
Mezhzvezdnaya sreda, 510

Michaux, C. M., 403, 404
Miczaika, G. R., 540
Middlehurst, B. M., 511, 539
Mihalis, D., 480, 504
Mikolaj Kopernik i jego epóka, 237
Milky Way, 501
Milky Way: An Elusive Road for Science, 299
Millman, P. M., 426
Mineralogical Abstracts, 16
Minnaert, M. G. J., 329
Minnis, C. M., 378
Misner, C. W., 443
Modern Astronomy, 319
Modern Astronomy: An Introduction, 169
Modern Astrophysics: A Memorial to Otto Struve, 454
Modern Cosmology, 528
Modern Optical Engineering: The Design of Optical Systems, 542
Modern Optics, 533
Moon: An International Journal of Lunar Studies, 128
Moon, 381
Moon in the Post-Apollo Era, 382
Moon Rocks and Minerals, 384
Moons and Planets, 354
Moore, P., 62, 95, 96, 153, 174, 197, 198, 228, 245, 246, 285, 358, 375, 392, 405, 406, 427
Morgenthaler, G. W., 551
Moroz, V. I., 407
Motion, Evolution of Orbits, and Origin of Comets, 419
Motivations, Tools and Theories of Pre-Modern Science, 276
Motz, L., 330
Mueller, I. I., 336
Muirden, J., 199, 229
Muller, P., 45
Murchie, G., 175
Muriel, A., 473
Music of the Spheres, 175
Mysteries of the Universe, 159

Nakayama, S., 304
Naked-Eye Astronomy, 198
Narratio prima, 266
National Geographic Society–Palomar Observatory Sky Survey, p. 57
National Research Council. Astronomy Survey Committee, 190
Nature of the Universe, 526
Nautical Almanac Office, U.S. Naval Observatory, 63, 64
Nebulae and Galaxies, 515
Nebulae and Interstellar Matter, 511
Nebular Variables, 494

Nébuleuses Galactiques et Matière Interstellaire, 506
Nebulose e gli Universi-isole, 515
Neely, H. M., 200
Neff, J., 328
Neighbors of the Earth, 359
Neutron Stars, Black Holes and Binary X-Ray Sources, 572
New and Full Moons: 1001 B.C. to A.D. 1651, 55
New Astronomies, 158
New Frontiers in Astronomy, 188
New Guide to the Planets, 358
New Horizons in Astronomy, 309
New Mars: The Discoveries of Mariner 9, 400
New Photographic Atlas of the Moon, 89
New Popular Star Atlas (Epoch 1950), 193
New Space Encyclopedia: A Guide to Astronomy and Space Exploration, 40
New Technical Books, 8
New Techniques in Space Astronomy, 550
New Universe, 173
Newton, R. R., 305
Nicolaus Copernicus and His Epoch, 237
Nicolson, I., 154, 408
Nietro, M. M., 306
Nilson, P., 97
Nine Planets, 409
1976-77 Graduate Programs in Physics, Astronomy, and Related Fields, 32
Nineteenth-Century Spectroscopy, 301
Nininger, H. H., 428
Non-Periodic Phenomena in Variable Stars, 491
Norton, A., 216
Norton, O. R., 176
Norton's Star Atlas and Reference Handbook (epoch 1950.0), 216
Nourse, A. E., 201, 409
Novotny, E., 481

Observation in Modern Astronomy, 535
Observational Astronomy for Amateurs, 232
Observatoires spatiaux, 555
Observatories of the World, 33
Observatory: A Review of Astronomy, 348
Observer's Handbook, 65
Obshchii Katalog Peremennykh Zvezd, 91
Omholt, A., 439
1001 Questions Answered about Astronomy, 66

298 / Author-Title Index

Optical Aurora, 439
Optical Production Technology, 538
Optical Telescope Technology, 541
Origin and Evolution of the Universe, 179
Origin of the Solar System, 351, 360
Original Theory or New Hypothesis of the Universe, 1750, 270
Origine et évolution des mondes, 179
Orthographic Lunar Atlas, 90
Oster, L., 319
Osterbrock, D. E., 512
Our Sun, 374
Our World in Space, 151
Out of the Zenith: Jodrell Bank, 1957-1970, 300
Outer Space Photography for the Amateur, 218
Outline of Astronomy, 340

Page, L. W., 217, 359, 360, 410, 463, 482, 505, 516, 552
Page, T., 33, 217, 359, 360, 410, 463, 482, 505, 516, 552
Palmer, H. P., 531
Palomar Sky Survey, p. 57
Pananides, N. A., 320
Pannekoek, A., 286
Papagiannis, M. D., 553
Parkes Catalogue of Radio Sources, p. 57
Paths of the Planets, 416
Paul, H. E., 218, 219
Payne-Gaposchkin, C., 321, 464
Pecker, J.-C., 554, 555
Peebles, P. J. E., 527
Peek, B. M., 411
Peltier, L. C., 177, 202
Percy, J. R., 65
Petrie, W., 440
Phillips, J. G., 144, 148
Photographic Lunar Atlas, 90
Photographic Study of the Brighter Planets, 415
Physical Characteristics of Comets, 430
Physical Cosmology, 527
Physical Foundations of General Relativity, 181
Physical Studies of Minor Planets, 420
Physics Abstracts, 17
Physics and Astronomy of the Moon, 383
Physics of Cosmic X-Ray, Gamma-Ray, and Particle Sources, 571
Physics of Planets, 407
Physics of Pulsars, 574
Physics of Stars and Stellar Systems, 467

Physics of Stellar Interiors, 488
Physics of the Solar Corona, 373, 376
Physics Today, 349
Pickering, E. C., 82
Pickering, J. S., 66
Pictorial Astronomy, 148
Pictorial Guide to the Moon, 380
Pictorial Guide to the Planets, 401
Pictorial Guide to the Stars, 459
Picture History of Astronomy, 285
Pikelner, S. B., 510
Pilcher, F., 67
Planet Jupiter, 397, 411
Planet Mercury, 413
Planet Saturn: A History of Observation, Theory and Discovery, 394
Planet Uranus, 395
Planet Venus, 406
Planetarium and Atmospherium: An Indoor Universe, 176
Planetarnye tumannosti, 508
Planetary and Space Science, 129
Planetary, Lunar and Solar Positions, 70
Planetary Nebulae, 508
Planets: Some Myths and Realities, 272
Planets, Stars, and Galaxies: An Introduction to Astronomy, 315
Podobed, V. V., 337
Point to the Stars, 194
Portrait of Isaac Newton, 264
Positional Astronomy, 335
Practical Work in Elementary Astronomy, 329
Primer for Star-Gazers, 200
Principles of Astrometry, 339
Principles of Astronomy, 332
Principles of Astronomy: A Short Version, 333
Principles of Celestial Mechanics, 446
Principles of Stellar Evolution and Nucleosynthesis, 474
Principles of Stellar Structure, 476
Pulsating Stars, 496, 575
Pulsiruyushchie zvezdy, 496

Quasars: Their Importance in Astronomy and Physics, 531
Quasi-Stellar Objects, 530
Quiet Sun, 369

RR Lyrae Stars, 499
Radiating Atmosphere, 438
Radiative Transfer, 472
Radio Astronomy, 560
Radio Astronomy: And How to Build Your Own Telescope, 226

Author-Title Index / 299

Radio Astronomy for Amateurs, 227
Radio Astronomy for the Amateur, 225
Radio Emission of the Sun and Planets, 567
Radio Exploration of the Sun, 564
Radio Science, 130
Radio Universe, 557
Radiotelescopes, 556
Rahe, J., 98
Raper, O., 400
Rectified Lunar Atlas, 90
Red Giants and White Dwarfs, 168
Redshift Controversy, 522
Referativnyi Zhurnal: Astronomiia, 18
Reference Catalogue of Bright Galaxies, 84
Relativity and Cosmology, 525
Revised New General Catalogue of Nonstellar Astronomical Objects, 104
Révolution astronomique, 242
Revolution in Astronomy, 165
Richardson, R. S., 247, 412, 429
Riddle of the Universe, 291
Riemer, M. F., 220
Roach, F. E., 441
Robinson, J. H., 68
Roemer, E., 423
Rohr, H., 155
Rohr, R. R. J., 235
Role of Schmidt Telescopes in Astronomy, 536
Ronan, C., 248, 249, 250, 251, 287, 288
Rose, W. K., 465
Rosen, E., 266
Roth, G. D., 221, 230
Routly, P. M., 504
Royal Astronomical Society. Monthly Notices, 131
Royal Astronomical Society of Canada. Journal, 133
Royal Astronomical Society. Quarterly Journal, 132
Ruffini, R., 572
Rükl, A., 99

STAR, 22
Sandage, A., 100, 517
Sandage, M., 517
Sandner, W., 413, 414
Santillana, G. de, 267
Saslaw, W. C., 178
Satellites of the Solar System, 414
Satterthwaite, G. E., 46
Schatzman, E. L., 179, 180

Scheifele, G., 448
Schmidt, M., 502
Schwarzschild, M., 483
Sciama, D. W., 181, 528
Science and Controversy: A Biography of Sir Norman Lockyer, 265
Science and Engineering Literature, 3
Science Citation Index, 19
Science of Astronomy, 323
Scientific and Technical Aerospace Reports, 22
Scientific, Engineering, and Medical Societies Publications in Print, 139
Scientific Instruments of the Seventeenth and Eighteenth Centuries, 296
Search the Solar System: The Role of Unmanned Interplanetary Probes, 362
Second Reference Catalogue of Bright Galaxies, 85
Seeing Beyond the Visible, 573
Seitter, W. C., 101
Selected Exercises in Galacitc Astronomy, 334
Sen, H. K., 479
Shadow of the Telescope: A Biography of John Herschel, 238
Shapere, D., 268
Shapley, H., 252, 289, 290
Shea, W. R., 269
Sheehy, E. P., 4
Shipman, H. L., 576
Shklovskii, I. S., 376, 484
Short History of Observatories, 274
Sidgwick, J. B., 231, 232
Simak, C. D., 182
Simonson, S. C., III, 559
Sinton, W. M., 540
Sky and Telescope, 350
Sky and Telescope Library of Astronomy, p. 88
Sky Observer's Guide: A Handbook for Amateur Astronomers, 195
Skyshooting: Photography for Amateur Astronomers, 215
Sleepwalkers: A History of Man's Changing Vision of the Universe, 281
Slettebak, A., 485
Slipher, E. C., 415
Smart, W. M., 291
Smith, A. G., 564
Smith, E. V. P., 331
Smith, W. J., 542
Sobolev, V. V., 466
Solar Activity Observations and Predictions, 372
Solar Atmosphere, 379
Solar Chromosphere, 365
Solar Physics, 134

Solar Radio Astronomy, 561
Solar Spectrum, 367
Solar System, 353, 356, 361
Solar System Astrophysics, 352
Solar Wind, 377
Solomon, J., 183
Sonnett, C. P., 377
Southern Durchmusterung, p. 56
Source Book in Astronomy: 1900-1950, 290
Source Book in Astronomy, 289
Soviet Astronomy, 135
Space Observatories, 555
Space Physics and Space Astronomy, 553
Space Science and Astronomy: Escape from Earth, 552
Space Science Reviews, 146
Spectral Line Formation, 458
Spectroscopic Astrophysics, 457
Spherical and Practical Astronomy as Applied to Geodesy, 336
Spherical Astronomy, 341
Spiral Structure of Our Galaxy, 500
Splendor in the Sky, 163
Standard Handbook for Telescope Making, 212
Star Atlas of Reference Stars and Nonstellar Objects, 102
Star Catalog, 103
Star Lovers, 247
Star Performance, 171
Stargazing with Telescope and Camera, 214
Starlight Nights: The Adventures of a Star-gazer, 177
Starlight: What It Tells Us about the Stars, 463
Stars, 205, 460
Stars and Clouds of the Milky Way, 505
Stars and Planets, 157
Stars and Stellar Systems, p. 89
Stars and the Milky Way System, 462
Stars in the Making, 464
Stars: Their Structure and Evolution, 489
Statistical Astronomy, 338
Steel, W. H., 543
Stein, R. F., 486
Stellar Atmospheres, 478, 480
Stellar Evolution, 473, 486
Stellar Interiors, 479
Stellar Rotation, 485
Stellar Spectroscopy: Normal Stars, 455
Stellar Spectroscopy: Peculiar Stars, 456

Stellar Structure, 468
Stelle e i planeti, 157
Stiefel, E. L., 448
Stockton, M. W., 94
Stonehenge Decoded, 277
Story of Jodrell Bank, 284
Stoy, R. H., 184
Strahlendes Weltall, 155
Strand, K. A., 69
Strickland, A. C., 378
Strohmeier, W., 497
Strong, J., 362
Structure and Evolution of the Galaxy, 503
Structure and Evolution of the Stars, 483
Structure of Space, 183
Structure of the Universe, 180
Struve, O., 292, 455, 456
Sulentic, J. W., 104
Sun, 371, 375
Sun and Its Influence, 368
Sun and the Amateur Astronomer, 206
Sundials: A Simplified Approach by Means of the Equatorial Dial, 233
Sundials: History, Theory and Practice, 235
Sundials: How to Know, Use and Make Them, 234
Sundials: Their Theory and Construction, 236
Suns, Myths and Men, 174
Sunspots, 366
Supernovae, 484
Supernovae and Supernova Remnants, 475
Supernovae and Their Remnants, 470
Survey of the Universe, 318
Swihart, T. L., 322, 487, 488
Szebehely, V., 449

Tables of Minor Planets, 67
Tables of the Characteristic Functions of the Eclipse and the Related Delta-Functions for Solution of Light Curves of Eclipsing Binary Systems, 58
Taschenbuch für der Planetenbeobachter, 221
Tayler, R. J., 489
Taylor, J., 577
Taylor, S. R., 384, 393
Technical Book Review Index, 9
Tektites, 417
Telescope and the World of Astronomy, 220

Telescope Handbook and Star Atlas, 213
Telescope Makers: From Galileo to the Space Age, 243
Telescopes, 539
Telescopes for Skygazing, 219
Telescopes: How to Make them and Use Them, 217
Telescopes in Space, 549
Terzian, Y., 513
Theory of Orbits: The Restricted Problem of Three Bodies, 449
Thermal Characteristics of the Moon, 385
Thewlis, J., 47
Thom, A., 307
Thompson, A. J., 222
Thompson, M. H., 316
Thorne, K. S., 443
Three Copernican Treatises, 266
Through Rugged Ways to the Stars, 252
Tifft, W. G., 104
Titus-Bode Law of Planetary Distances: Its History and Theory, 306
Tobias Mayer's Opera Inedita, 262
Tools of the Astronomer, 540
Translations Register-Index, 23
Tricker, R. A. R., 416
Trumpler, R. J., 338
Tsesevich, V. P., 498, 499
Tuckerman, B., 70

Underhill, A. B., 490
U.S. Observatories: A Directory and Travel Guide, 29
Universe, 149
Universe Unfolding, 317
Unveiling the Universe: The Aims and Achievements of Astronomy, 185
Uppsala General Catalogue of Galaxies, 97
Use of Physics Literature, 1

Valens, E. G., 444
Van de Kamp, P., 339
Van den Bergh, S., 519
Van Nostrand's Scientific Encyclopedia, 48
Vanderleen, W., 61
Variable Star Observer's Handbook, 209
Variable Stars, 495, 597
Variable Stars in Globular Clusters and in Related Systems, 492
Vehrenberg, H., 105, 203
Verschuur, G., 565, 566

Visitors from Afar: The Comets, 424
Vistas in Astronomy, 141
Vistas in Astronomy: An International Review Journal, 147
Voigt, H. H., 71, 340
Volkoff, I., 253
Vorontsov-Vel'yaminov, B. A., 467
Vzekhsvyatskii, S. D., 430

Walford, A. J., 5
Walker, J. C. G., 435
Wallenquist, A., 39
Wanderers in the Sky, 410
Warner, D. J., 254
Warnow, J. N., 256
Wasson, J. T., 431
Watchers of the Skies, 283
Watchers of the Stars: The Scientific Revolution, 246
Waugh, A. E., 236
Wavelength Tables, 93
Weaver, H. F. 338
Webb, The Rev. T. W., 223
Weekes, T. C., 578
Weigert, A., 49
Weiner, C., 256
What Star Is That?, 191
Wheeler, J. A., 443
Whipple, F. L., 318, 363
Whitaker, E., 86, 90
White, J. B., 277
White Dwarfs, 92
Whitney, C. A., 204, 293
Whitney's Star Finder: A Field Guide to the Heavens, 204
Wickramasinghe, N. C., 514
Widening Horizons: Man's Quest to Understand the Structure of the Universe, 282
Wilcox, J. M., 377
Wills, B. J., 521
Wills, D., 521
Wissenschaft von den Sternen, 460
Wonder and Glory: The Story of the Universe, 182
Wood, H., 185
Wood, J. A., 432
Woodbury, D. O., 294
Woods, J. A., 323
Woolard, E. W., 341
World Index of Scientific Translations, 23
Wright, F. W., 342
Wright, H., 255, 256
Wright, T., 270
Wurm, K., 98
Wyatt, S. P., 332, 333
Wyckoff, J., 195

X-Ray Astronomy, 570

Yale Zone Catalogues, p. 56
Yearbook of Astronomy, 62
Young, M. L., 34
Yü, C. S., 152

Zatmennye peremennye zvezdy, 498
Zebergs, V., 292
Zheleznyakov, V. V., 567
Zim, H. S., 205
Zimmerman, H., 49
Zirin, H., 379
Zvezdy tipa RR Liry, 499
Zwicky, F., 106

SUBJECT INDEX

Adams, John C., 289
Aeronautics/aerospace literature, 21 22
Amateur astronomy
  atlases, 193, 203, 210, 213, 216, 223
  binoculars, 192, 199
  biography, 177
  handbooks, 56, 62, 64-66, 68, 154
  journals, 122, 343, 350
  mathematics, 223
  Moon, 192
  photography, 207, 214, 215, 218
  planets, 221
  Sun, 206
  telescopes, 207, 208, 212-14, 217, 219, 222, 228, 231
  variable stars, 209
Aristarchus, 240, 282
Aristotle, 282, 287, 297, 299
Asteroids, 51, 67, 420
Astrodynamics, 123, 544, 545, 547
Astrometry, 335, 337, 339
Astronomical constants, 52, 68, 69
Astronomical formulae, 59
Astronomical gadgets, 171, 296
Astronomical libraries, 6, 34
Astronomical organizations, 26, 140
Astrophysics, 322, 327, 451, 454, 457, 465, 466
Atlases
  amateur astronomy, 193, 203, 210, 213, 216, 223
  aurorae, 88
  galaxies, 100
  general works, 81
  Moon, 73, 80, 86, 89, 90, 99
  sky, 76, 95, 96, 102, 105
  star, 74, 75, 78, 96
Atmospheres
  Earth, 433, 434, 436, 438
  planets, 435, 436
  popular works, 41

Atmospherium, 176
Aurorae
  atlases, 88
  popular works, 440
  technical works, 437, 439, 441

Baade, Walter F., 247
Bacon, Roger, 299
Banneker, Benjamin, 257
Barnard, E. E., 247
Binary stars, 58, 450, 452. *See also* Stars.
Binoculars (amateur astronomy), 192, 199
Black holes
  popular works, 162, 576, 577
  technical works, 569, 572
Bode, Johann Elert, 306
Bode's Law, 306
Bond, George, 279
Bond, William Cranch, 279
Borelli, Giovanni A., 242
Brahe, Tycho, 246-48, 281-83, 293

Catalogs
  emission line, 94
  galaxies, 84, 85, 97, 106
  general works, 54, 81
  Messier objects, 57
  sky, 77, 103, 104
  spectra, 82
  star, 53, 54, 77, 79, 87
  ultraviolet, 83
Celestial mechanics
  journals, 123
  technical works, 445-49
Clark, Alvan, 254
Collapsed objects, 572
Comets
  atlases, 98

Comets (cont'd)
  history, 298, 418
  popular works, 418, 424, 427, 429
  technical works, 419, 423, 430
Constellations (amateur astronomy)
  191, 194, 198, 200-204
Copernicus, 237, 242, 246, 248, 263, 266, 281-83, 287-89
Cosmology
  popular works, 520, 524-26
  technical works, 521, 522, 527, 528, 576

Doppler effect, 523
Double stars. See Binary stars.
Dwarf novae. See Variable stars.

Eclipses, 357
Einstein, Albert, 241, 247, 248
Electromagnetic spectrum, 93, 573
Encke, Johann Franz, 247
English literature and astronomy, 303
Ephemerides
  general works, 63, 64, 70
  Moon, 55
  Sun, 61
Eratosthenes, 240, 282
Euler, Leonhard, 261
Expanding Universe, 522

Films, 7
Flamsteed, John, 249, 260
Foreign language dictionaries, 35, 37, 47
Friedman, Herbert, 243

Galactic nebulae
  popular works, 506, 515
  technical works, 507, 511, 512
Galaxies
  atlases, 100
  catalogs, 84, 85, 97, 106
  popular works, 515, 516, 518
  technical works, 517, 519
Galileo, 243, 246, 248, 251, 259, 263, 267, 268, 269, 271, 281-83, 287-89, 293, 299
Gamma-ray astronomy, 571
Geological literature, 16, 19, 125, 126, 145
Goodricke, John, 247
Government R & D reports, 20, 22
Graduate study (in astronomy), 28, 32

Gravity
  popular works, 442, 444
  technical works, 443
Greek astronomy–History, 297, 305

H_{II} regions, 513
Hale, George Ellery, 243, 255, 256, 288, 294
Hall, Asaph, 247
Halley, Edmond, 247-50, 283, 289
Harrison, John, 294
Herschel, John, 238, 248
Herschel, William, 239, 243, 248, 249, 272, 288, 293, 299
Hevelius, Johannes, 253
High-energy particles, 571, 578
Hipparchus, 240
Horrocks, Jeremiah, 247
Hubble, Edwin, 299
Huggins, Sir William, 247

IQSY, 378
Infrared astronomy, 568
Instrumentation (technical works).
  See also Telescopes.
  optical, 533-43
  radio, 556, 562, 563
Interferometry, 543
International Years of the Quiet Sun, 378
Interstellar dust and gas
  popular works, 506
  technical works, 507, 510, 511, 514

Janssen, Zacharias, 247
Japanese astronomy–history, 304
Jodrell Bank, 284, 300
Journals
  historical, 346
  popular, pp. 172-76
  professional, pp. 70-84

Kepler, Johannes, 242, 243, 246, 248, 258, 281-83, 287-89, 293

Laboratory exercises, 312, 328, 329, 334
Leverrier, Urbain, J. J., 289
Light (radiation), 573
Lockyer, Sir Norman, 249, 265, 288
Lowell, Percival, 247

Magnetism, 509
Maskelyne, Nevil, 249

Mayer, Tobias, 261, 262
Messier objects, 57
Meteorology, 434
Meteors and meteorites
  handbooks, 60, 431
  popular works, 418, 421, 428
  technical works, 422, 425, 426, 432
  tektites, 417
Michelson, Albert A., 244
Milky Way
  history, 293, 299
  popular works, 501, 505
  spiral structure, 500
  technical works, 502-504
Minor planets. *See* Asteroids.
Moon
  atlases, 73, 80, 86, 89, 90, 99
  geology, 384, 391-93
  journals, 128
  popular works, 380, 392
  technical works, 381-83, 386-90
  thermal characteristics, 385
Mt. Palomar, 294

Nautical astronomy—history, 295
Navigation, 50, 295, 342
Newton, Isaac, 243, 246-49, 264, 282, 283, 287, 289, 293, 299
Nicholson, Seth Barnes, 247

Observatories
  directories, 29, 30, 33
  history, 274, 279, 280, 284, 294, 300, 307
  libraries, 6, 34
  publications, pp. 90-92
Optics (technical works), 533, 538, 541, 542

Photography. *See* Amateur astronomy.
Physics
  encyclopedias, 42, 47
  journals, 121, 349
  literature, 1, 3, 5, 14, 15, 17, 19
Pickering, Edward Charles, 279
Planetaria, 171, 176
Planetary nebulae
  popular works, 515
  technical works, 508, 511, 512
Planetoids. *See* Asteroids.
Planets
  journals, 127, 129
  Jupiter, 396-98, 403, 411
  Mars, 399, 400, 404, 405, 412
  Mercury, 413
  motion, 410, 416

Planets (cont'd)
  popular works, 401, 408-10
  satellites, 414
  Saturn, 394
  technical works, 407, 415
  terrestrial planets, 355
  Uranus, 395
  Venus, 402, 406
Plato, 297
Proceedings of meetings, pp. 84-88
Ptolemy, 240, 263, 283, 287
Publishers' series, pp. 88-89
Pulsars, 575

Quasars
  popular works, 529, 531
  technical works, 521, 530, 576

RR Lyrae stars. *See* Variable stars.
Radiative transfer, 472
Radio astronomy
  amateur, 225-27
  history, 278, 284, 300
  popular works, 557, 558, 566
  technical works, 559-61, 564, 565, 567
  telescopes, 278, 543, 556, 562, 563
Radio science, 130
Reber, Grote, 243
Relativity, 181, 525
Review literature, pp. 92-95
Roemer, Ole, 247
Rosse, Lord, 243, 245, 249, 293

Schmidt, Bernard, 243
Schmidt telescopes. *See* Telescopes.
Schwabe, S. Heinrich, 247
Smithson, James, 280
Solar system. *See also* Planets.
  astrophysics, 352
  journals, 127, 129
  origin of, 351, 360, 432
  popular works, 353, 354, 356, 359, 361-63
Space astronomy (technical works), 186, 541, 549-51, 553-55
Space exploration (popular works), 43, 151, 153, 350, 362, 401
Space science
  journals, 120, 129, 146
  popular works, 40, 43, 552
  technical works, 546, 548-51, 553
Spectroscopy. *See* Stars.
Spherical astronomy, 336, 341
Stars. *See also* Variable stars.
  atlases, 74, 75, 78, 96
  atmospheres, 476, 478, 480, 481, 487

Stars (cont'd)
  binary, 58, 450, 452
  catalogs, 53, 54, 77, 79, 87
  evolution, 469, 470, 473-75, 477,
    482-84, 486, 489, 490
  interiors, 472, 474, 476, 477, 479,
    481, 488
  mass loss, 453
  popular works, 459, 460, 463, 464, 482
  radial velocities, 72
  rotation, 485
  spectroscopy, 301, 455-58, 463
  structure, 468, 471, 472, 476, 477,
    483, 489
  supernovae, 470, 475, 484
  technical works, 462, 467
  white dwarfs, 92
Statistical astronomy, 338, 361
Stonehenge, 166, 277
Struve, Otto, 454, 457
Sun
  atmosphere, 379
  chromosphere, 365
  corona, 370, 373, 376
  journals, 134
  popular works, 371, 374, 375
  solar physics, 134, 302
  solar wind, 364, 370, 377
  spectrum, 367
  sunspots, 366
  technical works, 368, 369, 372, 378
Sundials, 233-36
Supernovae. *See* Stars.

Teaching of astronomy, 273
Technical reports, pp. 30-33
Tektites. *See* Meteors and Meteorites.
Telescopes. *See also* Amateur astronomy;
    Instrumentation; Radio astronomy.
  history, 271, 294, 296, 300, 532
  popular works, 532, 540
  Schmidt, 536
Thomson, J. J., 301

Variable stars. *See also* Stars.
  catalogs, 91
  dwarfs novae, 493
  eclipsing, 498
  intrinsic, 496
  popular works, 209, 495
  RR Lyrae stars, 499
  technical works, 491, 492, 494, 497
Von Fraunhofer, Joseph, 243
Von Wittenberg, Johann D. T., 306

Wavelength tables, 93
White dwarfs. *See* Stars.
Winlock, Joseph, 279
Wollaston, William Hyde, 301
Wright, Thomas, 270, 299

X-Ray astronomy, 570-72